Methods in Molecular Biology

Series Editor
John M. Walker
School of Life and Medical Sciences
University of Hertfordshire
Hatfield, Hertfordshire, UK

For further volumes:
http://www.springer.com/series/7651

For over 35 years, biological scientists have come to rely on the research protocols and methodologies in the critically acclaimed Methods in Molecular Biology series. The series was the first to introduce the step-by-step protocols approach that has become the standard in all biomedical protocol publishing. Each protocol is provided in readily-reproducible step-by step fashion, opening with an introductory overview, a list of the materials and reagents needed to complete the experiment, and followed by a detailed procedure that is supported with a helpful notes section offering tips and tricks of the trade as well as troubleshooting advice. These hallmark features were introduced by series editor Dr. John Walker and constitute the key ingredient in each and every volume of the Methods in Molecular Biology series. Tested and trusted, comprehensive and reliable, all protocols from the series are Indexed in PubMed.

Manuscript Instructions https://drive.google.com/file/d/1K4Az5vpAFVzYadJ3jKx2-QSo7Dqa_SNA/view?usp=sharing

Intestinal Stem Cells

Methods and Protocols

Second Edition

Edited by

Paloma Ordóñez-Morán

Translational Medical Sciences Unit, School of Medicine, Centre for Cancer Sciences, Biodiscovery Institute, University of Nottingham, Nottingham, UK

Humana Press

Editor
Paloma Ordóñez-Morán
Translational Medical Sciences Unit
School of Medicine
Centre for Cancer Sciences
Biodiscovery Institute
University of Nottingham
Nottingham, UK

ISSN 1064-3745 ISSN 1940-6029 (electronic)
Methods in Molecular Biology
ISBN 978-1-0716-5411-8 ISBN 978-1-0716-5412-5 (eBook)
https://doi.org/10.1007/978-1-0716-5412-5

This Humana imprint is published by the registered company Springer Science+Business Media, LLC, part of Springer Nature.
The registered company address is: 1 New York Plaza, New York, NY 10004, U.S.A.

Preface to the Second Edition

Since the publication of the first edition, the intestinal stem cell field has undergone transformative advances driven by revolutionary technologies that have expanded our ability to study stem cell biology with unprecedented resolution and physiological relevance. Single-cell and spatial omics have revealed striking heterogeneity within stem cell populations and their dynamic interactions with the niche during homeostasis, injury, and disease. Simultaneously, organoid technology has matured into sophisticated platforms that now support co-cultures with immune cells, endothelial networks, and stromal compartments, enabling studies of inflammatory responses, and therapeutic responses that closely recapitulate human pathophysiology.

This second edition builds upon the foundation established in the first by integrating these cutting-edge approaches. We have substantially expanded coverage of omics technologies, now including multi-omics integration, transcriptomics, and proteomics to dissect stem cell plasticity and regulatory networks. The applications of organoids have been comprehensively updated to reflect advancements in patient-derived models that are driving progress in cancer research and precision medicine. New sections address emerging methodologies such as computational modeling of stem cell dynamics and advanced imaging techniques for live tracking of lineage hierarchies.

The protocols remain rigorously optimized for accessibility, with step-by-step guidance suitable for both established investigators and those newly entering the field. Our goal remains to empower researchers with tools that bridge fundamental discovery with translational impact, reducing reliance on animal models while enhancing human relevance.

I extend my deepest gratitude to all contributors for their expertise and dedication, and to Dr. John Walker for continued guidance throughout this revision.

Nottingham, UK
February 2026

Paloma Ordóñez-Morán

Contents

Contributors

BRINDA BALASUBRAMANIAN • *Translational Medical Sciences Unit, School of Medicine, Centre for Cancer Sciences, Biodiscovery Institute, University of Nottingham, Nottingham, UK*

JOEP BEUMER • *Institute of Human Biology, Roche Pharma Research and Early Development, Roche Innovation Center Basel, Basel, Switzerland*

FRÉDÉRIC BRETAGNOL • *Service de chirurgie, Groupement Hospitalier de l'Ouest Lémanique (GHOL), Nyon, Switzerland*

BRAM VAN DEN BROEK • *Division of Cell Biology, The Netherlands Cancer Institute, Amsterdam, The Netherlands*

FABIO BUTTI • *Service de chirurgie, Groupement Hospitalier de l'Ouest Lémanique (GHOL), Nyon, Switzerland*

MICHELE CERVELLERA • *Tissue Engineering and Biomaterials Group, Department of Human Structure and Repair, Faculty of Medicine and Health Sciences, Ghent University, Ghent, Belgium*

CEWEN CHEN • *Institute for Chemical Reaction Design and Discovery (WPI-ICReDD), Hokkaido University, Kita-ku, Sapporo, Japan; Department of Cancer Pathology, Faculty of Medicine, Hokkaido University, Kita-ku, Sapporo, Japan*

HANS CLEVERS • *Hubrecht Institute, Royal Netherlands Academy of Arts and Sciences (KNAW) and UMC Utrecht, Utrecht, The Netherlands; Oncode Institute, Hubrecht Institute, Utrecht, The Netherlands; The Princess Maxima Center for Pediatric Oncology, Utrecht, The Netherlands; Oncode Institute, Hubrecht Institute, Royal Netherlands Academy of Arts and Sciences, Utrecht, The Netherlands; Institute of Human Biology, Roche Pharma Research and Early Development, Roche Innovation Center Basel, Basel, Switzerland*

ROEI COHEN-ALMAGOR • *Translational Medical Sciences Unit, School of Medicine, Centre for Cancer Sciences, Biodiscovery Institute, University of Nottingham, Nottingham, UK*

NICOLA COLUCCI • *Service de chirurgie, Groupement Hospitalier de l'Ouest Lémanique (GHOL), Nyon, Switzerland*

WILLIAM DALLEYWATER • *Translational Medical Sciences Unit, Centre for Cancer Sciences, Biodiscovery Institute, School of Medicine, University of Nottingham, Nottingham, UK; Cellular Pathology Department, Nottingham University Hospital, Nottingham, UK*

BERT DEVRIENDT • *Laboratory of Immunology, Department of Translational Physiology, Infectiology and Public Health, Faculty of Veterinary Medicine, Ghent University,, Merelbeke, Belgium*

RUSLAN I. DMITRIEV • *Tissue Engineering and Biomaterials Group, Department of Human Structure and Repair, Faculty of Medicine and Health Sciences, Ghent University, Ghent, Belgium; Ghent Light Microscopy Core, Ghent University,, Ghent, Belgium; Ghent Light Microscopy Core, Ghent University, Ghent, Belgium*

JUSTINE EPINEY • *Tumorshare, École Polytechnique Fédérale de Lausanne, Lausanne, Switzerland; Swiss Institute for Experimental Cancer Research (ISREC), EPFL, Lausanne, Switzerland; Division of Haematology and Central Haematology Laboratory, Lausanne University Hospital (CHUV), Lausanne, Switzerland*

YEGANEH FARSI • *Laboratory of Ocular Immunology, Transplantation and Regeneration, Schepens Eye Research Institute of Massachusetts Eye and Ear, Department of Ophthalmology, Harvard Medical School, Boston, MA, USA*
PIERRE FOURNIER • *Service de chirurgie, Groupement Hospitalier de l'Ouest Lémanique (GHOL), Nyon, Switzerland*
QIAOCHU FU • *Institute for Chemical Reaction Design and Discovery (WPI-ICReDD), Hokkaido University, Kita-ku, Sapporo, Japan; Department of Cancer Pathology, Faculty of Medicine, Hokkaido University, Kita-ku, Sapporo, Japan*
LOUIS GALL • *Centre for Systems Modelling & Quantitative Biomedicine (SMQB), University of Birmingham, Birmingham, UK*
CHAO GAO • *Institute for Chemical Reaction Design and Discovery (WPI-ICReDD), Hokkaido University, Kita-ku, Sapporo, Japan; Department of Cancer Pathology, Faculty of Medicine, Hokkaido University, Kita-ku, Sapporo, Japan*
MARIE-ISABELLE GARCIA • *Institut de Recherche Interdisciplinaire en Biologie Humaine et Moléculaire (IRIBHM), Faculty of Medicine, Université Libre de Bruxelles ULB, Brussels, Belgium*
MARIA EUGENIA GULINO • *Translational Medical Sciences, School of Medicine, Queen's Medical Centre, University of Nottingham, Nottingham, UK; Department of Internal Medicine, Yale School of Medicine, New Haven, USA*
LAURA GULL • *HistologiX, BioCity, Nottingham, UK*
STEFANO GUSSAGO • *Service de chirurgie, Groupement Hospitalier de l'Ouest Lémanique (GHOL), Nyon, Switzerland*
HIKARU HANYU • *Department of Integrated Medicine and Biochemistry, Keio University School of Medicine, Shinjuku-ku, Tokyo, Japan*
LULU HUANG • *Hubrecht Institute, Royal Netherlands Academy of Arts and Sciences (KNAW) and UMC Utrecht, Utrecht, The Netherlands; Oncode Institute, Hubrecht Institute, Utrecht, The Netherlands*
MASAMICHI IMAJO • *Institute for Chemical Reaction Design and Discovery (WPI-ICReDD), Hokkaido University, Kita-ku, Sapporo, Japan*
KEES JALINK • *Division of Cell Biology, The Netherlands Cancer Institute, Amsterdam, The Netherlands*
U KAICHENG • *Tri-Institutional Computational Biology & Medicine, Weill Cornell Medicine, New York, NY, USA*
JOE KHALLOUF • *Service de chirurgie, Groupement Hospitalier de l'Ouest Lémanique (GHOL), Nyon, Switzerland*
AMBER DE KINDER • *Tissue Engineering and Biomaterials Group, Department of Human Structure and Repair, Faculty of Medicine and Health Sciences, Ghent University, Ghent, Belgium*
THIBAUD KOESSLER • *Department of Oncology, University Hospital Geneva (HUG), Geneva, Switzerland*
MORGANE LEPROVOTS • *Institut de Recherche Interdisciplinaire en Biologie Humaine et Moléculaire (IRIBHM), Faculty of Medicine, Université Libre de Bruxelles ULB, Brussels, Belgium*
YASHWANT R. MAHIDA • *Translational Medical Sciences, School of Medicine, Queen's Medical Centre, University of Nottingham, Nottingham, UK*

MAGDA MARTINEZ-ESPUGA • *Department of Environmental and Chemical Engineering, Biodiscovery Institute, University of Nottingham, Nottingham, UK; Translational Medical Sciences Unit, School of Medicine, Centre for Cancer Sciences, Biodiscovery Institute, University of Nottingham, Nottingham, UK*

ADRIANA MARTINEZ-SILGADO • *Hubrecht Institute, Royal Netherlands Academy of Arts and Sciences (KNAW) and UMC Utrecht, Utrecht, The Netherlands; Oncode Institute, Hubrecht Institute, Utrecht, The Netherlands*

ALVARO MATA • *Department of Environmental and Chemical Engineering, Biodiscovery Institute, University of Nottingham, Nottingham, UK; School of Pharmacy, Biodiscovery Institute, University of Nottingham, Nottingham, UK*

ROMANE MEERTS • *Institut de Recherche Interdisciplinaire en Biologie Humaine et Moléculaire (IRIBHM), Faculty of Medicine, Université Libre de Bruxelles ULB, Brussels, Belgium*

MAX NOBIS • *Intravital Imaging Expertise Center, VIB Center for Cancer Biology, KU Leuven, Leuven, Belgium; Manchester Cell-Matrix Centre, Division of Cell Matrix Biology and Regenerative Medicine, School of Biological Sciences, Faculty of Biology Medicine and Health, The University of Manchester, Manchester, UK*

IRINA A. OKKELMAN • *Tissue Engineering and Biomaterials Group, Department of Human Structure and Repair, Faculty of Medicine and Health Sciences, Ghent University, Ghent, Belgium; Ghent Light Microscopy Core, Ghent University, Ghent, Belgium*

PALOMA ORDÓÑEZ-MORÁN • *Translational Medical Sciences Unit, School of Medicine, Centre for Cancer Sciences, Biodiscovery Institute, University of Nottingham, Nottingham, UK*

SHIVAM PATEL • *Translational Medical Sciences Unit, School of Medicine, Centre for Cancer Sciences, Biodiscovery Institute, University of Nottingham, Nottingham, UK*

CARMEN PIN • *Clinical Pharmacology and Quantitative Pharmacology, Clinical Pharmacology and Safety Sciences, R&D, AstraZeneca, Cambridge, UK*

ANA PIRES • *HistologiX, BioCity, Nottingham, UK*

FARAH QADERI • *Division of Gastrointestinal and Oncologic Surgery, Department of Surgery, Massachusetts General Hospital, Boston, MA, USA; Center for Engineering in Medicine and Surgery, Department of Surgery, Massachusetts General Hospital, Boston, MA, USA; Shriners Hospital for Children, Boston, MA, USA*

NIMA SAEIDI • *Division of Gastrointestinal and Oncologic Surgery, Department of Surgery, Massachusetts General Hospital, Boston, MA, USA; Center for Engineering in Medicine and Surgery, Department of Surgery, Massachusetts General Hospital, Boston, MA, USA; Shriners Hospital for Children, Boston, MA, USA; Harvard Stem Cell Institute, Cambridge, MA, USA*

SAMNA SAGADEVAN • *Translational Medical Sciences Unit, School of Medicine, Centre for Cancer Sciences, Biodiscovery Institute, University of Nottingham, Nottingham, UK*

PÉNÉLOPE SAINT-AMOUR • *Service de chirurgie, Groupement Hospitalier de l'Ouest Lémanique (GHOL), Nyon, Switzerland*

ALBERT SANTAMARIA-MARTÍNEZ • *Tumorshare, École Polytechnique Fédérale de Lausanne, Lausanne, Switzerland; Swiss Institute for Experimental Cancer Research (ISREC), EPFL, Lausanne, Switzerland*

TOSHIRO SATO • *Department of Integrated Medicine and Biochemistry, Keio University School of Medicine, Shinjuku-ku, Tokyo, Japan*

Mireia Sueca-Comes • *HistologiX, BioCity, Nottingham, UK*
Shinya Sugimoto • *Department of Integrated Medicine and Biochemistry, Keio University School of Medicine, Shinjuku-ku, Tokyo, Japan; Division of Gastroenterology, Department of Internal Medicine, Keio University School of Medicine, Shinjuku-ku, Tokyo, Japan*
Shinya Tanaka • *Department of Cancer Pathology, Faculty of Medicine, Hokkaido University, Kita-ku, Sapporo, Japan*
Teresa Torres • *Service de chirurgie, Groupement Hospitalier de l'Ouest Lémanique (GHOL), Nyon, Switzerland*
Aurélie Vuagniaux • *Service de chirurgie, Groupement Hospitalier de l'Ouest Lémanique (GHOL), Nyon, Switzerland*
Joost J. A. P. M. Wijnakker • *Oncode Institute, Hubrecht Institute, Royal Netherlands Academy of Arts and Sciences, Utrecht, The Netherlands*
Junhan Zhao • *Department of Pediatrics, the University of Chicago, Chicago, IL, USA; Department of Biomedical Informatics, Harvard Medical School, Boston, MA, USA*
Hang Zhou • *Tissue Engineering and Biomaterials Group, Department of Human Structure and Repair, Faculty of Medicine and Health Sciences, Ghent University, Ghent, Belgium*

Part I

Intestinal Stem Cells: Characterization and Multi-omics Profiling

Check for updates

Chapter 1

Intestinal Stem Cells: Regulation, Plasticity, and Therapeutic Potential

Samna Sagadevan, Roei Cohen-Almagor and Paloma Ordóñez-Morán

Abstract

Intestinal stem cells (ISCs) are central to the maintenance and regeneration of the gut epithelium, supporting its rapid turnover and adaptive capacity. Recent advances, including the identification of LGR5$^+$ stem cells, 3D in vitro organoid technology, and single-cell transcriptomics, have uncovered the remarkable heterogeneity and plasticity of ISCs and their niche. The ISC niche comprises dynamic interactions among epithelial, stromal, immune, and microbial components, enabling responses to injury and disease. Notably, both active and reserve stem cell pools, as well as progenitor and differentiated cells, can contribute to regeneration. These insights are shaping novel strategies for treating gastrointestinal diseases and cancers through personalized medicine.

Keywords Intestinal stem cells, Niche, Differentiation, Cellular plasticity, Organoids, Wnt/Notch/BMP signaling

1 Introduction

The intestinal epithelium is one of the most rapidly self-renewing tissues in the body, undergoing complete turnover every 3–5 days. This renewal is driven by intestinal stem cells (ISCs) residing at the crypt base, which are essential for maintaining gut barrier integrity, nutrient absorption, and immune defense [1]. ISCs operate within a complex and dynamic niche that includes neighboring epithelial and stromal cells, extracellular matrix components, and signals from the gut microbiota [2, 3]. Over the past decade, landmark discoveries, such as the identification of LGR5 as a marker of active ISCs and the development of intestinal 3D in vitro organoids, have transformed our understanding of epithelial biology [4]. These advances have revealed that ISCs are not a homogeneous population; instead, they exhibit significant heterogeneity and plasticity, enabling robust responses to environmental insults and injury [5, 6]. Importantly, the ISC compartment is not fixed. Reserve stem cells and even differentiated

Paloma Ordóñez-Morán (ed.), *Intestinal Stem Cells*, Methods in Molecular Biology, vol. 3059,
https://doi.org/10.1007/978-1-0716-5412-5_1,

cells can revert to a stem-like state under specific conditions [7]. In parallel, increasing evidence points to the influence of microbial and glial interactions in modulating ISC behavior and epithelial homeostasis [8, 9].

As research tools and models improve, our understanding of stem cell dynamics, niche interactions, and their therapeutic potential continues to advance. These insights are reshaping strategies for gastrointestinal disease management. This review examines recent advances in ISC biology and its niche within a rapidly evolving field, where ongoing discoveries consistently refine and expand current paradigms.

2 Types and Hierarchy of ISCs

ISCs are organized in a complex and dynamic hierarchy within the intestinal crypts, with multiple distinct populations contributing to tissue maintenance (Fig. 1). Recent research has refined our understanding of ISC types and their hierarchical relationships, revealing both structured organization and significant plasticity (Table 1).

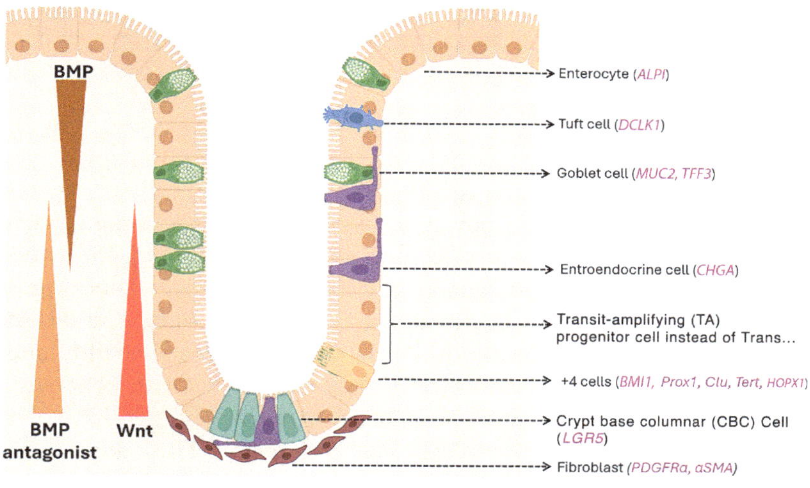

Fig. 1 Schematic of colon crypt architecture and signaling gradients. The intestinal epithelium is organized into crypts, with stem cells ($LGR5^+$) residing at the crypt base, supported by surrounding niche cells ($PDGFR\alpha^+$ fibroblasts and αSMA^+ myofibroblasts). Proliferating transit-amplifying cells occupy the mid-crypt region, differentiating into secretory lineages (Goblet cells: MUC2, TFF3; Enteroendocrine cells: CHGA) and absorptive enterocytes (ALPI) as they migrate toward the lumen. BMP signaling is highest at the top of the crypt, promoting differentiation, whereas Wnt signaling is enriched at the crypt base, maintaining stemness. BMP antagonists at the crypt base counteract BMP activity to sustain the stem cell niche. Key markers for specific cell types are indicated on the right. Figure created with BioRender.com

Table 1

Overview of intestinal stem cell (ISC)/Transit-amplifying (TA) progenitors: as LGR5- crypt base columnar cells (CBCs), label retaining cells, revival stem cells, and TA progenitor populations, summarizing their anatomical location within the crypt, proliferative activity, contribution to epithelial homeostasis and regeneration, and associated molecular markers as reported in the indicated references

ISC/TA type	Location	Activity	Role in homeostasis/regeneration	Key markers	References
$LGR5^+$ CBCs	Crypt base	Highly cycling	Daily renewal	LGR5	[1, 10–12]
+4/Label retaining cells	+4 position	Quiescent	Injury response	BMI1, HOPX	[10, 11, 13, 14]
Revival stem cells	Crypt, rare	Quiescent	Regeneration post-injury	CLU	[12, 15]
TA progenitors	Above ISCs	Rapidly cycling	Rapid proliferation and differentiation	Various (including Prom1/CD133, ARID3A, METTL3, ID1, PCNA, and ZNF277)	[1, 4, 16–22]

2.1 LGR5⁺ Crypt Base Columnar (CBC) Cells

These are the primary, actively cycling ISCs located at the crypt base. They drive epithelial renewal and can generate both new stem cells and TA. $LGR5^+$ CBCs lineage priming and cell-cycle exit are influenced by polarity cues and signaling pathway switches [1, 10–12, 16]. Moreover, multiple subtypes of $LGR5^+$ ISCs have now been identified, including $LGR5^+Mex3a^+$ slowly cycling cells and $LGR5^+p27^+$ quiescent cells, each with distinct proliferative capacities and roles in regeneration [2, 23].

2.2 Quiescent/Label retaining (+4) Cells

Positioned just above the crypt base, these cells are more quiescent and serve as a reserve pool. They can be activated during injury or stress to replenish the active stem cell compartment. These quiescent cells may express both stem cell and secretory lineage genes, indicating a transitional state [10, 11, 13, 14, 16].

2.3 Revival Stem Cells (revSCs)

Identified as CLU^+ cells, these are rare under normal conditions but expand after injury to regenerate the ISC pool and the epithelium [12, 15].

2.4 Hierarchical Organization

Stem cell compartments are typically structured according to a hierarchical organization, wherein a small population of multipotent stem cells gives rise to increasingly specialized progenitor and differentiated cells. This framework helps maintain tissue homeostasis and enables efficient regeneration following injury.

Traditionally, ISCs were thought to follow a linear hierarchy, with quiescent stem cells replenishing the active, cycling population, similar to models in other tissues like the hematopoietic system [13]. New evidence challenges the rigid linear model, suggesting that the ISC hierarchy is more fluid. ISCs can transition between active and quiescent states, and lineage priming can occur directly from ISCs through unipotent intermediates. The precise relationships and regulatory mechanisms between these populations remain under active investigation [13, 16].

3 The Intestinal Stem Cell Niche

The bottom of the crypt contains a specialized microenvironment that surrounds and regulates ISCs, playing a crucial role in maintaining the balance between stem cell self-renewal, differentiation, and regeneration. This niche is composed of a diverse array of supporting cells and molecular signals that together orchestrate ISC behavior (Fig. 2).

3.1 Key Cellular Components Regulating ISC Function

In the small intestine, Paneth cells are located adjacent to ISCs at the crypt base. They secrete essential factors such as Wnt3, EGF, and Notch ligands, which are critical for ISC maintenance, proliferation, and fate determination [24, 25]. Paneth cells also contribute to the antimicrobial defense of the crypt.

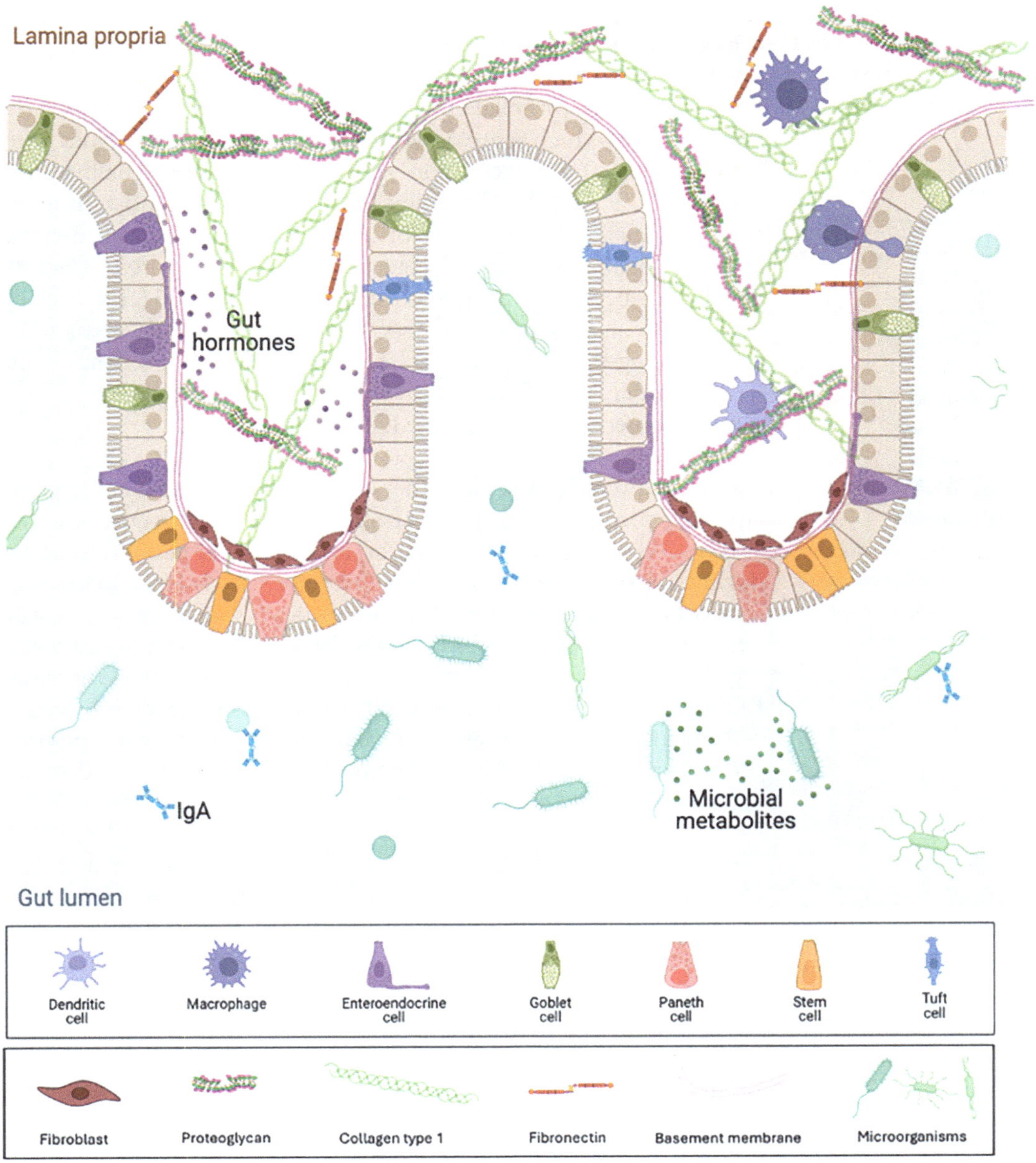

Fig. 2 Molecular and environmental cues governing ISC maintenance and differentiation. This diagram illustrates the complex cellular and molecular interactions within the intestinal crypt and lumen. It shows how various cell types, including Paneth cells and other niche cells, secrete factors that influence ISC maintenance and differentiation. The figure also highlights the crucial role of environmental cues from the gut lumen, such as microbial metabolites and immune signals, in regulating these processes. Figure created with BioRender.com

Subepithelial mesenchymal cells, including myofibroblasts and specialized CD34^{+} Gp38^{+} mesenchymal cells, secrete R-spondins, BMP antagonists (e.g., Gremlin1), and Wnt ligands, which modulate Wnt signaling gradients and help maintain the balance between

ISC self-renewal and differentiation [26–28]. These mesenchymal cells are rapidly activated after injury, upregulating niche factors and cytokines to support tissue repair [28].

Other non-epithelial cells that regulate ISCs function are the immune cells, which sense tissue damage and release cytokines (e.g., Interleukin (IL)-22, Interferon-γ, IL-1) and activate signaling pathways (STAT1, STAT3), boosting their proliferation and aiding epithelial repair [2, 29–32]. Endothelial cells help maintain the ISC niche by ensuring proper blood supply and delivering growth factors, which are especially important during regeneration [33]. Enteric glial cells, particularly GFAP$^+$ subsets, respond to injury by expanding and releasing Wnt ligands, which are essential for ISC activation and tissue regeneration. Their heterogeneity allows for tailored responses during homeostasis and disease [34–36].

3.2 ECM Regulation of ISC Activity

The ECM provides a scaffold that organizes ISCs and their progeny, ensuring proper tissue architecture and renewal. Its components, such as collagens and glycoproteins, are also dynamically remodeled during injury or infection, which alters the signals ISCs receive and supports regeneration [37]. Research studies reveal that ECM stiffness regulates stem cell lipid metabolism via the Lipin-1 and Sterol regulatory element-binding proteins (SREBP) pathways. Increased ECM stiffness leads to changes in lipid synthesis, which is linked to cell survival and proliferation. This demonstrates that the physical properties of the ECM can influence ISC behavior by modulating key metabolic and signaling pathways [38].

3.3 Microbial Metabolites and ISC Modulation

Recent studies highlight the role of gut microbiota and their metabolites in modulating ISC activity, suggesting that the microbiome is an integral component of the niche [39]. In Drosophila, an RNAi screen identified that extracellular adenosine, a metabolite released by various cell types, activates the adenosine receptor (AdoR) on ISCs, promoting their proliferation. After tissue damage, reduced activity of the enzyme Adgf-A (which degrades adenosine) leads to higher adenosine levels and increased ISC activity. While some niche cell development is independent of the microbiota, microbial metabolites can still modulate ISC function and the broader niche environment, especially during injury or inflammation [40].

3.4 Gut Hormones

Gut hormones are key niche signals in the intestine, integrating inputs from nutrients, the microbiota, and the epithelium to regulate ISC behavior and maintain homeostasis. Secreted by enteroendocrine cells in response to nutrient and luminal stimuli, their release can be precisely studied using models like the isolated perfused rat small intestine [41]. These hormones not only support epithelial renewal and barrier integrity but also interact with the nervous and immune systems to shape systemic physiology [42–44]. Microbial signals influence gut hormone secretion, making these hormones adaptive regulators of environmental and microbial changes [43, 45]. For instance, microbial metabolites like valeric acid stimulate enteric neurons to produce serotonin (5-HT), which prompts macrophages to release prostaglandin E2 (PGE2), activating Wnt/β-catenin signaling in ISCs to support self-renewal. Hormones like thyroid hormone (T3) also modulate ISC function by regulating metabolic and developmental pathways [46, 47].

4 Regulation of ISC Function

4.1 Core Signaling Pathways

ISC function is orchestrated by a sophisticated network of intrinsic signaling pathways and extrinsic cues from the surrounding niche, ensuring the balance between self-renewal, proliferation, and differentiation. Among the core pathways, Wnt/β-catenin signaling stands out as a principal driver of ISC self-renewal and proliferation. Pioneer studies in both Drosophila and mammals have shown that Wnt signals, often provided by neighboring niche cells such as the circular muscle or mesenchymal cells, are essential for maintaining the stem cell pool. Disruption of Wnt signaling leads to ISC quiescence and differentiation, while overactivation can result in excessive proliferation and tumorigenesis. Notably, Wnt signaling also interacts with Notch to coordinate cell fate decisions, with Notch activity directing progenitors toward absorptive or secretory lineages, thus ensuring balanced differentiation within the intestinal epithelium [3, 48, 49].

The bone morphogenetic protein (BMP) pathway provides inhibitory cues that restrict ISC expansion and promote differentiation, acting as a counterbalance to Wnt activity. This interplay prevents uncontrolled proliferation and maintains epithelial homeostasis [3]. Epidermal growth factor (EGF) signaling, delivered by niche cells such as the visceral muscle, is crucial for maintaining ISC proliferative capacity, especially during regeneration after injury. EGF signals converge with stress pathways, such as JNK, to adjust ISC proliferation in response to environmental conditions [50].

Recent research has highlighted the importance of the Hippo/YAP pathway in epithelial repair. Activation of YAP, particularly in response to injury or metabolic cues, is necessary for effective regeneration. Bile acids, for example, can activate ISCs via the TGR5 receptor (also known as GPBAR1), leading to YAP activation and promoting epithelial renewal and repair after injury [51].

4.2 Additional Regulatory Layers

Metabolic and epigenetic regulation affects ISC identity and function, which are shaped by metabolic state and epigenetic landscapes. For example, the nascent polypeptide–associated complex (NAC) regulates mitochondrial metabolism, which is essential for maintaining stemness [52]. Transcription factors such as HNF4A and HNF4G control fatty acid oxidation, supporting ISC renewal [53]. Epigenetic and metabolic shifts allow ISCs to adapt to physiological needs or damage [52, 53]. Also, ISCs can integrate redox signals and stress responses to maintain homeostasis. The unfolded protein response (UPR) and redox regulators such as Nrf2 and Keap1 control ISC proliferation and protect against oxidative damage, with dysregulation leading to hyperproliferation or degeneration [54, 55].

5 Plasticity and Regeneration

The intestinal epithelium is renowned for its remarkable regenerative capacity, a feature largely underpinned by the plasticity of both ISCs and their differentiated progeny. Following injury from sources such as radiation, inflammation, or chemical insult, not only ISCs but also progenitor and even mature cells, can revert to a stem-like state through dedifferentiation. This process is orchestrated by injury-specific signals from the niche, which reprogram these cells to repopulate the ISC pool and restore epithelial integrity [2, 23, 56].

5.1 Dedifferentiation of Mature Cells

In response to severe epithelial injury, committed progenitor cells, including absorptive and secretory precursors, and even fully differentiated cells such as Paneth, tuft, and enteroendocrine cells, can dedifferentiate, re-acquiring stem-like properties and contributing to the regeneration of the intestinal epithelium [2, 57–59]. This reversion to a stem-like state is orchestrated by the reactivation of key stemness-associated transcription factors, particularly ASCL2, which plays a central role in re-establishing the ISC compartment. During regeneration, ASCL2 is re-expressed prior to LGR5, following a defined transcriptional trajectory that reflects progressive reprogramming toward stemness [57, 58]. This phenomenon of cellular plasticity is not exclusive to the intestine. Dedifferentiation as a regenerative strategy has also been observed in other organs, such as the liver, lung, and skin, highlighting its evolutionary conservation as a robust mechanism for tissue repair following damage [60, 61].

5.2 Paligenosis

In addition to dedifferentiation, paligenosis has emerged as a distinct cellular response through which mature intestinal cells regain proliferative capacity in response to injury. Unlike general dedifferentiation, paligenosis follows a stepwise, transcriptionally defined program that includes suppression of mature cell identity, activation of autophagy, and reactivation of genes associated with cell cycle re-entry [62]. In the intestinal epithelium, this process has been observed in Paneth cells and other secretory lineages, which can undergo paligenosis when canonical ISC populations (e.g., LGR5$^+$ cells) are depleted or damaged. These paligenotic cells do not immediately express classical stem cell markers like LGR5, but instead transiently re-express early stress and stemness-related factors, ultimately supporting epithelial regeneration [63].

5.3 Emergence of Revival Stem Cells

The identification of CLU⁺ revSCs has revealed a transient, damage-induced population that plays a crucial role in crypt restoration. These cells emerge specifically during regeneration and are capable of replenishing the ISC compartment when the primary stem cell pool is compromised [2, 23].

6 ISCs in Aging and Disease

Aging leads to profound changes in both ISCs and their surrounding niche, resulting in diminished regenerative capacity and increased vulnerability to disease. Aged ISCs display altered gene expression, reduced responsiveness to niche-derived signals, and impaired proliferative ability, which together contribute to compromised epithelial barrier function and heightened inflammation [64–68]. The niche itself is remodeled with age, including shifts in mesenchymal and immune cell populations and changes in the extracellular matrix, further reducing the tissue's ability to recover from damage [64, 66, 69].

6.1 Mechanisms of ISC Aging

Mitochondrial Dysfunction:

Aging ISCs exhibit impaired mitochondrial function, leading to decreased energy production, increased oxidative stress, and reduced stemness. Disruption of mitochondrial metabolism, such as through loss of Fut2-mediated fucosylation, accelerates ISC aging and impairs tissue homeostasis [67, 68].

Altered Adhesion and Clonal Dynamics:

Aged ISCs show reduced expression of adhesion molecules, resulting in weaker attachment to the matrix and accelerated clonal drift, which may compromise crypt stability and function [69].

Metabolic and Epigenetic Changes:

Age-associated metabolic shifts, such as changes in folate and asparagine metabolism, impact ISC proliferation and homeostasis. Interventions like asparagine supplementation or lipoic acid administration can restore ISC function and delay aging phenotypes by promoting autophagy and endocytosis [70, 71].

Over-Proliferation and Dysregulation:

In some models, aging leads to ISC over-proliferation and tissue dysplasia, driven by metabolic and signaling changes, such as increased expression of the white gene in Drosophila [70].

6.2 Disease Associations

Increased Disease Susceptibility:

The decline in ISC function with age is linked to increased intestinal permeability, chronic inflammation, and a higher risk of age-related intestinal diseases such as inflammatory bowel disease (IBD), colorectal cancer, diverticulosis, and microscopic colitis [64, 65, 67, 68].

Colorectal Cancer Initiation and Progression:

ISCs are implicated in the development of colorectal cancer. Oncogenic mutations, especially in the Wnt/β-catenin pathway, can transform ISCs or their progenitors into cancer stem-like cells, which drive tumor initiation, growth, and relapse [73, 74, 75]. These cancer stem cells share self-renewal and multipotency features with normal ISCs, making them resistant to therapy and responsible for disease recurrence [72, 73, 74].

The tumor microenvironment, including fibroblasts, immune cells, and the gut microbiota, further influences ISC behavior and cancer risk. Chronic inflammation, dysbiosis, and altered niche signals can disrupt ISC regulation, promoting malignant transformation [76, 77]. Additionally, factors like MUC1-C and metabolic changes enhance ISC "stemness" and support tumor progression [77, 78, 79].

7 Experimental Models and Techniques

The study of ISCs has advanced rapidly due to innovative experimental models and analytical techniques that allow for precise investigation of stem cell behavior and their interactions with the niche (Fig. 3).

7.1 Organoid Cultures

ISCs can be isolated and grown into 3D organoids that mimic the native intestinal epithelium, including crypt-villus architecture and multilineage differentiation. These organoids are used to study stem cell dynamics, disease modeling, drug responses, and regenerative processes. Organoids derived from $LGR5^{+}$ cells can self-organize and even fuse to form macroscopic tubes, further recapitulating in vivo structures [4, 39, 79, 80].

7.2 Genetic Lineage Tracing

In vivo genetic lineage tracing in animal models enables researchers to track the fate of individual cells over time. This technique has clarified the origins, plasticity, and regenerative potential of various intestinal cell types, revealing how differentiated cells can revert to a stem-like state during injury and repair [23].

7.3 Single-Cell RNA Sequencing

Single-cell RNA Sequencing (scRNAseq) allows researchers to profile gene expression at the single-cell level, revealing cellular heterogeneity, rare ISC subpopulations, and transitional states within the intestinal epithelium. This technology has enabled the construction of high-resolution cellular atlases, identification of novel cell types, and tracking of dynamic gene expression changes during homeostasis and regeneration. It is also instrumental in studying cell lineage, differentiation trajectories, and disease mechanisms [53, 81–84]. Computational methods for scRNAseq data include clustering,

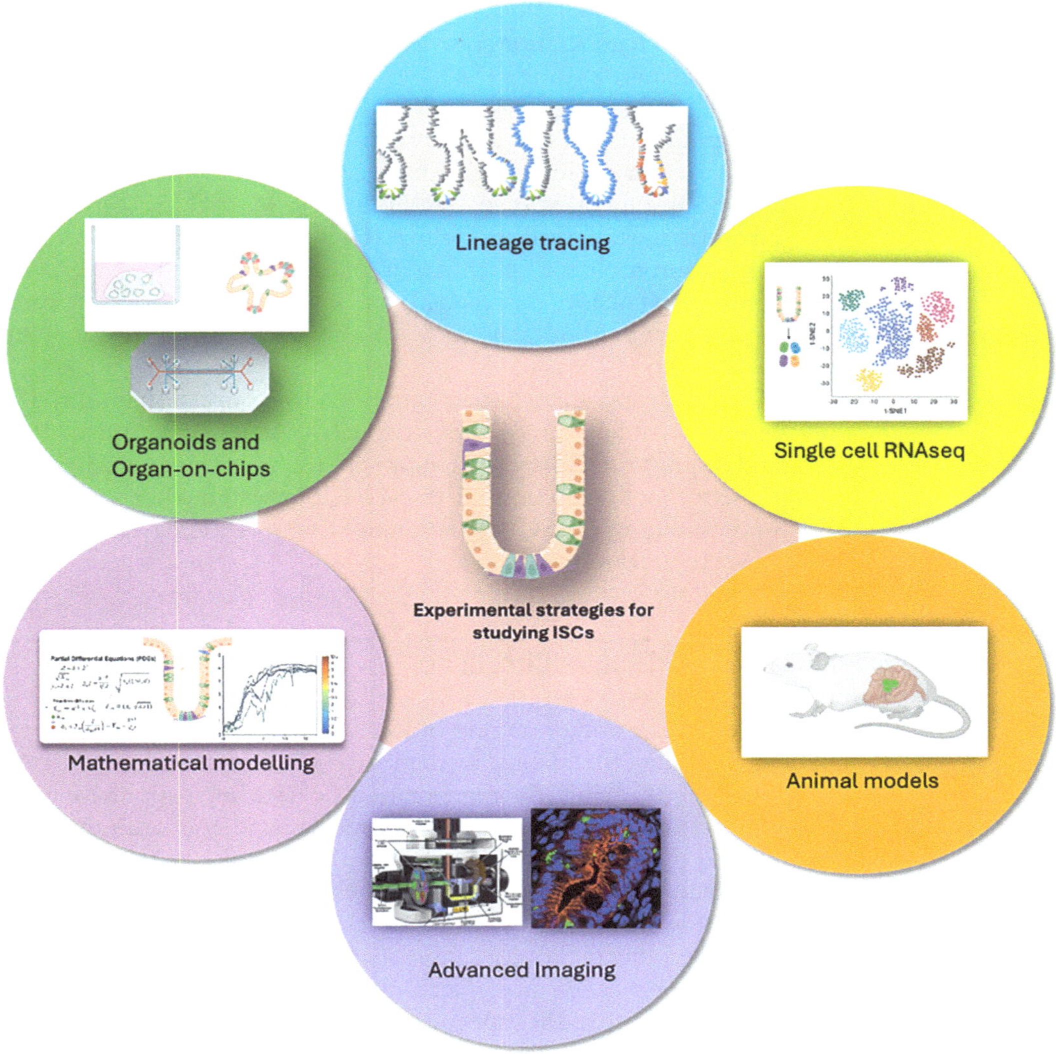

Fig. 3 Experimental strategies for studying intestinal stem cells (ISCs). This figure summarizes various cutting-edge approaches used to investigate the complex biology of ISCs. These strategies include in vitro models like organoids and organ-on-chips, advanced genetic techniques such as lineage tracing and single-cell RNA sequencing, in vivo animal models, sophisticated advanced imaging techniques, and computational approaches like mathematical modeling. Together, these methods provide a comprehensive toolkit for understanding ISC maintenance, differentiation, and their role in health and disease. Figure created with BioRender.com

normalization, dimensionality reduction, and trajectory inference, all of which are crucial for interpreting the vast and complex data generated [53, 83–86].

7.4 Advanced Imaging and Animal Models

High-resolution imaging techniques, such as confocal and live-cell microscopy, are used to visualize ISC behavior and niche interactions in real time. These approaches are often combined with genetically engineered animal models, particularly mice, to simulate disease, injury, and regeneration, providing platforms for preclinical testing and mechanistic studies [84].

7.5 Computational and Statistical Modeling

Computational models and advanced statistical designs are increasingly used to optimize experimental approaches and analyze scRNAseq and imaging data. Methods include machine learning for clustering, statistical modeling for differential expression, and simulation tools for benchmarking and hypothesis testing. These approaches enhance the reliability and interpretability of ISC research by addressing technical noise, batch effects, and biological variability [53, 82–86].

These experimental tools have provided critical insights into ISC biology, enabling the study of stem cell dynamics, niche interactions, and responses to injury or therapy with unprecedented precision. As these technologies continue to evolve, they will further deepen our understanding of ISC regulation and support the development of novel therapeutic strategies.

8 Therapeutic Implications and Future Directions

A deepening understanding of ISC biology and niche regulation is transforming therapeutic strategies for gastrointestinal diseases and cancer. Enhancing ISC regeneration or restoring supportive niche signals offers hope for mitigating tissue damage from chemotherapy, radiation, or chronic inflammatory diseases like IBD, potentially improving mucosal healing and patient outcomes [4]. Modulating key pathways such as Wnt and Notch, reactivating quiescent ISCs, and promoting dedifferentiation of mature cells are promising regenerative approaches under active investigation [87]. In colorectal cancer, targeting ISC-specific pathways may help prevent recurrence and overcome drug resistance by eradicating cancer stem-like cells [25]. The integration of patient-derived organoid systems into personalized medicine pipelines enables ex vivo drug testing and the development of tailored therapies, increasing the likelihood of effective, individualized treatment [4].

Overall, these advances have broad implications: they highlight the importance of ISC plasticity and the ability of differentiated cells to contribute to regeneration, especially when the primary stem cell pool is compromised. They also underscore the impact of aging and disease on ISC function and the niche, pointing to the need for interventions that restore or maintain stem cell health in older or diseased tissues [87]. Continued research into ISC heterogeneity, niche dynamics, and plasticity is expected to yield further breakthroughs in regenerative medicine and oncology, ultimately paving the way for more effective and personalized therapies for a range of intestinal disorders [4, 25, 87].

9 Conclusion

ISCs are fundamental for preserving the structure and function of the gut epithelium, driving its rapid renewal and enabling adaptation to physiological challenges and injury [5, 1, 4]. Their activity is orchestrated by a complex interplay of intrinsic genetic programs and extrinsic signals from the surrounding niche, which includes neighboring cells, the extracellular matrix, and the gut microbiota [5, 3, 6, 8, 88]. ISCs display remarkable heterogeneity and plasticity, allowing differentiated or quiescent cells to revert to a stem-like state during regeneration, a process that is especially important when the primary stem cell pool is compromised [2, 4]. Aging and disease can disrupt ISC function and niche integrity, leading to impaired regeneration, increased inflammation, and heightened risk of disorders such as IBD and colorectal cancer [5, 2, 4].

Technological advances, including organoid cultures, genetic lineage tracing, and single-cell RNA-seq, have revolutionized ISC research, providing unprecedented insight into stem cell dynamics, niche interactions, and cellular diversity. These tools have also accelerated translational research, supporting the development of personalized medicine approaches such as patient-derived organoid drug testing [4]. Therapeutically, strategies that enhance ISC regeneration, restore niche signals, or target cancer stem cell pathways are showing promise for improving outcomes in gastrointestinal diseases and cancer [5, 4]. The crosstalk between ISCs and the gut microbiota is emerging as a key area for future interventions, with the potential to modulate epithelial homeostasis and disease susceptibility [3, 6, 8, 88].

In summary, ongoing research into ISC biology, niche regulation, and their interactions with environmental factors is paving the way for innovative therapies in regenerative medicine and oncology. As our understanding deepens, the translation of these discoveries into clinical practice holds great promise for improving gut health and treating a wide range of intestinal disorders.

Acknowledgments

This work was supported by MRC AIM iCASE Grant Ref: MR/W007002/1.

References

1. Van der Flier LG, Clevers H (2009) Stem cells, self-renewal, and differentiation in the intestinal epithelium. Annu Rev Physiol 71:241–260
2. Kurokawa K, Hayakawa Y, Koike K (2020) Plasticity of intestinal epithelium: stem cell niches and regulatory signals. Int J Mol Sci 22:357
3. Ma N, Chen X, Johnston LJ, Ma X (2022) Gut microbiota–stem cell niche crosstalk: a new territory for maintaining intestinal homeostasis. iMeta 1:36

4. Beumer J, Clevers H (2020) Cell fate specification and differentiation in the adult mammalian intestine. Nat Rev Mol Cell Biol 22:39–53
5. Choi J, Augenlicht LH (2024) Intestinal stem cells: guardians of homeostasis in health and aging amid environmental challenges. Exp Mol Med 56:1–15
6. Ahmad Sophien AN, Jusop AS, Tye G et al. (2023) Intestinal stem cells and gut microbiota therapeutics: hype or hope?. Front Med 10:1223456
7. Cui Y, Chang JT (2016) The intestinal crypt, a prototype stem cell compartment. Cell 154:274–284
8. Luo H, Li M, Wang F et al. (2022) The role of intestinal stem cell within gut homeostasis: focusing on its interplay with gut microbiota and the regulating pathways. Int J Biol Sci 18:5185–5206
9. Wu H, Mu C, Xu L et al. (2024) Host-microbiota interaction in intestinal stem cell homeostasis. Gut Microbes 16:2353399
10. Yousefi M, Li L, Lengner C (2017) Hierarchy and plasticity in the intestinal stem cell compartment. Trends Cell Biol 27:753–764
11. Li N, Yousefi M, Nakauka-Ddamba A et al. (2014) Single-cell analysis of proxy reporter allele-marked epithelial cells establishes intestinal stem cell hierarchy. Stem Cell Reports 3:876–891
12. Ayyaz A, Kumar S, Sangiorgi B et al. (2019) Single-cell transcriptomes of the regenerating intestine reveal a revival stem cell. Nature 569:121–125
13. Smith N, Gallagher A, Wong M (2016) Defining a stem cell hierarchy in the intestine: markers, caveats and controversies. J Physiol 594:4861–4871
14. Bankaitis E, Ha A, Kuo C, Magness S (2018) Reserve stem cells in intestinal homeostasis and injury. Gastroenterology 155:1348–1361
15. Qin X, Rodriguez FC, Sufi J et al. (2023) An oncogenic phenoscape of colonic stem cell polarization. Cell 186:5554–5568
16. Böttcher A, Büttner M, Tritschler S et al. (2021) Non-canonical Wnt/PCP signalling regulates intestinal stem cell lineage priming towards enteroendocrine and Paneth cell fates. Nat Cell Biol 23:23–31
17. Snippert H, Van Es J, Van den Born M et al. (2009) Prominin-1/CD133 marks stem cells and early progenitors in mouse small intestine. Gastroenterology 136:2187–2194
18. Xie G, Peng Z, Liang J et al. (2022) Zinc finger protein 277 is an intestinal transit-amplifying cell marker and colon cancer oncogene. JCI Insight 7:150894
19. Geng H, Subramanian S, Bu H, Tan X (2023) The proliferation and migration features of Lgr5+ stem cells and transit amplifying cells in the intestine are dynamically changed during the postnatal development. Physiology 38:5731206
20. Zhang N, Yantiss R, Nam H et al. (2014) ID1 is a functional marker for intestinal stem and progenitor cells required for normal response to injury. Stem Cell Reports 3:716–724
21. Danan C, Naughton K, Hayer K et al. (2023) Intestinal transit-amplifying cells require METTL3 for growth factor signaling and cell survival. JCI Insight 8:171657
22. Angelis N, Baulies A, Kucharska A et al. (2023) ARID3A coordinates the proliferation-differentiation switch of transit-amplifying cells in the intestine. bioRxiv 25:559311
23. Sun X, Yang Q, Rogers CJ et al. (2023) Intestinal stem cell niche: the crosstalk between stem cells and their microenvironment. Front Cell Dev Biol 11:1122334
24. Takahashi T, Takase Y (2025) The intestinal stem cell niche: generation and utilization of intestinal organoids. Organoids 4:6
25. Sailaja B, He X, Li L (2016) The regulatory niche of intestinal stem cells. J Physiol 594:4827–4836
26. McCarthy N, Kraiczy J, Shivdasani R (2020) Cellular and molecular architecture of the intestinal stem cell niche. Nat Cell Biol 22:1033–1041
27. Meran L, Baulies A, Li V (2017) Intestinal stem cell niche: the extracellular matrix and cellular components. Stem Cells Int 2017:7970385
28. Stzepourginski I, Nigro G, Jacob J et al. (2017) CD34+ mesenchymal cells are a major component of the intestinal stem cells niche at homeostasis and after injury. Proc Natl Acad Sci U S A 114:E506–E513
29. Takashima S, Sharma R, Chang W et al. (2025) STAT1 regulates immune-mediated intestinal stem cell proliferation and epithelial regeneration. Nat Commun 16: 55227
30. Cox C, Storm E, Kapoor V et al. (2021) IL-1R1–dependent signalling coordinates epithelial regeneration in response to intestinal damage. Sci Immunol 6:abe8856
31. Lindemans C, Calafiore M, Mertelsmann A et al. (2015) Interleukin-22 promotes intestinal stem cell-mediated epithelial regeneration. Nature 528:560–564
32. Guenin-Macé L, Konieczny P, Naik S (2022) Immune-epithelial cross talk in regeneration and repair. Annu Rev Immunol 41:207–228
33. Palikuqi B, Rispal J, Klein O (2021) Good neighbors: the niche that fine tunes mammalian intestinal regeneration. Cold Spring Harb Perspect Biol 13:a040865
34. Baghdadi M, Ayyaz A, Coquenlorge S et al. (2021) Enteric glial cell heterogeneity regulates

intestinal stem cell niches. Cell Stem Cell 28:1592–1606
35. Baghdadi M, Kim T (2023) The multiple roles of enteric glial cells in intestinal homeostasis and regeneration. Semin Cell Dev Biol 136:81–88
36. Progatzky F, Shapiro M, Chng S et al. (2021) Regulation of intestinal immunity and tissue repair by enteric glia. Nature 599:125–130
37. Tomlin H, Piccinini A (2018) A complex interplay between the extracellular matrix and the innate immune response to microbial pathogens. Immunology 155:186–193
38. Romani P, Brian I, Santinon G et al. (2019) Extracellular matrix mechanical cues regulate lipid metabolism through Lipin-1 and SREBP. Nat Cell Biol 21:338–347
39. Duckworth C (2021) Identifying key regulators of the intestinal stem cell niche. Biochem Soc Trans 49:2163–2176
40. Xu C, Franklin B, Tang H et al. (2019) An in vivo RNAi screen uncovers the role of AdoR signalling and adenosine deaminase in controlling intestinal stem cell activity. Proc Natl Acad Sci U S A 117:464–471
41. Kuhre R, Holst J (2019) Mechanisms underlying gut hormone secretion using the isolated perfused rat small intestine. J Vis Exp 144:58533
42. Gribble F, Reimann F (2019) Function and mechanisms of enteroendocrine cells and gut hormones in metabolism. Nat Rev Endocrinol 15:226–237
43. Sun L, Li J, Nie Y (2020) Gut hormones in microbiota–gut–brain cross-talk. Chin Med J 133:826–833
44. Cho H, Lim J (2024) The emerging role of gut hormones. Mol Cells 47:100126
45. Zhao X, Qiu Y, Liang L, Fu X (2025) Interkingdom signaling between gastrointestinal hormones and the gut microbiome. Gut Microbes 17:2456592
46. Zhu P, Lu T, Wu J et al. (2022) Gut microbiota drives macrophage-dependent self-renewal of intestinal stem cells via niche enteric serotonergic neurons. Cell Res 32:555–569
47. Giolito M, Plateroti M (2022) Thyroid hormone signaling in the intestinal stem cells and their niche. Cell Mol Life Sci 79:4503
48. Lin G, Xu N, Xi R (2008) Paracrine Wingless signalling controls self-renewal of Drosophila intestinal stem cells. Nature 455:1119–1123
49. Tian A, Benchabane H, Wang Z, Ahmed Y (2016) Regulation of stem cell proliferation and cell fate specification by Wingless/Wnt signalling gradients enriched at adult intestinal compartment boundaries. PLoS Genet 12:e1005822
50. Biteau B, Jasper H (2011) EGF signalling regulates the proliferation of intestinal stem cells in Drosophila. Development 138:1045–1055
51. Sorrentino G, Perino A, Yildiz E et al. (2020) Bile acids signal via TGR5 to activate intestinal stem cells and epithelial regeneration. Gastroenterology 159:2101–2114
52. Ramalho S, Alkan F, Prekovic S et al. (2025) NAC regulates metabolism and cell fate in intestinal stem cells. Sci Adv 11:e9750
53. Chen L, Vasoya R, Toke N et al. (2019) HNF4 regulates fatty acid oxidation and is required for renewal of intestinal stem cells in mice. Gastroenterology 158:985–999
54. Wang L, Zeng X, Ryoo H, Jasper H (2014) Integration of UPRER and oxidative stress signalling in the control of intestinal stem cell proliferation. PLoS Genet 10:e1004568
55. Hochmuth C, Biteau B, Bohmann D, Jasper H (2011) Redox regulation by Keap1 and Nrf2 controls intestinal stem cell proliferation in Drosophila. Cell Stem Cell 8:188–199
56. Spit M, Koo BK, Maurice MM (2018) Tales from the crypt: intestinal niche signals in tissue renewal, plasticity and cancer. Open Biol 8:180120
57. Liu Y, Chen Y (2020) Intestinal epithelial plasticity and regeneration via cell dedifferentiation. Cell Regen 9:10
58. Murata K, Jadhav U, Madha S et al. (2020) Ascl2-dependent cell dedifferentiation drives regeneration of ablated intestinal stem cells. Cell Stem Cell 26:377–390
59. Goyal S, Guo C, Ranger A et al. (2024) Bacterial ADP-heptose initiates a revival stem cell program in the intestinal epithelium. bioRxiv 15:575749
60. Johansson B (2007) Regeneration and plasticity in the brain and spinal cord. J Cereb Blood Flow Metab 27:1417–1430
61. Galliot B, Ghila L (2010) Cell plasticity in homeostasis and regeneration. Mol Reprod Dev 77:837–855
62. Brown J, Cho C, Mills J (2021) Paligenosis: cellular remodelling during tissue repair. Annu Rev Physiol 83:365–389
63. Mei X, Gu M, Li M (2020) Plasticity of Paneth cells and their ability to regulate intestinal stem cells. Stem Cell Res Ther 11:349
64. Jasper H (2020) Intestinal stem cell aging: origins and interventions. Annu Rev Physiol 82:203–226
65. Park J, Sung M, Na H (2025) Drosophila model systems reveal intestinal stem cells as key players in aging. Ann N Y Acad Sci 1547:88–99
66. Rodriguez-Fernandez I, Tauc H, Jasper H (2020) Hallmarks of aging Drosophila intestinal stem cells. Mech Ageing Dev 190:111285
67. Duan C, Wang Z, Wu J et al. (2024) Fut2 deficiency promotes intestinal stem cell aging by damaging mitochondrial functions via down-regulating α1,2-fucosylation of Asah2 and Npc1. Research (Wash D C) 7:343

68. Quan H, Lu Y, Lin Y et al. (2025) Alternate day fasting enhances intestinal epithelial function during aging by regulating mitochondrial metabolism. Aging Cell 24:e70052
69. Hageb A, Thalheim T, Nattamai K et al. (2022) Reduced adhesion of aged intestinal stem cells contributes to an accelerated clonal drift. Life Sci Alliance 5:e202201408
70. Sasaki A, Nishimura T, Takano T et al. (2021) White regulates proliferative homeostasis of intestinal stem cells during ageing in Drosophila. Nat Metab 3:546–557
71. Luo T, Zhao L, Feng C et al. (2025) Asparagine prevents intestinal stem cell aging via the autophagy–lysosomal pathway. Aging Cell 24:e14423
72. Merlos-Suárez A, Barriga F, Jung P et al. (2011) The intestinal stem cell signature identifies colorectal cancer stem cells and predicts disease relapse. Cell Stem Cell 8: 511–524
73. Vermeulen L, Snippert H (2014) Stem cell dynamics in homeostasis and cancer of the intestine. Nat Rev Cancer 14:468–480
74. Shi G, Li Y, Shen H et al. (2024) Intestinal stem cells in intestinal homeostasis and colorectal tumourigenesis. Life Med 3:lnae042
75. Marzano M, Fosso B, Piancone E et al. (2021) Stem cell impairment at the host–microbiota interface in colorectal cancer. Cancers 13:996
76. Pani G (2023) Fusobacterium & Co. at the stem of cancer: microbe–cancer stem cell interactions in colorectal carcinogenesis. Cancers 15:2583
77. Li W, Zhang N, Jin C et al. (2020) MUC1-C drives stemness in progression of colitis to colorectal cancer. JCI Insight 5:137112
78. Sedlak J, Yilmaz Ö, Roper J (2022) Metabolism and colorectal cancer. Annu Rev Pathol 17:463–486
79. Morgan R, Mortensson E, Williams A (2018) Targeting LGR5 in colorectal cancer: therapeutic gold or too plastic?. Br J Cancer 118:1410–1418
80. Yin Y, De Jonge H, Wu X, Yin Y (2019) Mini-gut: a promising model for drug development. Drug Discov Today 24:178–184
81. Jovic D, Liang X, Zeng H et al. (2022) Single-cell RNA sequencing technologies and applications: a brief overview. Clin Transl Med 12: 694
82. Bacher R, Kendziorski C (2016) Design and computational analysis of single-cell RNA-sequencing experiments. Genome Biol 17: 63
83. Andrews T, Kiselev V, McCarthy D, Hemberg M (2020) Tutorial: guidelines for the computational analysis of single-cell RNA sequencing data. Nat Protoc 16:1–9
84. AlJanahi A, Danielsen M, Dunbar C (2018) An introduction to the analysis of single-cell RNA-sequencing data. Mol Ther Methods Clin Dev 10:189–196
85. Petegrosso R, Li Z, Kuang R (2019) Machine learning and statistical methods for clustering single-cell RNA-sequencing data. Brief Bioinform 21:1209–1223
86. Kharchenko P (2021) The triumphs and limitations of computational methods for scRNAseq. Nat Methods 18:723–732
87. Pentinmikko N, Katajisto P (2020) The role of stem cell niche in intestinal aging. Mech Ageing Dev 191:111330
88. Markandey M, Bajaj A, Ilott N et al. (2021) Gut microbiota: sculptors of the intestinal stem cell niche in health and inflammatory bowel disease. Gut Microbes 13:1990827

Check for updates

Chapter 2

In Silico Single-Cell Frame work for Modeling Intestinal Stem and Transit-Amplifying Progenitor Cells Dynamics

Brinda Balasubramanian

Abstract

Single-cell RNA sequencing (scRNA-seq) has revolutionized the ability to resolve cellular heterogeneity within complex tissues, enabling the identification of discrete cell states. Here, we present an in silico analytical pipeline designed to characterize intestinal stem cells (ISC), transit-amplifying (TA) progenitors, and BEST4^{+} enterocyte precursors from human scRNA-seq datasets, with a focus on inflammatory contexts such as inflammatory bowel disease (IBD). The pipeline integrates dataset acquisition, quality control, normalization, dimensionality reduction, unsupervised clustering, and cell type annotation using a reference cell atlas. We implemented iterative subsetting and re-clustering of ISC and TA compartments to identify inflammation-associated subpopulations and epithelial biomarkers. While demonstrated in the context of IBD, this computational framework is broadly applicable to other tissues and pathological conditions where stem/progenitor dynamics underpin disease progression and tissue repair.

Keywords Intestine, Pipeline, scRNA-seq, Stem cells, Progenitors, Inflammation

1 Introduction

The intestinal epithelium is a rapidly renewing tissue sustained by LGR5+ crypt base columnar intestinal stem cells (ISCs) that generate short-lived transit-amplifying (TA) progenitors and all differentiated lineages. Foundational lineage-tracing work established Lgr5 as a defining ISC marker and demonstrated the multipotency and self-renewal properties of crypt base columnar cells, providing a functional framework for stem-progenitor hierarchies [1].

The ex vivo derivation of long-term expanding "mini-gut" organoids from single LGR5^{+} ISCs cemented the concept of an ISC-centric epithelial renewal system and enabled mechanistic studies of niche signaling, including Wnt/RSPO, Notch, EGF, and BMP axes that jointly control proliferation and lineage allocation [2].

While bulk RNA-seq transcriptomics has provided valuable insights into intestinal biology and aberrations in diseases, it averages expression across diverse cell type populations and obscures cellular

Paloma Ordóñez-Morán (ed.), *Intestinal Stem Cells*, Methods in Molecular Biology, vol. 3059,
https://doi.org/10.1007/978-1-0716-5412-5_2,

heterogeneity. Single-cell RNA sequencing (scRNA-seq) overcomes this limitation, enabling resolution of stem/progenitor cell states, lineage hierarchies, and disease-specific programs.

Early atlases of the small intestine revealed unexpected diversity among enterocytes, enteroendocrine cells, and tuft cells, and identified transcriptional programs that adaptively change during infection [3].

Here, we present a reproducible in silico workflow to identify, annotate, and investigate SCs, TA, and BEST4^{+} enterocyte precursors from human scRNA-seq dataset in health and inflammation, as shown in a previous study (Balasubramanian et al.) [4]. The pipeline standardizes quality control, normalization, dimensionality reduction, and unsupervised clustering, and reference-based cell type annotation (SingleR) to ensure consistent labeling of epithelial subtypes using a curated epithelial database (Burclaff et al.) [5]. Finally, Iterative subsetting and re-clustering of stem/progenitor pools enables detection of proliferative, immune-enriched, and regenerative states. This computational framework is designed to be modular and extensible. Each step functions as an independent module that can be adapted or replaced without disrupting the entire pipeline.

This workflow is demonstrated using a subtype of IBD, ulcerative colitis (UC), the (SCP259) dataset contains epithelial cells from IBD non-inflamed (IBD normal) and inflamed tissues, and from healthy individual biopsies (UC = 18, healthy = 12 and CD = 46, healthy = 25) cohort [6]. Finally, we provide practical guidance on parameterization of clustering resolution and visualization strategies (UMAPs, marker heatmaps, dot plots, and volcano plots) to facilitate transparent reporting and reusability. In contexts where stem/progenitor dynamics underlie disease initiation, progression, or treatment response, from chronic inflammation to neoplasia, this single-cell computational approach provides a principled route to dissect cellular states, lineage transitions, and pathway drivers at high resolution in other tissues and diseases.

2 Materials

2.1 Software

1. R (v4.3+).
2. Seurat (v5.x) and SeuratObject packages for preprocessing, clustering, UMAP.
3. SingleR (Bioconductor) for reference-based annotation.
4. ggplot2, dplyr, tidyverse for plotting and data handling.
5. (*Optional*) EnhancedVolcano, pheatmap/ComplexHeatmap, viridis for figures.

2.2 Dataset

1. Discovery human UC scRNA-seq datasets
   Intra- and Inter-cellular Rewiring of the Human Colon during Ulcerative Colitis; download epithelial matrices and metadata from Single Cell Portal accession: SCP259.
2. Reference atlas
   A proximal-to-distal survey of healthy adult human small intestine and colon epithelium by single-cell transcriptomics; GSE185224 Gene Expression Ominibus (GEO) accession: GSE185224
3. Hardware
   a. Workstation or server with ≥16 GB RAM (Higher recommended for large cohorts).
   b. Optional GPU is not required (UMAP runs on CPU).
4. Validation dataset (*optional*).
   - CD—*Crohn's disease single-cell atlas spanning ileum and colon; download*; epithelial matrices and metadata Single Cell Portal accession: SCP1884.

3 Methods

3.1 Environment Setup

```
# R >= 4.3
library(Seurat# v5+
library(SingleR)
library(ggplot2)
library(dplyr)
set.seed(1234)
```

3.2 Load the Epithelial Reference Atlas for Standardized and Robust Cell Type Annotation (see Note 1)

```
# Load saved Seurat object for the epithelial
reference atlas (GSE185224)
ref <- readRDS("GSE185224/GSE185224.rds")
# Get raw counts and log-normalize (Seurat v5
'layer' semantics)
ref.counts <- GetAssayData(ref, layer = "counts")
ref.counts <- LogNormalize(ref.counts)
rownames(ref.counts) <- ref@assays[["RNA"]]@meta.
features[["feature_name"]]
ref.labels <- ref@meta.data$type # cell-type labels
from atlas
```

3.3 Load SCP259 Epithelial Matrix and Metadata

1. Read the Matrix Market file and create a Seurat object (*see* **Note 2**).
2. Attach study metadata (all.meta2.txt), which contains condition labels.

```
mtx <- "SCP259/Epi.matrix.mtx"
features <- "SCP259/Epi.genes.tsv" # gene symbols
column 1
cells <- "SCP259/Epi.barcodes2.tsv"
sce259 <- ReadMtx(mtx = mtx, features = features,
cells = cells, feature.column = 1)
obj259 <- CreateSeuratObject(counts = sce259,
project = "SCP259_Epi"
min.cells = 5, min.features = 200)
# Optional: add published metadata (condition/
health, sample IDs, etc.)
md.path <- "SCP259/all.meta2.txt"
md <- read.table(md.path, header = TRUE, sep =
"\t", stringsAsFactors = FALSE)
rownames(md) <- md$NAME; md$NAME <- NULL
obj259 <- AddMetaData(obj259, metadata = md)
```

3.4 Quality Control and Pre-Processing

1. Filter cells based on the following thresholds:
   a. Gene count per cell: exclude <200 or >6000 features.
   b. Mitochondrial gene content: exclude >5%–20% (*see* **Note 3**).
2. Normalize counts and select highly variable genes (HVGs), scale, and compute PCA and visualize via UMAP

```
obj259[["percent.mt"]] <-
PercentageFeatureSet(obj259, pattern = "^MT-")
obj259 <- subset(obj259, subset = nFeature_RNA >
200 & nFeature_RNA < 2500 & percent.mt < 5)
obj259 <- NormalizeData(obj259)
obj259 <- FindVariableFeatures(obj259, selection.
method = "vst", nfeatures = 2000)
obj259 <- ScaleData(obj259)
obj259 <- RunPCA(obj259)
obj259 <- FindNeighbors(obj259, dims = 1:20)
obj259 <- FindClusters(obj259)
obj259 <- RunUMAP(obj259, dims = 1:20)
```

3.5 Automated Epithelial Annotation with SingleR

```
# Extract counts from query (Seurat v5 layer
syntax)
q.counts <- GetAssayData(obj259, layer = "counts")
pred <- SingleR(test = q.counts, ref = ref.counts,
labels = ref.labels, de.method = "wilcox")
obj259$SingleR.labels <-
pred$labels[match(rownames(obj259@meta.data),
rownames(pred))]
```

(*see* **Note 4**).

3.6 Subset SC/TA/BEST4+ and Recluster

1. Subset by SingleR labels that correspond to stem/progenitor compartments.

2. Re-run normalization, HVGs, scaling, PCA, neighbor graph, UMAP, and test multiple clustering resolutions (0.1–2.0). For the purpose of this demo, we have used **1:20** for dims and resolutions around **1.2**. (Fig. 1) (*see* **Notes 5 and 6**).
3. Inspect UMAPs and canonical markers to validate stem/progenitor identities before downstream differential expression analyses using *FeaturePlot* (Fig. 2) (*see* **Note 7**).

```
stem_classes <- c("ISC", "BEST4+", "TA") # adapt
to label names based on the atlas used
# set Idents(obj259) <- obj259$SingleR.labels first
if needed
obj259_sub <- subset(obj259, idents = stem_classes)
obj259_sub <- NormalizeData(obj259_sub)
obj259_sub <- FindVariableFeatures(obj259_sub,
selection.method = "vst", nfeatures = 2000)
obj259_sub <- ScaleData(obj259_sub)
obj259_sub <- RunPCA(obj259_sub)
obj259_sub <- FindNeighbors(obj259_sub, dims =
1:20)
obj259_sub <- RunUMAP(obj259_sub, dims = 1:20)
obj259_sub <- FindClusters(obj259_sub, resolution
= 1.2) # adjust per dataset
FeaturePlot(obj259_sub, features =
c("LGR5","PCNA","BEST4"))
```

3.7 Identify Differential Gene Expression Within Inflammation State Using the Subset Data and Compare Inflamed vs Healthy Using the Health Variable from SCP259 Metadata

```
# Ensure the condition variable is present; this
metadata uses 'Health' with levels:
# Healthy / Non-inflamed / Inflamed (check exact
spelling in all.meta2.txt)
Idents(obj259_sub) <- obj259_sub$Health
# Inflamed vs Healthy within the subset
deg_inf_vs_healthy <- FindMarkers(obj259_sub,
ident.1 = "Inflamed", ident.2 = "Healthy"
test.use = "wilcox", only.pos = FALSE)
# Stringent subset (optional)
deg_inf_vs_healthy_sig <- deg_inf_vs_healthy %>%
dplyr::filter(p_val_adj < 0.05 & abs(avg_log2FC) >
1.5)
write.csv(deg_inf_vs_healthy, file = "SCP259_ISC_TA_
BEST4_inflamed_vs_healthy_DEGs.csv")
```

Visualize using volcano plot (Fig. 3):

```
library(EnhancedVolcano)
EnhancedVolcano(deg_inf_vs_healthy, lab =
rownames(deg_inf_vs_healthy)
x = "avg_log2FC", y = "p_val_adj"
pCutoff = 0.05, FCcutoff = 1.5)
# Heatmap of top markers(Figure 3)
topN <- 50
```

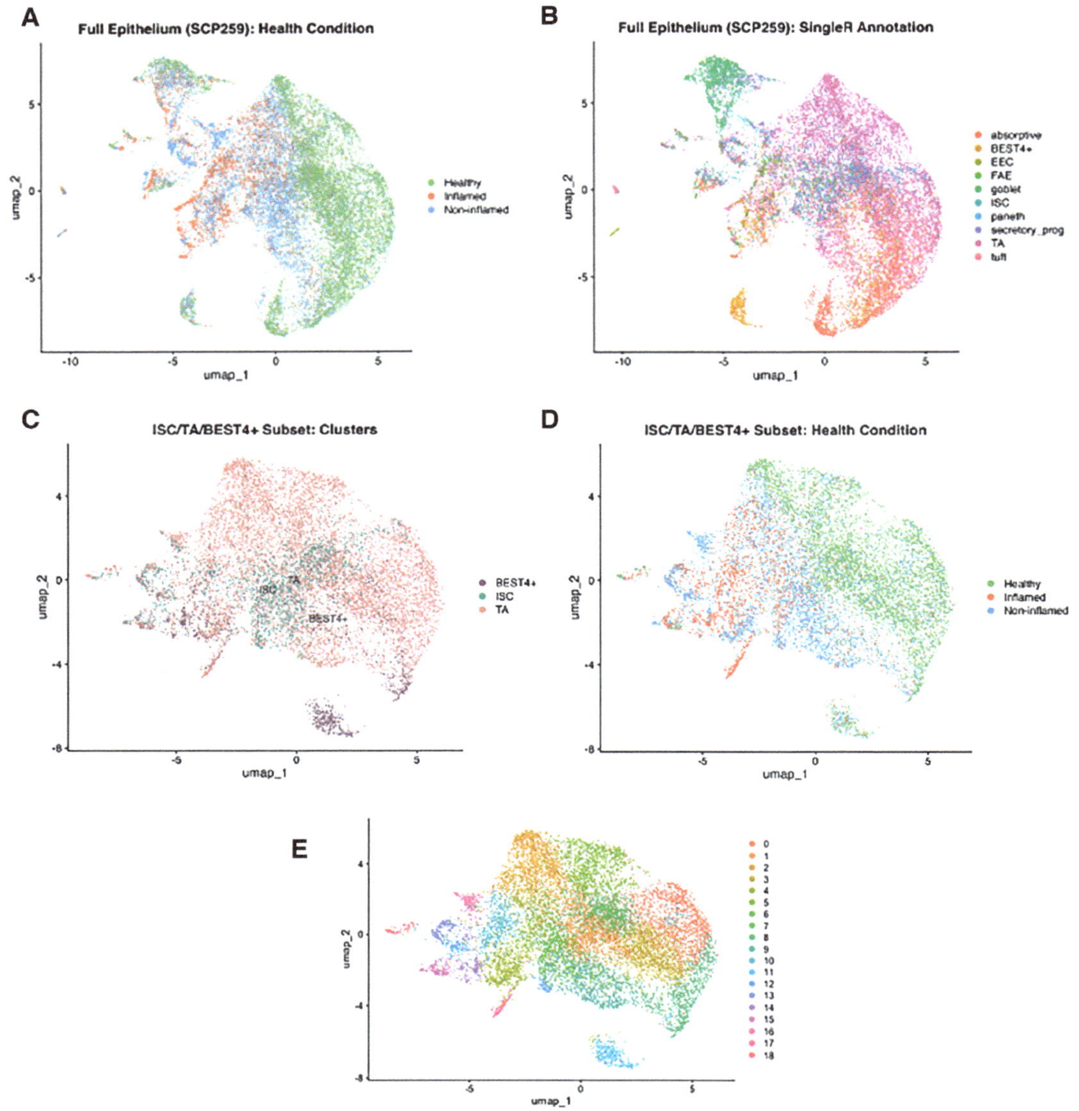

Fig. 1 UMAP of epithelium SCP259 annotated by reference-based labels. (**A**) Uniform Manifold Approximation and Projection (UMAP) embedding of epithelial cells from ulcerative colitis (UC) cohort (SCP259). (**B**) Cells are colored by SingleR annotation using the Burclaff et al. epithelial reference atlas (GSE185224). Major lineages include intestinal stem cells (ISC), transit-amplifying (TA) progenitors, absorptive enterocytes, goblet cells, and BEST4$^+$ enterocyte precursors. (**C**) UMAP of ISC, TA progenitors, and BEST4$^+$ enterocyte precursors extracted from SCP259. (**D**) Reclustering reveals proliferative and inflammation-enriched states within stem/progenitor pools. (**E**) Unique subpopulations of ISC, TA progenitors, and BEST4+ enterocyte precursors. (Figure reproduced with permission from [4])

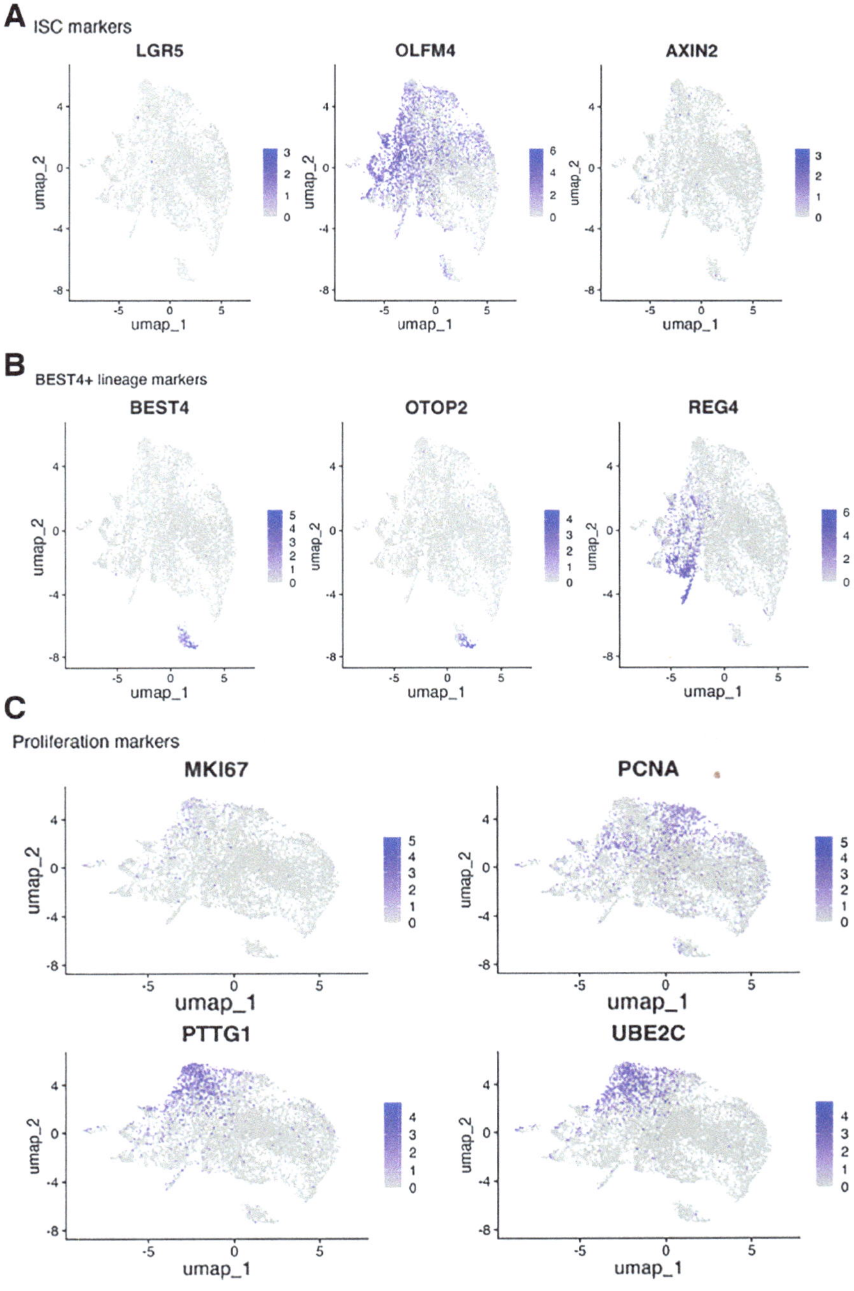

Fig. 2 Feature plots showing canonical expression of cell type markers in the reclustered subset. (**A**) ISC markers (LGR5, OLFM4, AXIN2). (**B**) BEST4$^+$ lineage markers (BEST4, OTOP2, REG4). (**C**) Cell cycle and proliferation markers (MKI67, PCNA, PTTG1, UBE2C) identify actively cycling progenitor clusters within ISC/TA compartments. (Figure reproduced with permission from [4])

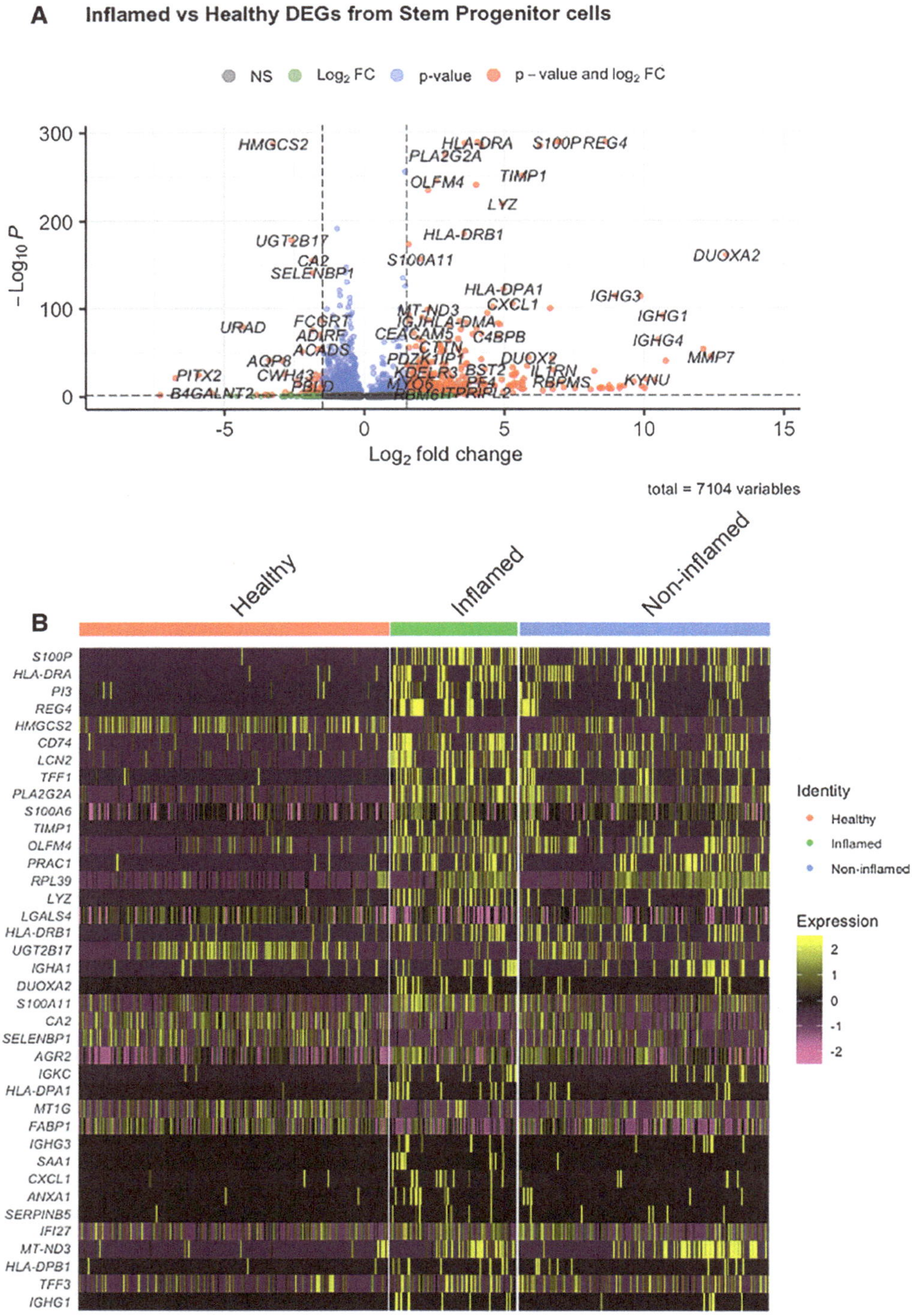

Fig. 3 Differentially expressed genes in the reclustered ISC/TA/BEST4+ subset (**A**) Volcano plot of ISC/TA/BEST4$^+$ subset comparing inflamed vs healthy conditions. Genes with $|\log_2 FC| > 1.5$ and adjusted $p < 0.05$ are highlighted. Top upregulated genes in inflamed progenitors include REG4, ANXA1, and TIMP1. (**B**) Heatmap of the top 50 genes distinguishing inflamed vs healthy ISC/progenitor cells. Rows represent genes; columns represent cells grouped by health status. Inflammation-associated clusters show strong expression of immune-modulatory and regenerative programs. (Figure reproduced with permission from [4])

```
top_markers <- deg_inf_vs_healthy %>% arrange(p_
val_adj) %>% head(topN) %>% rownames()
DoHeatmap(obj259_sub, features = top_markers,
group.by = "Health")
```

Final marker checks and visualization (*see* **Note 8**).

```
FeaturePlot(obj259_sub, features =
c("MKI67","PCNA","PTTG1","UBE2C"))
FeaturePlot(obj259_sub, features =
c("REG4","ANXA1","TIMP1","PI3"))
```

(*see* **Note 9**).

3.8 (Optional) Validate in SCP1884 (Crohn's Disease)

Repeat Sects. 3.3–3.7 for SCP1884 epithelial data, annotate with GSE185224, subset ISC/TA/BEST4$^+$, and compare signatures qualitatively (e.g., FeaturePlot of REG4/ANXA1/TIMP1/PI3).

- (Optional) Seurat provides an end-to-end scRNA-seq workflow, but this can be adapted for Scanpy to scale up (see **Note 10).**

4 Notes

1. Seurat v5 layers vs slots: The scripts use GetAssayData(…, layer="counts"). If you are an older Seurat, use slot = "counts" instead.
2. The feature file column: In ReadMtx, feature.column can be 1 or 2, depending on whether the gene table stores symbols or Ensembl IDs. Make sure to select the one suited for downstream analysis.
3. Mitochondrial content filter (percent.mt < 5) and nFeature bounds (200–2500) are conservative; adjust to the distribution per dataset (colon tissue often tolerates a slightly higher mitochondria percentage under inflammation).
4. SingleR label mapping: Ensure rownames(pred) align with cell barcodes of the Seurat object; the match() used in the script handles ordering. After labelling, visually verify with UMAP + known markers (LGR5/OLFM4 for ISCs; BEST4/OTOP2/REG4 for BEST4$^+$).
5. Clustering resolution: This script explored 0.1–2.0 and used ~1.2 for stem/progenitor granularity. If over-clustering (tiny clusters) occurs, lower the resolution; if under-clustering (heterogeneous clusters) occurs, increase the resolution or PCs.

6. Reclustering after subsetting increases sensitivity to proliferative vs inflammation-enriched progenitor states that may be diluted in full epithelium embeddings.
7. Inspect UMAP colored by SingleR labels and key markers (e.g., LGR5/OLFM4/AXIN2 for ISC; BEST4/OTOP2/REG4 for BEST4^{+} lineage) (Fig. 2)
8. Biological interpretation anchors:
   - ISCs: LGR5, OLFM4, AXIN2.
   - BEST4^{+} lineage: BEST4, OTOP2, often REG4 region-specific markers.
   - Proliferation: MKI67, PCNA, PTTG1, UBE2C (cell cycle programs).
9. Reproducibility: Save intermediate Seurat objects (saveRDS()), export DEG tables (CSV), and record sessionInfo().
10. Scaling up: For very large matrices, consider SCTransform and/or using **Scanpy** with **BBKNN**. The results are compatible with downstream SingleR or map-labels in Scanpy. Seurat provides an end-to-end scRNA-seq workflow, but this can be adapted for Scanpy.

Acknowledgments

Work supported by MRC grant MR/W01825X/1.

References

1. Barker N et al (2007) Identification of stem cells in small intestine and colon by marker gene Lgr5. Nature 449(7165):1003–1007
2. Sugimoto S, Sato T (2017) Establishment of 3D intestinal organoid cultures from intestinal stem cells. Methods Mol Biol 1612:97–105
3. Haber AL et al (2017) A single-cell survey of the small intestinal epithelium. Nature 551(7680):333–339
4. Balasubramanian B et al (2025) Intestinal stem and progenitor cells exhibit distinct adaptive responses to inflammatory stress in IBD. Stem Cell Res Ther 17(1):71
5. Burclaff J et al (2022) A proximal-to-distal survey of healthy adult human small intestine and colon epithelium by single-cell transcriptomics. Cell Mol Gastroenterol Hepatol 13(5):1554–1589
6. Smillie CS et al (2019) Intra- and inter-cellular rewiring of the human colon during ulcerative colitis. Cell 178(3):714–730.e22

Check for updates

Chapter 3

Using In Silico Pseudobulk Differential Expression and Trajectory Analysis of Single Cell RNA Sequencing to Reveal Stem Cell State Transitions in Colonic Inflammation

William Dalleywater

Abstract

Colonic inflammation induces profound alterations in the intestinal mucosa that are evident at both macroscopic and microscopic levels, including disruption of epithelial barrier integrity, crypt fission, and immune cell infiltration. Recent advances in single-cell and spatial RNA sequencing have greatly expanded our understanding of the cellular diversity of the colonic mucosa, enabling pathological features to be directly linked to underlying cellular and molecular mechanisms. This chapter demonstrates the application of single-cell RNA sequencing (scRNA-seq) analysis techniques to interrogate stem cell dynamics in the context of inflammatory bowel disease. Using publicly available datasets, the chapter provides a step-by-step workflow implemented in Python, covering data access, loading, integration of multiple datasets, and initial preprocessing. Cells are mapped to large single-cell atlas references to infer cell identities, followed by a pseudobulk analysis strategy to assess inflammation-associated changes in cell phenotypes. Finally, trajectory inference approaches are applied to explore potential mechanisms governing the specification and modulation of key cell types during inflammation. Accompanied by an online resource containing fully annotated scripts, this chapter offers guided instruction in contemporary scRNA-seq analysis workflows. It is intended as an accessible introduction for researchers seeking to develop practical skills in single-cell data analysis that can be readily applied to their own biological questions.

Keywords Bioinformatics, Single cell RNA-sequencing, Inflammation, Stem cells

1 Introduction

The colonic mucosa has a unique architecture, comprising epithelial cells arranged into crypts that allow niches to be established, which maintain the important and diverse functions of these cells. In broad terms, the base of the crypt contains intestinal epithelial stem cells (IESCs), and above these cells, there is a transit amplifying (TA) zone of uncommitted daughter cells from asymmetric stem cell division [1–3]. These TA cells are guided to their ultimate fate as they progress towards the crypt apex, specializing to one of many diverse colonic

Paloma Ordóñez-Morán (ed.), *Intestinal Stem Cells*, Methods in Molecular Biology, vol. 3059,
https://doi.org/10.1007/978-1-0716-5412-5_3,

Table 1
Recommended computing specifications

Component	Minimum specification
CPU	8 cores
RAM	32 GB
Storage	1 TB SSD
GPU	Optional, but all packages in the script below run using a CPU
Operating system	Linux

cell types, including colonocytes, goblet cells, neuroendocrine cells, tuft cells, and BEST4+ cells among others [4–6].

This system relies on a finely balanced, intricate network of signals, mainly orchestrated by surrounding stromal cell populations [7–9]. While the roles of Wnt and BMP signaling pathways are well-established—Wnt signals promoting an IESC phenotype and BMP favoring differentiation—there are many contributions from auxiliary signaling molecules—including pathway amplifiers (e.g., R-spondins) and inhibitors (e.g., GREM1, Noggin, sFRP1) and other signaling pathways, including Notch, TGFβ, Hippo, as well as effects of extracellular matrix and mechanotransduction [1, 3, 10–13].

These signaling networks break down during disease states such as neoplasia and inflammation [10, 14, 15]. The effect of inflammation is morphologically evident, as crypts become distorted by crypt fission, are depleted in certain cell types, such as goblet cells, and colonic crypts often inappropriately contain Paneth cells, which are usually found only in the small intestine and very proximal colon [6]. However, understanding what drives these changes at the molecular level has previously been challenging.

Single-cell RNA sequencing allows useful insights to be formed into underlying molecular cell biology, particularly in understanding transitions in cell states in cells with disease state annotations [4–6, 16]. In this chapter, trajectory modeling techniques will be demonstrated [17–19], showing how changes in cell states during inflammatory bowel disease, as a common exemplar of colonic inflammation, can be inferred. These powerful techniques have the potential to assist in supporting hypothesis generation for validating in silico findings using *in vitro* and *in vivo* models.

2 Materials

2.1 Single-Cell Datasets (for the Script Below—SCP1884 [20] and SCP259 [21] from the Single Cell Portal Were Used). See Note 1

2.1.1 Access to Computer or Server with Minimum Specification (See Table 1)

Linux (e.g., Ubuntu).

Conda implementation (e.g., Anaconda, Miniconda). *See* **Note 2**.

Python 3.2.

Jupyter Labs. *See* **Note 3**.

Python packages:

Scanpy 1.10.1 [17], Anndata 0.10.7, Umap 0.5.6, Numpy 1.26.4, Scipy 1.13.0, Pandas 2.2.2, Scikit-learn 1.4.2, statsmodels 0.14.2, igraph 0.11.4, pynndescent 0.5.12, bbknn [22], Celltypist [23], scFates [24], and decoupler [25].

3 Methods

3.1 Preparation of Dataset—Loading Matrix and Metadata

1. Install and import the packages into a Jupyter notebook (e.g., `import scanpy as sc`).
2. Load the dataframe in mtx format and show dimensions

```
adata = sc.read_mtx(data_path + 'CO_EPI.scp.
matrix.mtx')
adata.shape
Out: (28663,97788)
```

3. The dataframe rows (observations) and columns (variables) may be in the wrong orientation—gene names/ids need to be variables, while cell (barcode) names/ids need to be observations. In the example above, the gene names (28663) have been imported as observations and the barcode names (97788) as variables. To fix this, transpose the dataset:

```
adata = adata.transpose()
adata.shape
Out: (97788,28663)
```

4. The matrix market (.mtx) file contains no information about cell and gene names, thus these need to be imported separately

and merged. Calling gene_data can be used to check that the data are as expected.

```
gene_data = pd.read_csv(data_path + 'CO_EPI.scp.
features2.tsv')
epi_barcodes = pd.read_csv(data_path + 'CO_EPI.
scp.barcodes.tsv')
gene_data
gene_name
0AL627309.1
1AP006222.2
2RP4-669L17.10
```

5. First, merge the gene data ids with the data matrix, this is achieved by setting the imported gene names as an index for the variable column. This will also remove any data points in the matrix file with empty/null gene names and create unique pseudo-ids if necessary, where duplicate gene names are identified.

```
gene_data = gene_data[gene_data.gene_name.
notnull()]
notNa = gene_data.index
notNa = notNa.to_list()
adata = adata[:,notNa]
adata.var = gene_data
adata.var.set_index('gene_name', inplace=True)
adata.var.index.name = None
adata.var_names_make_unique()
```

6. Next, merge the barcodes. At the same time, cell metadata can also be imported into the dataset, and this should be prepared first. At this point, column names in the metadata can be relabeled as desired using *meta_data[new_name] = meta_data[old_name]*.

```
meta_data = pd.read_table(data_path + 'scp_
metadata_combined.v2.txt')
meta_data["barcodes"] = meta_data["NAME"]
meta_data['Health'] = meta_data['Type']
meta_data['disease'] =
meta_data['disease__ontology_label']
meta_data['Cluster'] = meta_data['Celltype']
meta_data['Location'] =
meta_data['organ__ontology_label']
meta_data['Sample'] = meta_data['biosample_id']
```

7. Unwanted metadata columns should also be removed. Note that the above script adds new columns replicating the old columns, thus the old columns should also be removed at the same time.

```
meta_data = meta_data.drop(columns=
["NAME","Site","Type","Celltype","organ__
ontology_label","biosample_id","n_
genes","n_counts","Chem","donor_
id","Layer","sex","species","species__
ontology_label","library_preparation_
protocol","library_preparation_protocol__ontology_
label","organ","disease","disease__ontology_
label"])
```

8. Next, add the barcode name as the index within both the barcode data file and the metadata file. The barcode columns can then be removed from both files.

```
epi_barcodes.index = epi_barcodes["barcodes"]
meta_data.index = meta_data["barcodes"]
epi_barcodes = epi_barcodes.drop(columns=
["barcodes"])
meta_data = meta_data.drop(columns=["barcodes"])
```

9. Next, merge the barcode data file and metadata files:

```
cell_meta = pd.concat([epi_barcodes, meta_data],
axis=1, join="inner")
```

10. Integrate the metadata file with the matrix:

```
adata.obs = cell_meta
adata.obs_names_make_unique()
```

11. Correct loading of the gene ids and cell ids can be shown as follows:

```
sc.get.var_df(adata)
Out:
AL627309.1
AP006222.2
RP4-669L17.10
sc.get.obs_df(adata)
Out:
N105446_L-ATTGTTCCAAACGTGG
N105446_L-TCGACGGGTGAGACCA
N105446_L-AGTAACCGTTAAGGGC
```

12. The complete dataset can be saved in anndata (h5ad) format for easy future loading:

```
adata.write_h5ad(data_path + 'SCP1884.epi.h5ad')
```

3.2 Merging and Annotating Datasets

1. Load the anndata file from the previously saved location

```
adata = sc.read_h5ad(data_path + 'SCP1884.epi.
h5ad')
```

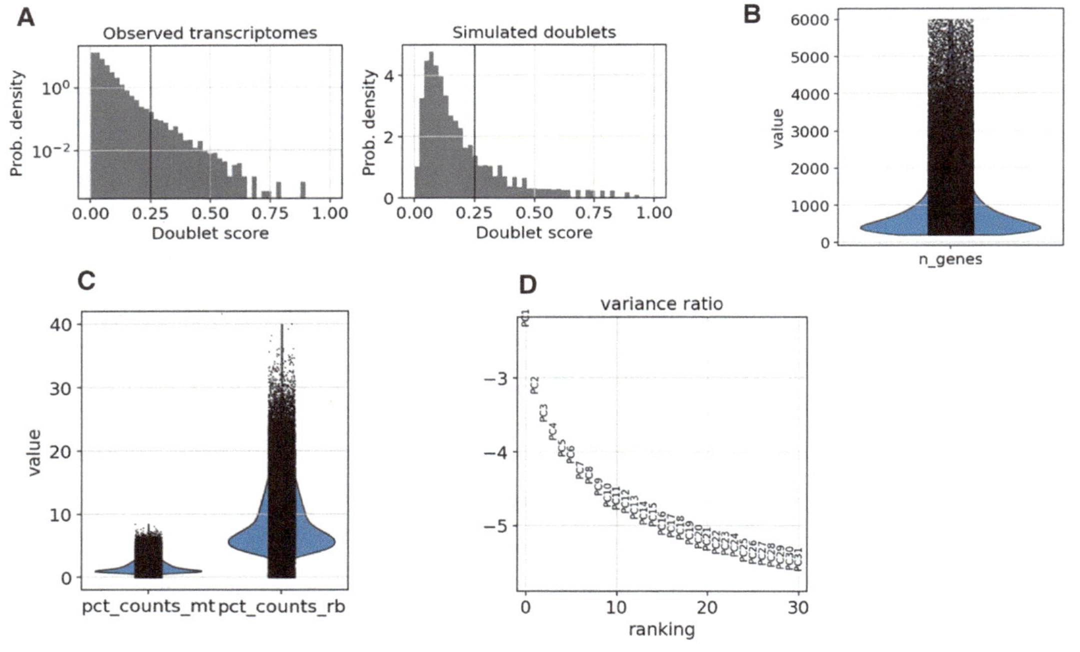

Fig. 1 An overview of QC metrics for the SCP1884 dataset. (**A**) Scrublet doublet prediction modeling, showing the likelihood of a cell being a doublet. (**B**) Violin plot showing distribution of gene counts among cells. (**C**) Violin plot showing mitochondrial (pct_counts_mt) and ribosomal (pct_counts_rb) among cells. (**D**) Elbow plot showing variance ratios of principal components (PC) identified through PCA

2. Run the dataset through Scrublet to identify likely doublet cells. The result can be visualized at the end (Fig. 1A)

```
import scrublet as scr
scrub = scr.Scrublet(adata.X,expected_doublet_
rate=0.061)
scrub.scrub_doublets(min_counts=2,min_cells=
3,min_gene_variability_pctl=85,n_prin_comps=30)
scrub.call_doublets(threshold=0.25)
scrub.plot_histogram()
adata.obs['doublet_scores'] = scrub.
doublet_scores_obs_
adata.obs['predicted_doublets'] = scrub.
predicted_doublets_
```

3. Perform QC steps to filter cells and genes based on the minimum number of genes expressed / cells containing genes:

```
sc.pp.filter_cells(adata, min_genes=200)
sc.pp.filter_genes(adata, min_cells=5)
```

4. Next, calculate mitochondrial and ribosomal mRNA reads per cell. The result can be visualized on violin plots (Fig. 1B, C).

```
adata.var['mt'] = adata.var_names.str.
startswith('MT-')
adata.var['rb'] = adata.var_names.str.
startswith(('RPS','RPL'))
sc.pp.calculate_qc_metrics(adata, qc_vars=['mt'],
percent_top=None, log1p=False, inplace=True)
sc.pp.calculate_qc_metrics(adata, qc_vars=['rb'],
percent_top=None, log1p=False, inplace=True)
sc.pl.violin(adata,['n_genes'])
sc.pl.violin(adata,['pct_counts_mt','pct_counts_
rb'])
```

5. Save the file after QC metrics have been calculated, but before any filtering steps are performed. The steps above should be repeated for the other dataset.

```
adata.write_h5ad(data_path + 'SCP1884.epi.
qc.h5ad')
```

6. Ideally, set up a new Jupyter notebook for the combined processing, importing all the required packages in the first step. Next, import both datasets into the notebook and use the concatenate function to merge the two datasets. If successful, the output should show an increased number of observations (cells) and a slightly reduced number of variables (gene count), as the procedure will keep only genes present in both datasets.

```
adata = sc.read_h5ad(data_path_1 + 'SCP1884.epi.
qc.h5ad')
adata2 = sc.read_h5ad(data_path_2 + 'SCP259.epi.
qc.h5ad')
adata.obs['dataset'] = "CD_SCP1884"
adata2.obs['dataset'] = "UC_SCP259"
adata_comb = adata.concatenate(adata2,batch_key=
'dataset')
adata.shape
Out:
(220351,18771)
```

7. Filter out low-quality reads based on the previously computed QC metrics:

```
adata = adata[adata.obs['n_genes'] < 6000,:]
adata = adata[adata.obs['pct_counts_mt'] < 20,:]
adata = adata[adata.obs['predicted_doublets'].
isin([False])]
```

8. Perform normalization 10,000 reads per cell and log transform the data. Calculate and select highly variable genes.

```
sc.pp.normalize_total(adata, target_sum = 1e4)
sc.pp.log1p(adata)
sc.pp.highly_variable_genes(adata, min_mean=
0.0125, max_mean=3, min_disp=0.5)
adata.var.highly_variable.value_counts()
sc.pl.highly_variable_genes(adata)
adata = adata[:, adata.var['highly_variable']]
```

9. Regress out the effect of total count, scale the data to a maximum gene expression of ten reads per gene and perform a principal component analysis (PCA). The final command produces an elbow plot (Fig. 1D)

```
sc.pp.scale(adata, max_value=10)
sc.tl.pca(adata, svd_solver='arpack')
sc.pl.pca_variance_ratio(adata, log=True)
```

10. Next, compute neighbors and project onto a uniform manifold approximation and projection (UMAP). Number of neighbors (n_neighbors) controls the local vs. global structure of the data representation (small n_neighbors preserves local cell communities while larger n_neighbors shows the bigger picture view)—this value should be balanced based on prior knowledge of how different the cell populations are and how much resolution in differentiating similar cells is required. The number of principal components (from the prior PCA) to be used (n_pcs) can also be adjusted based on the elbow plot.

```
sc.pp.neighbors(adata, n_neighbors=5, n_pcs=17,
knn=True)
sc.tl.umap(adata)
```

11. Run Leiden clustering to determine cell communities that are similar enough to be considered a specific cell type. The resolution parameter affects the stringency of cluster identification, a higher number will create more clusters.

```
sc.tl.leiden(adata, resolution=0.3)
```

12. Finally, the UMAP figures can be plotted. The color parameter can be used to highlight a specific cell annotation, such as leiden cluster, disease state, or original dataset. The results are shown in Fig. 2.

```
sc.pl.umap(adata, color=
['Sample','leiden','Health','dataset'],legend_loc=
'on data',size=40)
```

13. The UMAP for the datasets shows some segregation based on the source of the original dataset (i.e., a batch effect). This can be mitigated using batch-balanced k-nearest neighbors (bbknn). Here, the batch effect is being accounted for per subject (i.e., the donor) and based on cell type identity (Leiden cluster).

```
bbknn.ridge_regression(adata,batch_key=
['Subject'], confounder_key=['leiden'])
sc.tl.pca(adata)
bbknn.bbknn(adata,batch_key='Subject')
```

14. The UMAP can now be re-computed, accounting for the batch effect.

```
sc.pp.neighbors(adata, n_neighbors=5, n_pcs=17,
knn=True)
sc.tl.umap(adata)
sc.tl.leiden(adata, resolution=0.3)
```

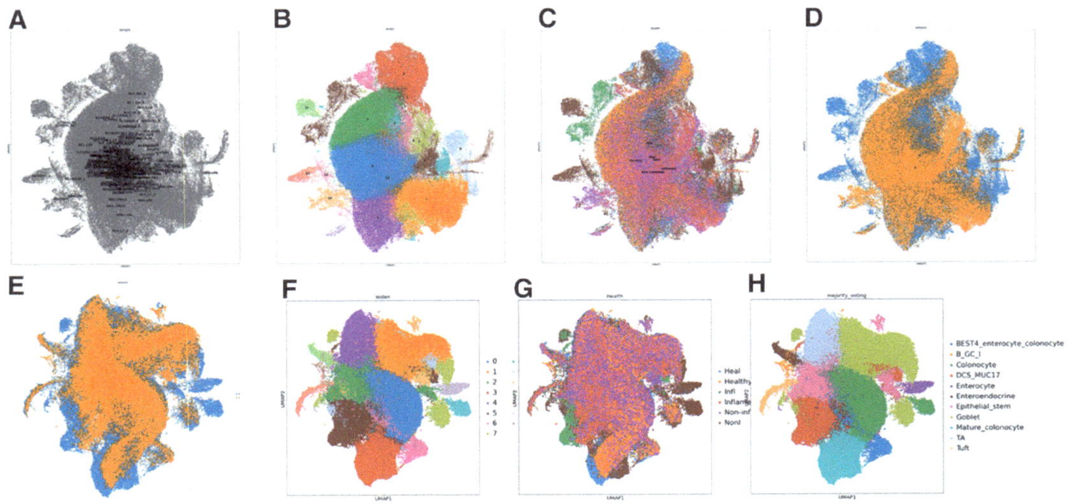

Fig. 2 Data structure of integrated datasets before and after batch correction. (**A**) UMAP showing individual sample labels, demonstrating some outlying samples. (**B**) Identified Leiden clusters in the uncorrected combined dataset. (**C**) Distribution of cell populations according to health status. (**D**) Distribution of cell populations according to dataset origin (prior to batch correction). (**E**) Distribution of cell populations after bbknn (batch correction algorithm). (**F**) Re-identification of Leiden clusters after bbknn, note the number of clusters has reduced from 16 to 8, showing better data integration. (**G**) Labeling of cell populations according to health status. (**H**) Automated annotation of cell populations using the CellTypist package

15. Re-plot the UMAP (Fig. 2E). Here, it can be seen that the dataset batch effect is reduced compared to the uncorrected data (Fig. 2D):

```
sc.pl.umap(adata, color=['Health','dataset'])
```

16. Next, CellTypist will be used to classify the cell types automatically. The custom package (full_gi.pkl - https://www.gutcellatlas.org/pangi.html) will be used for classification. Finally, plot the predictions on a UMAP (Fig. 2F–H) (leiden clusters [Fig. 2F] and health status [Fig. 2G] are also shown for comparison).

```
model = models.Model.load(model='full_gi.pkl')
predictions = celltypist.
annotate(adata,model,majority_voting=True,over_
clustering='leiden')
predictions.predicted_labels
adata_p = predictions.to_adata()
sc.pl.umap(adata,color=
['leiden','Health','majority_voting'],size=
50,legend_fontsize=20)
```

17. Some automated cell labels may not be correct and can be checked using the following commands that find marker genes. The results can be seen in Fig. 3A.

```
sc.tl.rank_genes_groups(adata, method="wilcoxon",
groupby="cell_label")
sc.pl.rank_genes_groups_dotplot(adata, swap_axes=
True, dendrogram=False, n_genes=5,standard_scale=
'var')
```

18. Incorrect cell labels can be corrected using the following command. In the example dataset, a few categories were merged (e.g., Goblet_1, Goblet_2, Goblet_3) or modified (Stem_2 to Paneth):

```
adata.obs['cell_label'] = adata.obs['cell_label'].
replace('old_category,'new_category')
```

19. The cell types according to the health status can be plotted using the crosstab function (Fig. 3B):

```
tmp = pd.crosstab(adata.obs['cell_label'],adata.
obs['Health'], normalize='columns').T.plot(kind=
'bar', stacked=True, figsize=(10,10))
tmp.legend(title='CellType vs Health', bbox_to_
anchor=(1.5, 1.02),loc='upper right')
```

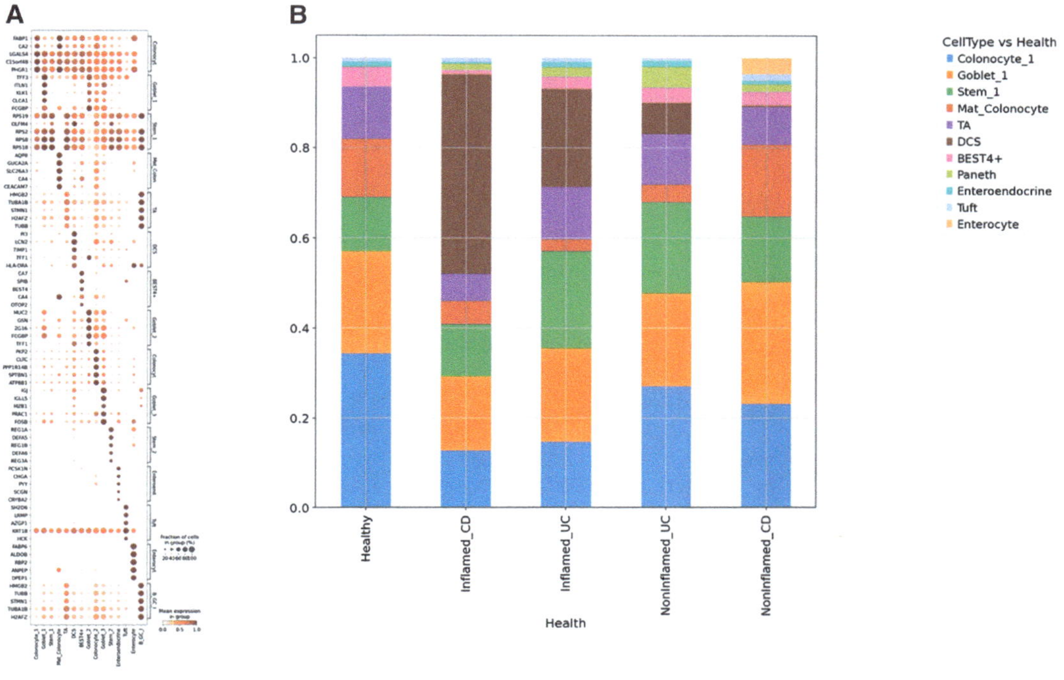

Fig. 3 Evaluation of cell annotation. (**A**) Automated top 5 cell marker identification for each of the CellTypist-annotated cell populations. This allows manual determination of the quality of annotation. For example, Goblet_2 cells have well-described MUC2 and TFF1 among their marker genes, suggesting accurate annotation. On the other hand, Stem_2 includes DEFA5/6, which are regarded as Paneth cell markers, required correction to this annotation. (**B**) A crosstabulation barplot can be created to show the distribution of cell populations according to a known annotation, e.g., health status

20. The steps so far can be saved:

```
adata.write_h5ad(data_path + 'combined.epi.anno.
h5ad')
```

3.3 Performing "Pseudobulk" Cell State Differential Expression Analysis

In the example below, shifts in the deep crypt secretory cell (DCS) population will be explored in inflamed and non-inflamed conditions. To do so, cells in each condition will be grouped together and treated as a pseudobulk dataset to perform differential expression analysis using tools familiar to bulk RNA sequencing, for example, DeSeq2.

1. In a new notebook, import the standard packages. Also import the decoupler package:

```
import decoupler as dc
```

2. Load the saved dataset from above.
3. Prepare the dataset for loading into the decoupler:

```
adata.X = np.round(adata.X)
adata.X = np.absolute(adata.X)
adata.layers['counts'] = adata.X
```

4. Load the dataset into the decoupler:

```
pdata = dc.get_pseudobulk(
adata
sample_col='Sample'
groups_col='cell_label'
layer='counts'
mode='sum'
min_cells=0
min_counts=0)
```

5. Next, performing normalization, scaling, and PCA:

```
pdata.layers['counts'] = pdata.X.copy()
sc.pp.normalize_total(pdata, target_sum=1e4)
sc.pp.log1p(pdata)
sc.pp.scale(pdata,max_value=10)
sc.tl.pca(pdata)
dc.swap_layer(pdata, 'counts', X_layer_key=None,
inplace=True)
```

6. Prepare the dataset for analysis by selecting the metadata associations with principal components that need to be examined:

```
dc.get_metadata_associations(
pdata
obs_keys = ['Health', 'cell_label', 'psbulk_n_
cells', 'psbulk_counts']
obsm_key='X_pca'
uns_key='pca_anova'
inplace=True)
```

7. Here, the DCS cell type will be explored. Select this using the following command:

```
dcs = pdata[pdata.obs['cell_label'] == 'DCS'].
copy()
```

8. Optional: to increase the statistical stringency of the analysis, genes can be filtered according to their total number of counts. This can be optimized first by plotting (Fig. 4A) with the intended parameters before filtering.

```
dc.plot_filter_by_expr(dcs, group='Health', min_
count=1, min_total_count=5)
genes = dc.filter_by_expr(dcs, group='Health',
min_count=1, min_total_count=5)
dcs = dcs[:, genes].copy()
dcs.X = np.round(dcs.X)
```

9. Next, differential expression analysis will be performed. To do so, DeSeq tools will be used. In this example, non-inflamed

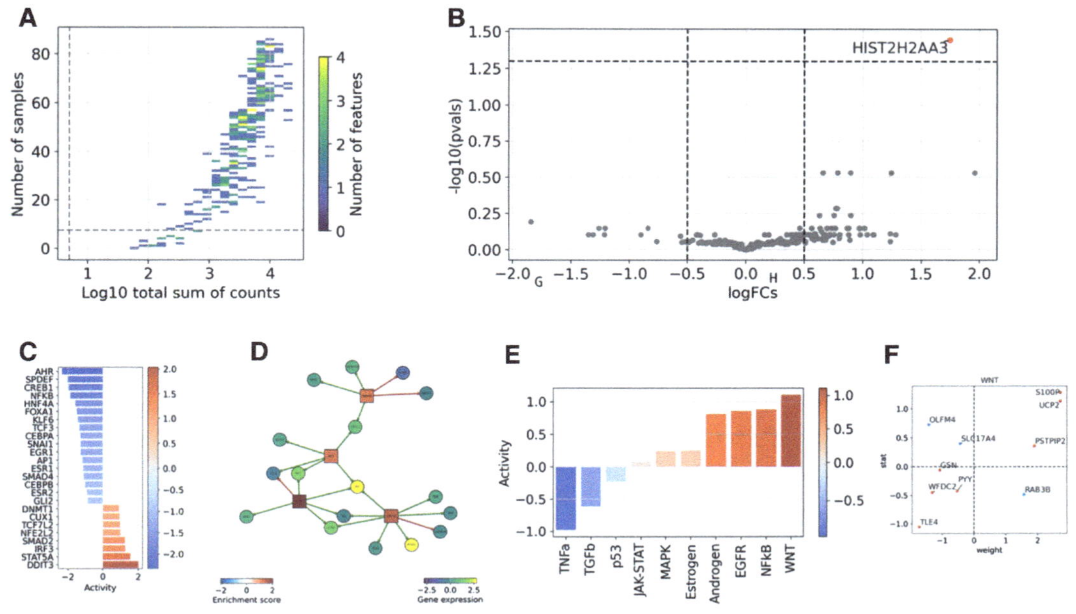

Fig. 4 Differential expression analysis of the DCS population in non-inflamed and inflamed conditions. (**A**) Expression filtering plot used to evaluate the effect of applying count thresholds. (**B**) Volcano plot showing differential expression of genes activated in DCS cells in inflammation vs. non-inflamed colon. (**C**) Differential transcription factor activity plot. (**D**) Network plot formed from the transcription factors with the highest activity (DDIT3, IRF3, SMAD2, and STAT5A). (**E**) Differential pathway activity plot. (**F**) Scatter plot showing contributions from genes in the WNT signaling pathway to the weighting in differential pathway activity

samples will be compared with inflamed. All non-inflamed (either Crohn's disease or ulcerative colitis) will be compared with all inflamed. First, these categories need to be combined, then the DeSeq object created.

```
dcs.obs['Health_Any'] = dcs.obs['Health']
dcs.obs['Health_Any'] = dcs.obs['Health_Any'].
replace('Inflamed_CD','Inflamed')
dcs.obs['Health_Any'] = dcs.obs['Health_Any'].
replace('Inflamed_UC','Inflamed')
dcs.obs['Health_Any'] = dcs.obs['Health_Any'].
replace('NonInflamed_CD','NonInflamed')
dcs.obs['Health_Any'] = dcs.obs['Health_Any'].
replace('NonInflamed_UC','NonInflamed')
from pydeseq2.dds import DeseqDataSet,
DefaultInference
from pydeseq2.ds import DeseqStats
inference = DefaultInference(n_cpus=8)
dds = DeseqDataSet(
adata=dcs
design_factors='Health_Any'
ref_level=['Inflamed', 'NonInflamed']
refit_cooks=True
```

```
inference=inference)
dds.deseq2()
stat_res = DeseqStats(
dds
contrast=["Health_Any", 'Inflamed','NonInflamed']
inference=inference)
results_df = stat_res.results_df
```

10. A volcano plot can now be made to evaluate genes that are differentially expressed in DCS cells in the non-inflamed and inflamed conditions (Fig. 4B).

```
dc.plot_volcano_df(
results_df
x='log2FoldChange'
y='padj'
top=20
figsize=(8, 4))
```

11. Transcription factor activity in the two conditions can now be evaluated. Visualize the activity in a barplot (Fig. 4C).

```
mat = results_df[['stat']].T.rename(index=
{'stat': 'DCS'})
net = dc.get_collectri(organism='human', split_
complexes=False)
tf_acts, tf_pvals = dc.run_ulm(mat=mat, net=net)
dc.plot_barplot(
acts=tf_acts
contrast='DCS'
top=25
vertical=True
figsize=(3, 6))
```

12. A network plot (Fig. 4D) can be generated using transcription factors selected from the barplot, to visualize connections in regulated genes between the various factors.

```
dc.plot_network(
net=net
obs=mat
act=tf_acts
n_sources=['DDIT3', 'STAT5A', 'IRF3', 'SMAD2']
n_targets=25
node_size=100
figsize=(7, 7)
c_pos_w='darkgreen'
c_neg_w='darkred'
vcenter=True)
```

13. Pathway analysis can next be performed and results visualized as a barplot (Fig. 4E).

```
progeny = dc.get_progeny(top=500)
pathway_acts, pathway_pvals = dc.run_mlm(mat=mat,
net=progeny)
dc.plot_barplot(
acts=pathway_acts
contrast='DCS'
top=25
vertical=False
figsize=(6, 3))
```

14. The effect of individual genes influencing the pathway weightings (e.g., WNT) can be assessed (Fig. 4F):

```
dc.plot_targets(
data=results_df
stat='stat'
source_name='WNT'
net=progeny
top=15)
```

3.4 Performing Cell State Trajectory Analysis

In the final section, the analysis of cell differential expression will be extended to explore the transition between stem cells and DCS cells.

1. Work in a new notebook, load the combined dataset, and import standard packages. Import scFates needed for this workflow using the command below.

```
import scFates as scf
```

2. Define the cell lineages of interest by selecting them from the main dataset:

```
lineages = adata[adata.obs['cell_label'].
isin(['Stem_1','DCS'])]
stem_dcs = lineages
```

3. Allow scFates to try to identify a trajectory through the data and then plot the results (Fig. 5A):

```
scf.tl.curve(stem_dcs,Nodes=30,use_rep="X_
pca",ndims_rep=2)
scf.pl.graph(stem_dcs,basis="pca")
```

4. Marker gene labels can be used to identify the cell types of interest and assess suitability to act as a root cell marker. In the example below, stem cell markers *EPHB2* (Fig. 5B) and *STMN1* (Fig. 5C) were evaluated:

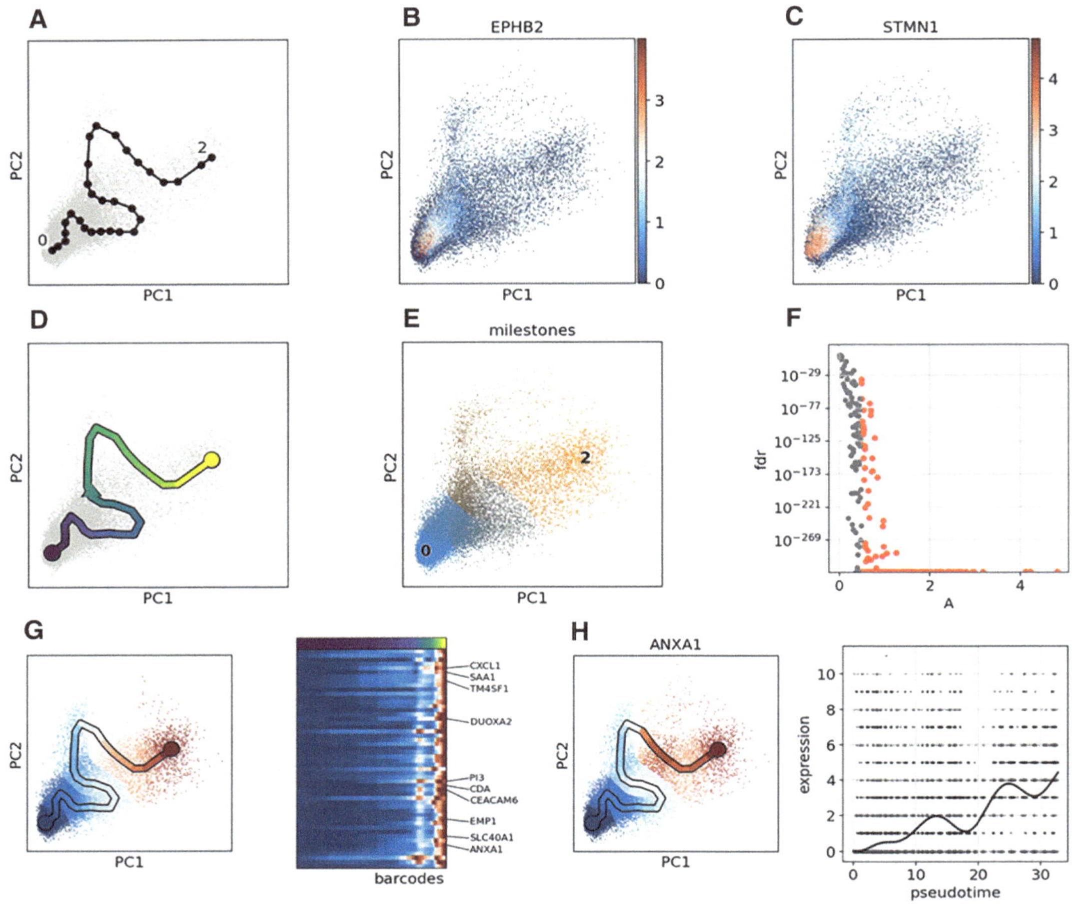

Fig. 5 Trajectory analysis of progression from Stem to DCS state using scFates. (**A**) Trajectory super-imposed over a PCA plot showing the path of cells from the progenitor (Stem) state at 0 to progeny (DCS) state at 2. (**B**/**C**) Expression of known stem cell marker genes (EPHB2/STMN1) is shown on the PCA plot to evaluate the utility to mark the root/progenitor state for further analysis. (**D**) After setting STMN1 as the progenitor marker gene, pseudotime is estimated along the trajectory. (**E**) Cells falling within each milestone (blue: progenitor/ orange: progeny) are shown on the PCA plot. (**F**) Plot to show genes significantly associated with the trajectory. (**G**) Plot showing the late part of the trajectory with a heatmap showing expression of genes that may be controlling or associated with the cell fate decision. (**H**) Projection of the trend of one gene (ANXA1) associated with the late part of the trajectory demonstrated an upward trend over pseudotime

```
sc.pl.pca(stem_dcs,color="EPHB2",cmap="RdBu_r")
sc.pl.pca(stem_dcs,color="STMN1",cmap="RdBu_r")
```

5. Choose *STMN1* as the root marker:

```
scf.tl.root(stem_dcs,"STMN1")
```

6. Perform pseudotime analysis and plot trajectories (Fig. 5D) and milestones (Fig. 5E).

```
scf.tl.pseudotime(stem_dcs,n_jobs=20,n_map=
100,seed=42)
scf.pl.trajectory(stem_dcs,basis="pca",arrows=
True,arrow_offset=3)
scf.pl.milestones(stem_dcs,basis="pca",annotate=
True)
```

7. Analyze which cells are progenitors, progenies, and bridge states.

```
scf.tl.linearity_deviation(stem_dcs,
 start_milestone="0",
 end_milestone="2",

n_jobs=20,plot=True,basis="pca")
```

8. Next genes (features) associated with the trajectory will be analyzed in the following lines. Significantly associated features will be shown on the association plot (Fig. 5F). Here, 194 significant features were identified. These features will then be fitted to the trajectory, and cells will be clustered based on their positions on the trajectory.

```
scf.tl.test_association(stem_dcs,n_jobs=20)
scf.tl.test_association(stem_dcs,reapply_filters=
True,A_cut=.5)
scf.pl.test_association(stem_dcs)
scf.tl.fit(stem_dcs,n_jobs=20)
scf.tl.cluster(stem_dcs,n_neighbors=50,metric=
"correlation")
```

9. The clusters can be inspected to determine their position along the trajectory. To view all the clusters, run the following command. An example of one of the cluster diagrams is shown in Fig. 5G.

```
for c in stem_dcs.var["clusters"].unique():
scf.pl.trends(stem_dcs,features=stem_dcs.
var_names[stem_dcs.var.clusters==c],basis=
"pca",ordering="quantile",return_genes=True)
```

10. A single gene trend within a cluster can be visualized using the next command. This can be useful for demonstrating changing gene expression in pseudotime. A plot for ANXA1 (a gene marking the end of the trajectory / DCS cells) is shown in Fig. 5H.

```
scf.pl.single_trend(stem_dcs,"ANXA1",basis=
"pca",color_exp="k")
```

4 Notes

1. Datasets can be downloaded from the Broad Institute Single Cell Portal. A free account is required to do so. Ensure that the following files are downloaded: barcode file (.tsv), gene file (.tsv), matrix file (.mtx), and any metadata (.txt).
2. Setting up Python and Conda environments is beyond the scope of this book chapter, but numerous tutorials are available online if needed.
3. Use of Jupyter Labs is not essential—the code can be written in a script file or direct into the command line, but Jupyter Labs notebooks are a convenient format for containing the code and text and graphical outputs.

References

1. Barker N (2014) Adult intestinal stem cells: critical drivers of epithelial homeostasis and regeneration. Nat Rev Mol Cell Biol 15(1):19–33
2. Gehart H, Clevers H (2019) Tales from the crypt: new insights into intestinal stem cells. Nat Rev Gastroenterol Hepatol 16(1):19–34
3. Meran L, Baulies A, Li VSW (2017) Intestinal stem cell niche: the extracellular matrix and cellular components. Stem Cells Int 2017:7970385
4. Elmentaite R, Ross ADB, Roberts K, James KR, Ortmann D, Gomes T et al (2020) Single-cell sequencing of developing human gut reveals transcriptional links to childhood Crohn's disease. Dev Cell 55(6):771–783.e5
5. Elmentaite R, Kumasaka N, Roberts K, Fleming A, Dann E, King HW et al (2021) Cells of the human intestinal tract mapped across space and time. Nature 597(7875):250–255
6. Oliver AJ, Huang N, Bartolome-Casado R, Li R, Koplev S, Nilsen HR et al (2024) Single-cell integration reveals metaplasia in inflammatory gut diseases. Nature 635(8039):699–707
7. Brügger MD, Valenta T, Fazilaty H, Hausmann G, Basler K (2020) Distinct populations of crypt-associated fibroblasts act as signaling hubs to control colon homeostasis. PLOS Biol 18(12):e3001032
8. Dutton LR, Hoare OP, McCorry AMB, Redmond KL, Adam NE, Canamara S et al (2019) Fibroblast-derived Gremlin1 localises to epithelial cells at the base of the intestinal crypt. Oncotarget 10(45):4630
9. Fawkner-Corbett D, Antanaviciute A, Parikh K, Jagielowicz M, Gerós AS, Gupta T et al (2021) Spatiotemporal analysis of human intestinal development at single-cell resolution. Cell. 184(3):810–826.e23
10. Piersma B, Bank RA, Boersema M (2015) Signaling in fibrosis: TGF-β, WNT, and YAP/TAZ converge. Front Med 2:59
11. Blache P, van de Wetering M, Duluc I, Domon C, Berta P, Freund J-N et al (2004) SOX9 is an intestine crypt transcription factor, is regulated by the Wnt pathway, and represses the CDX2 and MUC2 genes. J Cell Biol 166(1):37–47
12. Jho E-h, Zhang T, Domon C, Joo C-K, Freund J-N, Costantini F (2002) Wnt/β-Catenin/Tcf signaling induces the transcription of Axin2, a negative regulator of the signaling pathway. Mol Cell Biol 22(4):1172–1183
13. Orzechowska-Licari EJ, Bialkowska AB, Yang VW (2023) Sonic hedgehog and WNT signaling regulate a positive feedback loop between intestinal epithelial and stromal cells to promote epithelial regeneration. Cell Mol Gastroenterol Hepatol 16(4):607–642
14. He S, Lei P, Kang W, Cheung P, Xu T, Mana M et al (2023) Stiffness restricts the stemness of the intestinal stem cells and skews their differentiation toward goblet cells. Gastroenterology 164(7):1137–1151.e15
15. Zidar N, Langner C, Jerala M, Boštjančič E, Drobne D, Tomažič A. Pathology of fibrosis in Crohn's disease—contribution to understanding its pathogenesis. Front Med 7:167
16. Kinchen J, Chen HH, Parikh K, Antanaviciute A, Jagielowicz M, Fawkner-Corbett D et al (2018) Structural remodeling of the human colonic mesenchyme in inflammatory bowel disease. Cell 175(2):372–386.e17

17. Wolf FA, Angerer P, Theis FJ (2018) SCANPY: large-scale single-cell gene expression data analysis. Genome Biol 19(1):15
18. Stassen SV, Yip GGK, Wong KKY, Ho JWK, Tsia KK (2021) Generalized and scalable trajectory inference in single-cell omics data with VIA. Nat Commun 12(1):5528
19. Van den Berge K, Roux de Bézieux H, Street K, Saelens W, Cannoodt R, Saeys Y et al (2020) Trajectory-based differential expression analysis for single-cell sequencing data. Nat Commun 11(1):1201
20. Kong L, Pokatayev V, Lefkovith A, Carter GT, Creasey EA, Krishna C et al (2023) The landscape of immune dysregulation in Crohn's disease revealed through single-cell transcriptomic profiling in the ileum and colon. Immunity 56(2):444–458.e5
21. Smillie CS, Biton M, Ordovas-Montanes J, Sullivan KM, Burgin G, Graham DB et al (2019) Intra- and inter-cellular rewiring of the human colon during ulcerative colitis. Cell 178(3):714–730.e22
22. Polański K, Young MD, Miao Z, Meyer KB, Teichmann SA, Park J-E (2020) BBKNN: fast batch alignment of single cell transcriptomes. Bioinformatics 36(3):964–965
23. Xu C, Prete M, Webb S, Jardine L, Stewart BJ, Hoo R et al (2023) Automatic cell-type harmonization and integration across human cell Atlas datasets. Cell 186(26):5876–5891.e20
24. Faure L, Soldatov R, Kharchenko PV, Adameyko I (2023) scFates: a scalable python package for advanced pseudotime and bifurcation analysis from single-cell data. Bioinformatics 39(1):btac746
25. Badia-i-Mompel P, Vélez Santiago J, Braunger J, Geiss C, Dimitrov D, Müller-Dott S et al (2022) decoupleR: ensemble of computational methods to infer biological activities from omics data. Bioinform Adv 2(1):vbac016

Check for updates

Chapter 4

Cross-Species Translation of Drug-Induced Toxicity in the Small Intestine Using Mathematical Modeling

Louis Gall and Carmen Pin

Abstract

We previously detailed a quantitative systems toxicology modeling framework for early prediction of chemotherapeutic-induced clinical intestinal adverse effects. The model describes stem and progenitor cell dynamics in the small intestinal epithelium and integrates epithelial-related processes, such as transcriptional profiles, citrulline kinetics, and the probability that the organism will have diarrhea. This protocol describes a step-by-step guide to the examples detailed in the original manuscript, namely 5-FU- and doxorubicin-induced intestinal toxicity.

Keywords Mathematical modeling, Clinical translation, Intestinal epithelium, Drug-induced toxicity, Quantitative systems toxicology, Dose regimens, Cross-species translation

1 Introduction

The functional integrity of the intestinal epithelium requires tight coordination between cell production, migration, and shedding along the crypt-villus axis [1, 2]. Dysregulation of any one of these processes, e.g., due to disease or drug insult, can result in a loss of barrier integrity, susceptibility to infection, allergies, diarrhea, malabsorption, inflammatory bowel disease, and cancer [3–5].

Chemotherapeutic agents act by disrupting cellular division, thereby inhibiting the proliferation of both malignant and healthy cells, aiming to suppress tumor progression without causing too many harmful side effects to the patient. In the gut, inhibiting cell division can compromise the integrity of the intestinal epithelium, resulting in adverse gastrointestinal (GI) effects such as diarrhea [5, 6]. Chemotherapy-induced diarrhea remains a significant challenge in the development of safe and tolerable anticancer drugs, representing a leading cause of treatment interruption during both preclinical and clinical studies. The incidence of chemotherapy-induced diarrhea has been reported to be as high as 80% [5]. To address this issue, we have

Paloma Ordóñez-Morán (ed.), *Intestinal Stem Cells*, Methods in Molecular Biology, vol. 3059,
https://doi.org/10.1007/978-1-0716-5412-5_4,

developed a mathematical modeling framework capable of predicting clinical toxicity based on preclinical data, to assist in the development of safer, more tolerable chemotherapeutics [7].

The model is used by first integrating experimental data from mice (such as drug plasma concentrations and epithelial cell population counts) into a mouse-specific mathematical model of the gastrointestinal epithelium, coupled with a drug- and species-specific pharmacokinetic (PK) model. This integrated model is used to quantify the drug's dose–effect relationship on epithelial cell populations in mice. The established dose-effect relationship is then used in an analogous coupled human GI epithelium and drug PK model. This human model predicts clinical outcomes, such as the incidence of diarrhea associated with a given drug and dosing regimen, extrapolated from preclinical mouse data.

This protocol will focus on using the model for clinical translation of the known compounds, 5-FU and doxorubicin, that were used to demonstrate the framework's capabilities in the original manuscript. We will outline the process of taking preclinical data and using a quantitative systems toxicology (QST) framework to predict drug-induced human toxicity. We will not detail how the experimental data were obtained, nor how the mathematical model was designed. PK models were obtained using Simcyp [8, 9], and their construction will also not be detailed.

2 Materials

2.1 Hardware and Software

- MATLAB (tested on versions 2019b to 2025a) [10, 11].
- Simcyp [8, 9] was used for animal and human PK simulations.

2.2 Code Availability

Model codes reproducing the previously published figures [7] are available on BioModels [12, 13] as MODEL2212120003 available at [14].

3 Methods

3.1 Small Intestine Epithelial Mathematical Model

The small intestinal epithelium is captured as a system of ordinary differential equations that describe the dynamics of the multiple heterogeneous epithelial cell types across the crypt-villus axis. In total, the model considers eight cell populations or compartments: proliferative stem and progenitor cells in the crypt that give rise to a differentiated cell population that transits from the crypt to the villus (as two distinct populations), as well as apoptotic and arrested cells in both the crypt and villus (*see* **Note 1**). Figure 1 in the original

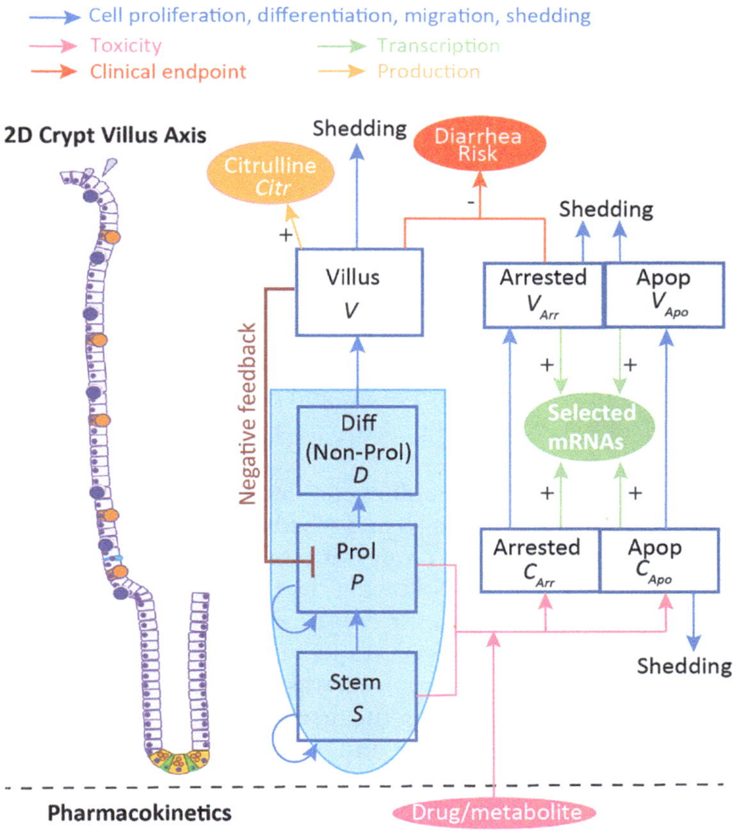

Fig. 1 Schematic of the intestinal epithelial model to describe the effect of drugs perturbing proliferative cells. The epithelium comprises three compartments in the crypt to describe stem (S) cells, proliferative progenitors (P), and differentiated (D) non-proliferative cells, and a villus (V) compartment with epithelial cells. Villus cells feedback on the proliferation of crypt cells. Drug-induced arrested and apoptotic proliferative cells are described by additional compartments in the crypt (CArr and CApo, respectively) and eventually on the villus (VArr and VApo, respectively) after cell migration. Transcription of selected genes is increased in arrested and apoptotic cells. Citrulline (Citr) is produced by mature epithelial cells on the villus and decreases upon villus injury. The integrity of the villus, quantified here by the number of cells, impacts on the risk of diarrhea, which increases as the cell number decreases. Plus/minus signs describe the direct/inverse relationship between quantities. Without loss of generality and in alignment with the data, the model refers to two-dimensional longitudinal crypt villus strips. Figure and caption reproduced from "A dynamic model of the intestinal epithelium integrates multiple sources of preclinical data and enables clinical translation of drug-induced toxicity" by Gall et al., published in CPT: Pharmacometrics and Systems Pharmacology (2023) 12(10), available under CC BY-NC-ND 4.0

manuscript [7], reproduced here for convenience, depicts the compartments of the epithelial model and their relationship with one another.

The model simulates temporal changes in these cell populations and, based on cell dynamics, predicts citrulline in plasma, transcriptomic profiles of the p53 pathway, and the risk of diarrhea associated with a given treatment regimen (*see* **Note 2**). The framework includes both mouse and human parameterizations, which can be used in tandem for cross-species pre-clinical translation. The mouse and human models have the same compartmental structure, with species-specific parameter values (cell numbers, proliferation rates, transfer rates, etc.). In mice, parameters were determined primarily from experimental data. For the human model, we used a mixture of human ileum values from the literature and values extrapolated from those in mice. The parameter values for both mouse and human can be found in the original manuscript [7]. A MATLAB script that simulates this epithelial model for both mice and humans with 5-FU and doxorubicin can be downloaded from BioModels [14].

3.2 Experimental Data

5-FU and metabolites concentrations were obtained from a previous study [15]. Dosing and schedules were the following: 20 or 50 mg/kg, b.i.d. for 4 days. Data was recorded at (approximately) 6, 24, 96, and 144 h following the first dose.

Doxorubicin: two dosing schedules were considered: 5 or 10 mg/kg intravenous (i.v.) bolus every 24 h for 2 days, followed by 2 days of recovery. Data was recorded at (approximately) 6, 24, 72, and 96 h following the first dose.

For each dose and timepoint, three vehicle and four to six untreated animals were included (*see* **Notes 3 and 4**). The tissue collection, determination of tissue drug and citrulline concentrations, histopathology, histomorphometry, immunohistochemistry, and stool consistency assessment in the mice were performed as outlined in the original study [15] (*see* **Notes 5 and 6**).

3.3 Quantification of Drug Effect in Mice

3.3.1 Mouse Plasma Concentration Profiles

5-FU: a PK model was constructed detailing the plasma concentration of 5-FU and relevant metabolites, implemented from mouse experiments.

The plasma concentration of 5-FU in patients was simulated using Simcyp [8, 9], while 5-FU metabolism and kinetics were assumed to mirror the dynamics observed in mice.

Doxorubicin: a two-compartment linear PK model was parameterized using mouse data as described in the manuscript [7].

Human plasma concentration–time profiles following dosing were generated using Simcyp [8, 9].

We adopted the free-drug hypothesis [16] and incorporated species-specific binding protein considerations, as appropriate.

3.3.2 Mechanisms of Toxicity

5-FU and doxorubicin are cytotoxic chemotherapies and were assumed to induce both cell cycle arrest and/ or apoptosis of proliferative cells, which comprise two populations in our model: stem cells and proliferative progenitors.

*3.3.3 Dose-Effect Relationship (See **Note 7**)*

5-FU: a Hill function was used to capture the dose-effect relationship linking drug concentration and the rate of cell arrest and apoptosis. FUTP was assumed to be the metabolite responsible for toxicity in the epithelium, causing RNA damage that leads to cell cycle arrest and apoptosis, in alignment with previous reports [17].

Doxorubicin: a linear relationship between drug concentration and induction of arrest/apoptosis was used due to limited data availability.

3.3.4 Determination of Parameter Values

The parameters of the mouse GI epithelium model were determined from experimental data for each drug and published sources listed in the original manuscript. These include parameters describing the PK and drug metabolism, epithelial homeostasis, and the dose-effect relationship for each compound. These parameter values, and the relevant equations, are presented in the original manuscript. The coupled GI epithelium and PK models were implemented in MATLAB.

3.4 Translation to Humans

3.4.1 Human Plasma Concentration Profiles

The plasma profile concentrations of both 5-FU and doxorubicin in patients were simulated using Simcyp [8, 9]. The concentrations of the 5-FU metabolites in human plasma were modeled by assuming that their associated rates were equal in mouse and human. Like in the mouse, these PK models were implemented in MATLAB and coupled with the corresponding human epithelial model. The supplementary material of the original manuscript contains the full mathematical models.

The dosing regimens considered in humans were:

5-FU:

- 46h i.v. infusion resulting in AUC 18 or 28 mg·h/L every 3 weeks.
- 450 mg/day for 5 days every 4 weeks.

Doxorubicin: 75 mg/m^2 i.v. bolus every 3 weeks.

3.4.2 Shared Mechanisms of Toxicity and Dose-Effect Relationships

The dose-effect relationships for both 5-FU and doxorubicin were assumed to act in the same manner in mice and humans, with identical parameters (see **Note 8**). To account for species-specific proliferation, the total effect was scaled by the division time (see equations S10–S18 in the supplementary material of the original manuscript).

3.4.3 Simulations

For each clinical regimen listed above, the coupled epithelium and PK model predicts stem/progenitor/villus cell counts, arrested/apoptotic compartments, plasma citrulline, and predicted diarrhea probability over time. MATLAB scripts that perform these simulations for 5-FU and doxorubicin in patients can be found on BioModels [14].

4 Notes

1. The model makes extensive use of steady-state values derived from the literature, with other parameters derived using the steady-state approximation (all derivatives equal to 0) and experimentally derived data. The manuscript captures all parameters for the GI epithelium model and their various sources and/or justifications.
2. Diarrhea incidence is reported in the literature as a single % for the entirety of the treatment duration. Our model produced predictions over time, with the same temporal resolution as the simulations. A parameterization was deemed a good fit when the reported values were in the range predicted by the model.
3. For integration with dynamic mathematical models, longitudinal data are required with a resolution relevant to the timescales of the disease.
4. Transcriptional data of the p53 pathway were selected by comparatively analyzing mice colon and small intestine, as well as mouse and human small intestinal and colon organoids exposed to 5-FU. To enable cross-species translation, intensity measurements in the mouse small intestinal mucosa were transformed into Z-scores. The linear relationship between p53 pathway gene transcriptional Z-scores and the proportion of arrested/apoptotic cells was assumed to be conserved across species.
5. Citrulline kinetics were modeled by considering citrulline production by enterocytes and systemic clearance, which results in a steady-state level that reduces when the villus is damaged. In mice, steady-state plasma citrulline levels were measured in 5-FU–treated mice, and the two other parameters were calculated to maintain this steady-state level. In patients, two parameter values (steady-state plasma concentration and elimination rate of citrulline) were taken from publicly available sources [18, 19] and used to determine the citrulline production rate.
6. Diarrhea risk is calculated using a nonlinear logistic regression where the predictor is villus cell number (as a proxy of

epithelial integrity). These numbers were calibrated using data obtained from mice exposed to 5-FU.

7. This assumption, that the drug acts through cell cycle arrest and apoptosis, captures the typical cellular response to chemotherapeutics. Different classes of drugs will exhibit different effects, and domain knowledge is vital here.
8. The assumption that the dose-effect relationship parameters can be directly translated between mice and humans should be applied with caution. In our original manuscript, we demonstrated that our modeling framework failed to predict the clinical safety risks of gefitinib [7]. Gefitinib selectively inhibits EGFR in humans [20], and it showed no detectable epithelial toxicity in mice, even at exposures exceeding those that cause diarrhea in patients. With no observable epithelial toxicity in mice, the model cannot predict gastrointestinal damage in humans.

References

1. Gehart H, Clevers H (2019) Tales from the crypt: new insights into intestinal stem cells. Nat Rev Gastroenterol Hepatol 16:19. https://doi.org/10.1038/s41575-018-0081-y
2. Barker N et al (2007) Identification of stem cells in small intestine and colon by marker gene Lgr5. Nature 449:1003
3. Van Der Heijden M, Vermeulen L (2019). Stem cells in homeostasis and cancer of the gut. Mol Cancer18:66. https://doi.org/10.1186/s12943-019-0962-x
4. Henderson P, Van Limbergen JE, Schwarze J, Wilson DC (2011) Function of the intestinal epithelium and its dysregulation in inflammatory bowel disease. Inflamm Bowel Dis 17:382. https://doi.org/10.1002/ibd.21379
5. Stein A, Voigt W, Jordan K (2010) Review: chemotherapy-induced diarrhea: pathophysiology, frequency and guideline-based management. Ther Adv Med Oncol 2:51. https://doi.org/10.1177/1758834009355164
6. McQuade RM et al. (2016) Gastrointestinal dysfunction and enteric neurotoxicity following treatment with anticancer chemotherapeutic agent 5-fluorouracil. Neurogastroenterol Motil 28:1861–1875
7. Gall L et al (2023) A dynamic model of the intestinal epithelium integrates multiple sources of preclinical data and enables clinical translation of drug-induced toxicity. CPT Pharmacometrics Syst Pharmacol 12:1511
8. Jamei M et al (2009) The Simcyp® population-based ADME simulator. Expert Opin Drug Metab Toxicol 5:211. https://doi.org/10.1517/17425250802691074
9. Jamei M et al (2013) The Simcyp population based simulator: architecture, implementation, and quality assurance. In Silico Pharmacol 1:9
10. The Mathworks Inc. Download and Install MATLAB. https://uk.mathworks.com/help/install/ug/install-products-with-internet-connection.html
11. The Mathworks Inc (2022) MATLAB. Preprint at https://www.mathworks.com
12. Glont M et al (2018) BioModels: expanding horizons to include more modelling approaches and formats. Nucleic Acids Res 46:D1248
13. Malik-Sheriff RS et al (2020) BioModels-15 years of sharing computational models in life science. Nucleic Acids Res 48:D407
14. Gall L, Pin C (2022) Gall&Pin2022 - ODE model of the intestinal epithelium. https://www.ebi.ac.uk/biomodels/MODEL2212120003#Overview
15. Jardi F et al (2023) Mouse organoids as an in vitro tool to study the in vivo intestinal response to cytotoxicants. Arch Toxicol 97:235
16. Smith DA, Di L, Kerns EH (2010) The effect of plasma protein binding on in vivo efficacy: misconceptions in drug discovery. Nat Rev Drug Discov 9:929–939. https://doi.org/10.1038/nrd3287
17. Pritchard M, Watson AJM, Potten CS, Jackman AL, Hickman JA (1997) Inhibition by uridine but not thymidine of p53-dependent intestinal apoptosis initiated by 5-fluorouracil: evidence

for the involvement of RNA perturbation. Proc Natl Acad Sci U S A 94:1795
18. Fragkos KC, Forbes A (2018) Citrulline as a marker of intestinal function and absorption in clinical settings: a systematic review and meta-analysis. United Eur Gastroenterol J 6:181. https://doi.org/10.1177/2050640617737632
19. Moinard C et al (2008) Dose-ranging effects of citrulline administration on plasma amino acids and hormonal patterns in healthy subjects: the Citrudose pharmacokinetic study. Br J Nutr 99:855
20. Herbst RS, Fukuoka M, Baselga J (2004) Gefitinib - a novel targeted approach to treating cancer. Nat Rev Cancer 4:956. https://doi.org/10.1038/nrc1506

Check for updates

Chapter 5

A Modelling Approach to Resolve Gene Temporal Dynamics in Intestinal Inflammation

Shivam Patel, Brinda Balasubramanian and Paloma Ordóñez-Morán

Abstract

The intestinal epithelium is maintained by stem cells at the crypt base, however, acute or chronic inflammation can severely disrupt stem cell function and tissue homeostasis. To characterize mucosal inflammation and tissue damage, we evaluated time-course gene expression changes using a dextran sulfate sodium (DSS)-induced mouse model of colitis. By applying normalization, Z-score transformation, and spline curve modeling, we trace the temporal expression dynamics of key intestinal genes. This computational approach offer insights into the molecular responses during inflammation and may help identifying new biomarkers for chronic inflammatory diseases.

Keywords Stem cells, Inflammation, DSS mouse model, Transcriptomics, Spline curve modeling

1 Introduction

Intestinal epithelium is maintained by the population of rapidly self-renewing intestinal stem cells (ISCs) located at the base of crypts [1]. The ability of tissue regeneration and stem cell differentiation process can be impaired due to environmental cues such as inflammation [2]. Chronic intestinal inflammation characterized by ulceration plays an important role in the pathogenesis of Inflammatory Bowel Disease (IBD) [3]. By understanding gene expression dynamics of intestinal inflammation-associated genes, novel insights for IBD progression and therapeutic strategies can be determined.

In mouse models, acute or chronic inflammation can be induced by chemical agents such as dextran sulfate sodium (DSS) [4, 5]. This mimics colitis by triggering ulceration, epithelial damage, and immune activation. In the last few years, transcriptomic studies based on DSS mouse models have emerged to study temporal expression profiling [6], IBD patient stratification [7], and colitis-associated cancer [8].

Paloma Ordóñez-Morán (ed.), *Intestinal Stem Cells*, Methods in Molecular Biology, vol. 3059,
https://doi.org/10.1007/978-1-0716-5412-5_5,

Publicly available transcriptomic datasets serve as a valuable resource to analyze gene expression changes in response to inflammation stimuli. One such dataset, GSE22307, consists of microarray expression profiles of DSS-treated mice at various but continuous time points [9]. Here, using this experimental design as leverage, expression trajectories of candidate genes over the course of increased inflammation were studied. Similarly, these gene expression datasets are useful to track ISC gene expression correlation with other biological processes and regulatory pathways.

In this method, we employed a computational correlation approach to compare candidate genes with previously derived inflammation gene signature using normalized-transformed data of DSS-treated mouse models at continuous timepoints. To do this, we standardized gene expression values by applying Z-score transformation and defined a spline curve function to model gene expression dynamics over a period of 6 days. We provide a systematic framework for evaluating gene expression in inflammatory conditions and for identifying potential biomarkers or therapeutic targets.

2 Materials

2.1 Hardware and Software

64-bit computer running either on Windows OS or MacOS with 4 GB of RAM (8 GB preferred).

Software: R (Version 4.4.1) can be downloaded and installed in Windows or MacOS accordingly (https://cran.r-project.org), the same should be done for the R Studio IDE (Version) (https://posit.co/download/rstudio-desktop/).

2.2 Installation of Dependencies

The following R packages must be installed using the code shown in Fig. 1:

1. GEOquery [10].
2. limma [11].
3. dplyr [12].
4. tidyverse [13].
5. ggplot2 [14].
6. ggpubr [15].

```
if (!requireNamespace("BiocManager", quietly = TRUE)) install.packages("BiocManager")
BiocManager::install(c("GEOquery", "limma"))
install.packages(c("dplyr", "tidyverse", "ggplot2", "ggpubr", "stringr", "gridExtra"))
```

Fig. 1 Code to install all dependencies

7. stringr [16].
8. gridExtra [17].

2.3 Code Availability

Code used in this methodology is available on GitHub https://github.com/shivam97scripts/Spline_curve_analysis/blob/9cffaec0a93443786d3560bdb83f804e2c658b37/GSE22307_pub.R
https://github.com/shivam97scripts/Spline_curve_analysis.git

3 Methods

3.1 Preparation of R Environment

1. In R Studio source panel, load the following R packages: GEOquery, ggplot2, dplyr, ggpubr and stringr using library() function (*see* **Note 1**).

3.2 Import Dataset from NCBI and Data Gene Expression Manipulation

1. Import Microarray dataset of DSS-treated mouse with accession number GSE22307 from Gene Expression Omnibus (GEO) database into a new list named as "gse" with the help of getGEO("GSE22307") function (*see* **Note 2**).
2. From "gse" retrieve ExpressionSet object into a new variable which contains raw microarray expression data, phenotype data (i.e., treatment time of DSS), and feature data (i.e., probeIDs and gene symbols). Name this retrieved object to eSet.
3. Isolate raw microarray expression data into a new matrix raw_data by executing exprs(eSet) function (*see* **Note 3**).
4. Apply the Quantile method of normalization to raw_data using normalizeBetweenArrays(raw_data) function. Resultant data from this would be stored in new matrix normalize_data.
5. By applying log2(noramlize_data+1) function to stabilize the variation in new data and to avoid having zero values in normalized expression matrix (*see* **Note 4**).
6. Extract phenotype data using pData(eSet) and feature data using fData(eSet) in two separate data frame variables with names pheno_data and feature_data, respectively.
7. Following feature data extraction, execute probe_ID<-feature_data$ID and geneSymbol<-feature_data$'Gene Symbol' to obtain values for annotation process.
8. Convert probeIDs in normalize_data to gene symbols by carrying out following command, rownames(normalize_data) < - geneSymbol[match(rownames(normalize_data), probe_ID)]

3.3 Prepare Phenotype Data

1. Extract two columns "title," "geo_accession," and "time point (day):ch1" with row names in pheno_data in a new data frame col_data.

2. Using mutate() function of dplyr package, group them 23 samples into either "Untreated" or Number + "Days of DSS treatment" according to values in "time point (day):ch1" column.
3. Convert time points in numeric values from characters, run as.numeric(col_data$ 'time point (day):ch1') (*see* **Note 5**).

3.4 Quality Check of Expression Data

1. To check the quality of raw expression data, run boxplot(log2(raw_data+1)) and observe the distribution across the n=23 samples in Fig. 2 (*see* **Notes 6 and 7**)
2. Next, generate a boxplot using the same function as in the previous step, replacing raw_data with normalize_data (*see* **Note 8**).

3.5 Z-Score Transformation and Extracting Inflammation Signature

1. Calculate Z-score by applying scale() function, centering and scaling expression data, followed by truncating values to a range of ±2 standard deviation (*see* **Note 9**).
2. Reshape normalized Z-score matrix into the new data frame using the gather() function, while extracting sample-wise expression values, merge phenotype data (i.e., col_data) into the same data frame based on the same GEO sample IDs.
3. By importing 26 gene symbols (Table 1), filter expression data to retain genes only present in the inflammation signature, use %in% operator to make Inflam_expression data frame (*see* **Note 10**).

3.6 Create a Spline Curve Function and Test Gene Candidate

1. Define Inf_dynamics function using code in Fig. 2 having ggplot() and geom_smooth to plot an exrpression dynamics graph of inflammation gene signature and test gene expression

```
# Create spline curve

Inf_dynamics<-function(g1){
  {test_expression<-data_long[data_long$gene == g1,]}
  ggplot() +
  geom_smooth(data = test_expression,
              aes(x = day_of_DSS_treatment, y = Normalized_expressoin),
              formula = y ~ s(x, bs = "cs", k = 4),
              method = "gam",
              color = "darkgreen",
              fill = "lightgreen",
              se = TRUE)+
  geom_smooth(data = Inflam_expression,
              aes(x = day_of_DSS_treatment, y = Normalized_expressoin),
              formula = y ~ s(x, bs = "cs", k = 4),
              method = "gam",
              color = "red",
              fill = "pink",
              se = TRUE)+
  theme_classic()+
  annotate("text",x =6, y = 2.7, label = "Inflammation gene expression", color = "red", hjust = 1)+
  annotate("text",x =6, y = 2.1, label = as.character(substitute(g1)), color = "darkgreen", hjust = 1)
}
```

Fig. 2 Script to define Inf_dynamics() function, which takes gene symbol, e.g., "Sox9" as input

Table 1
Gene markers used to characterize the inflammation process, indicating their primary biological role and supporting literature references

Genes	Role in inflammation	Reference
Aqp9	Regulates the transportation of water and glycerol in cells, and it is also expressed in CD (Crohn's disease) patients intestinal mucosa	[18, 19]
Clec5a	Peripheral mononuclear blood cells of CD patients show higher expression of CLEC5A compared to healthy human individuals	[20]
Csf3	Gene expressed in the inflammatory fibroblasts of Ulcerative colitis (UC) and CD patients	[21]
Cxcl5 *Cxcl10* *Cxcl9* *Cxcl11* *Cxcl6* *Cxcl8* *Cxcl1*	These genes are upregulated in intestinal organoids by the inflammatory cytokine INFγ or IL17A	[22–24]
Cxcl13	Gene associated with intestinal inflammatory response and correlated with IBD severity	[25]
Cxcl17	Mitigation of mycotoxin-induced inflammation by PI3K-AKT pathway modulation	[26]
Cxcr1 *Cxcr2*	Upregulated gene in colonic tissues of UC patients and correlated with neutrophil recruitment	[27]
Fpr1	Highly expressed gene in gut inflammation of UC and CD patients	[28]
Il17a	Pro-inflammatory gene encoding a cytokine that plays a key role in the pathogenesis of CD and UC	[29]
Il1a *Il1b*	Gene responsible for stromal-neutrophil interaction associated to resistance against IBD therapeutics	[30]
Il6	Pathways regulated by IL6 are involved in sustaining chronic inflammation in IBD	[31]
Mmp3	Elevated serum and mucosal levels of this gene correlate with mucosal inflammation, supporting its utility as a biomarker in UC	[32]
Inhba *Osm*	Genes associated with non-response to anti-TNF therapy in IBD patients	[33, 34]
Pdpn	Gene expressed by stromal cells during the IBD pathogenesis	[30]
Prok2	Gene that indicates neutrophil increase in UC and CD	[21]
Sele	Gene that encoded for a protein that facilitates initial leukocyte adhesion and recruitment to inflamed tissues	[35]
Serpine1	Upregulated gene in colonic tissues and associated with mucosal damage and disease severity	[36]

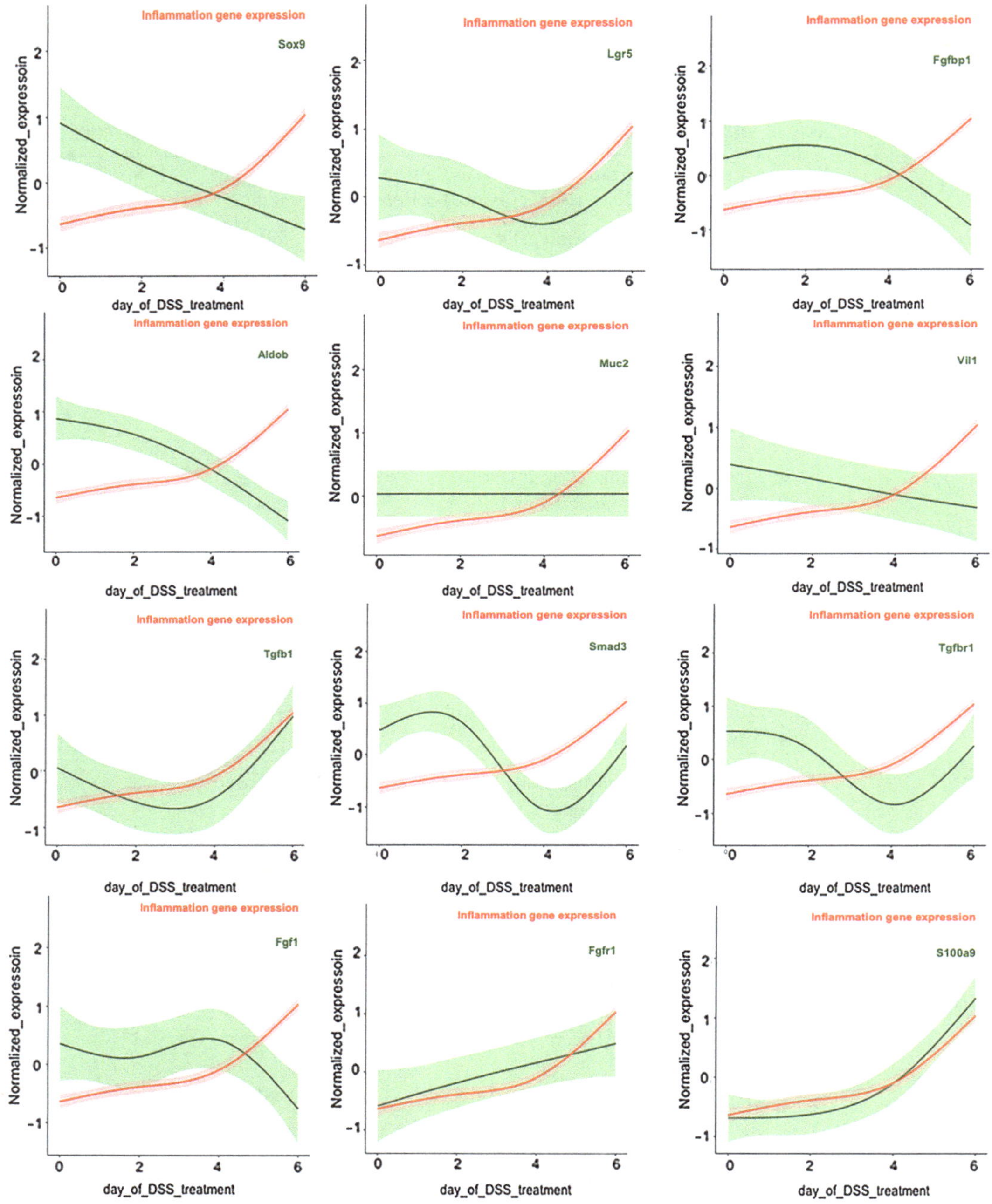

Fig. 3 Spline curve graphs showing change in gene expression (green) dynamics in comparison to inflammatory gene signature (red) of stem cell/progenitor genes: Sox9, Lgr5, Fgfbp1; differentiation genes:*Aldob*, *Muc2*, *Vil1*; *Tgfβ* pathway genes: *Tgfβ1*, *Smad3*, *Tgfbr1*; and FGF pathway genes: *Fgf1*,*Fgfr1*. *S100a9* was used as a positive control to establish inflammation correlation

Table 2
Examples of gene markers used to characterize intestinal epithelial cell populations, indicating their primary biological role and supporting literature references

Gene markers	Role	Reference
Lgr5	Receptor expressed by intestinal stem cells of both small intestine and colon	[37]
Sox9	Wnt responsive gene expressed by intestinal progenitor cell types	[38]
Aldob	Enterocyte differentiation marker in small intestine	[39]
Muc2	Gene expressed by mucin-secreting goblet cells in the intestine	[40]

value. Use formula y~s(x,bs="cs",k=4). Where x is the day of DSS treatment and Y is the Z-score value of normalized expression.

2. Execute Inf_dynamics("Sox9") with candidate gene (e.g., Sox9) to compare its dynamics compared to inflammation gene signature dynamics (Table 2) (*see* **Note 11**).
3. For the demonstration, we described a spline curve showing the dynamics of multiple genes either involved in cell signaling process or indicating cell type marker (Fig. 3) using Inf_dynamics() function.

4 Notes

1. In case of a missing source panel, open a new script from File -> New File -> R Script or press Ctrl+Shift+N on windows operating system or Command+Shift+N on MacOS.
2. Having specified GSEMatrix = TRUE ensures that the curated ExpressionSet Object is loaded into gse list.
3. raw_data matrix contains 45,101 rows as probeIDs (e.g., "1415670_at") and columns as 23 samples (e.g., "GSM555189"). In case of importing matrix from downloaded CSV file, use read.csv("file_name.csv", row.names=1) to ensure the same matrix format.
4. To remove NA value from matrix, use na.omit(normalize_data) function.
5. For later use, geom_smooth() function needs numeric values.
6. In raw_data matrix median of each sample is not similar.
7. Color palette storing variable indicating colors in boxplot according to sample classification.

8. A box plot of normalized data can show similar median range across all 23 samples when compared with raw data.
9. Remove any NA values resulted because of extreme outliers by passing z-score matrix through na.omit() function.
10. List containing capitalized gene symbol (e.g., "AQP9") to title case (e.g., "Aqp9") by executing str_to_title() from stringr library, as gene symbols of mouse datasets always have the first character capitalized, while following first letters are small case.
11. To make plot area showing more graphs in one frame, call individual variable storing graph generated by Inf_dynamics() function and then execute grid.arrange() function with all variables having graphs.

Acknowledgments

This work was supported by Medical Research Council (MRC) New Investigator Research Award under grant number (MR/W01825X/1) and GUTS UK (ECR2021_1).

References

1. Barker N (2014) Adult intestinal stem cells: critical drivers of epithelial homeostasis and regeneration. Nat Rev Mol Cell Biol 15(1):19–33. https://doi.org/10.1038/nrm3721
2. Beumer J, Clevers H (2024) Hallmarks of stemness in mammalian tissues. Cell Stem Cell 31(1):7–24. https://doi.org/10.1016/j.stem.2023.12.006
3. Khor B, Gardet A, Xavier RJ (2011) Genetics and pathogenesis of inflammatory bowel disease. Nature 474(7351):307–317. https://doi.org/10.1038/nature10209
4. Creyns B, Cremer J, Hoshino T, Geboes K, de Hertogh G, Ferrante M, Vermeire S, Ceuppens JL, Van Assche G, Breynaert C (2019) Fibrogenesis in chronic DSS colitis is not influenced by neutralisation of regulatory T Cells, of major T helper cytokines or absence of IL-13. Sci Rep 9(1):10064. https://doi.org/10.1038/s41598-019-46472-6
5. Hong D, Kim HK, Yang W, Yoon C, Kim M, Yang CS, Yoon S (2024) Integrative analysis of single-cell RNA-seq and gut microbiome metabarcoding data elucidates macrophage dysfunction in mice with DSS-induced ulcerative colitis. Commun Biol 7(1):731. https://doi.org/10.1038/s42003-024-06409-w
6. Fang K, Bruce M, Pattillo CB, Zhang S, Stone 2nd R, Clifford J, Kevil CG (2011) Temporal genomewide expression profiling of DSS colitis reveals novel inflammatory and angiogenesis genes similar to ulcerative colitis. Physiol Genomics 43(1):43–56. https://doi.org/10.1152/physiolgenomics.00138.2010
7. Czarnewski P, Parigi SM, Sorini C, Diaz OE, Das S, Gagliani N, Villablanca EJ (2019) Conserved transcriptomic profile between mouse and human colitis allows unsupervised patient stratification. Nat Commun 10(1):2892. https://doi.org/10.1038/s41467-019-10769-x
8. Liu M, Zhong XS, Krishnachaitanya SS, Ou R, Dashwood RH, Powell DW, Li Q (2024) Erlotinib suppresses tumorigenesis in a mouse model of colitis-associated cancer. Biomed Pharmacother 175:116580. https://doi.org/10.1016/j.biopha.2024.116580
9. Fang K, Bruce M, Pattillo CB, Zhang S, Stone 2nd R, Clifford J, Kevil CG (2011) Temporal genomewide expression profiling of DSS colitis reveals novel inflammatory and angiogenesis genes similar to ulcerative colitis. Physiol Genomics 43(1):43–56. https://doi.org/10.1152/physiolgenomics.00138.2010
10. Davis S, Meltzer PS (2007) GEOquery: a bridge between the gene expression omnibus (GEO)

and BioConductor. Bioinformatics (Oxford, England) 23(14):1846–1847. https://doi.org/10.1093/bioinformatics/btm254
11. Ritchie ME, Phipson B, Wu D, Hu Y, Law CW, Shi W, Smyth GK (2015) limma powers differential expression analyses for RNA-sequencing and microarray studies. Nucleic Acids Res 43(7):e47. https://doi.org/10.1093/nar/gkv007
12. https://cran.r-project.org/web/packages/dplyr/index.html
13. Wickham H, Averick M, Bryan J, Chang W, McGowan LD, François R, Grolemund G, Hayes A, Henry L, Hester J, Kuhn M, Pedersen TL, Miller E, Bache SM, Müller K, Ooms J, Robinson D, Seidel DP, Spinu V, Takahashi K, Vaughan D, Wilke C, Woo K, Yutani H (2019) Welcome to the tidyverse. J Open Source Softw 4(43):1686. https://doi.org/10.21105/joss.01686
14. Wickham H (2016) ggplot2: elegant graphics for data analysis. Springer-Verlag, New York. ISBN 978-3-319-24277-4. https://ggplot2.tidyverse.org
15. Kassambara A (2023) ggpubr: 'ggplot2' based publication ready plots. R package version 0.6.0. https://rpkgs.datanovia.com/ggpubr/
16. Wickham H (2023) stringr: simple, consistent wrappers for common string operations. R package version 1.5.1. https://github.com/tidyverse/stringr, https://stringr.tidyverse.org
17. Auguie B, Antonov A (2017) gridExtra: miscellaneous functions for "grid" graphics. R Package Version 2.3. https://CRAN.R-project.org/package=gridExtra
18. Sachdeva R, Priyadarshini P, Gupta S (2023) Aquaporins display a diversity in their substrates. J Membrane Biol 256(1):1–23. https://doi.org/10.1007/s00232-022-00257-7
19. Martin JC, Chang C, Boschetti G, Ungaro R, Giri M, Grout JA, Gettler K, Chuang LS, Nayar S, Greenstein AJ, Dubinsky M, Walker L, Leader A, Fine JS, Whitehurst CE, Mbow ML, Kugathasan S, Denson LA, Hyams JS, Friedman JR, Kenigsberg E (2019) Single-cell analysis of Crohn's disease lesions identifies a pathogenic cellular module associated with resistance to anti-TNF therapy. Cell 178(6):1493–1508.e20. https://doi.org/10.1016/j.cell.2019.08.008
20. Schäffler H, Rohde M, Rohde S, Huth A, Gittel N, Hollborn H, Koczan D, Glass Ä, Lamprecht G, Jaster R (2018) NOD2- and disease-specific gene expression profiles of peripheral blood mononuclear cells from Crohn's disease patients. World J Gastroenterol 24(11):1196–1205. https://doi.org/10.3748/wjg.v24.i11.1196
21. Garrido-Trigo A, Corraliza AM, Veny M, Dotti I, Melón-Ardanaz E, Rill A, Crowell HL, Corbí Á, Gudiño V, Esteller M, Álvarez-Teubel I, Aguilar D, Masamunt MC, Killingbeck E, Kim Y, Leon M, Visvanathan S, Marchese D, Caratù G, Martin-Cardona A, Salas A (2023) Macrophage and neutrophil heterogeneity at single-cell spatial resolution in human inflammatory bowel disease. Nat Commun 14(1):4506. https://doi.org/10.1038/s41467-023-40156-6
22. Pavlidis P, Tsakmaki A, Treveil A, Li K, Cozzetto D, Yang F, Niazi U, Hayee BH, Saqi M, Friedman J, Korcsmaros T, Bewick G, Powell N (2022) Cytokine responsive networks in human colonic epithelial organoids unveil a molecular classification of inflammatory bowel disease. Cell Rep 40(13):111439. https://doi.org/10.1016/j.celrep.2022.111439
23. Workman MJ, Troisi E, Targan SR, Svendsen CN, Barrett RJ (2020) Modeling intestinal epithelial response to interferon-γ in induced pluripotent stem cell-derived human intestinal organoids. Int J Mol Sci 22(1):288. https://doi.org/10.3390/ijms22010288
24. Lee C, Song JH, Cha YE, Chang DK, Kim YH, Hong SN (2022) Intestinal epithelial responses to IL-17 in adult stem cell-derived human intestinal organoids. J Crohn's Colitis 16(12):1911–1923. https://doi.org/10.1093/ecco-jcc/jjac101
25. Liu T, Liu Y, Liu CX, Jiang YM (2022) CXCL13 is elevated in inflammatory bowel disease in mice and humans and is implicated in disease pathogenesis. Front Immunol 13:997862. https://doi.org/10.3389/fimmu.2022.997862
26. Sun C, Shen H, Cai H, Zhao Z, Gan G, Feng S, Chu P, Zeng M, Deng J, Ming F, Ma M, Jia J, He R, Cao D, Chen Z, Li J, Zhang L (2021) Intestinal guard: human CXCL17 modulates protective response against mycotoxins and CXCL17-mimetic peptides development. Biochem Pharmacol 188:114586. https://doi.org/10.1016/j.bcp.2021.114586
27. Xv Y, Feng Y, Lin J (2024) CXCR1 and CXCR2 are potential neutrophil extracellular trap-related treatment targets in ulcerative colitis: insights from Mendelian randomization, colocalization and transcriptomic analysis. Front Immunol 15:1425363. https://doi.org/10.3389/fimmu.2024.1425363
28. McAllister MJ, Hall R, Whelan RJ, Fischer LJ, Chuah CS, Cartlidge PD, Drury B, Rutherford DG, Duffin RM, Cartwright JA, Dorward DA, Rossi AG, Ho GT (2024) Formylated peptide Receptor-1-mediated gut inflammation as a therapeutic target in inflammatory bowel disease. Crohn's Colitis 360 6(1):otae003. https://doi.org/10.1093/crocol/otae003
29. Petitpain N, D'Amico F, Yelehe-Okouma M, Jouzeau JY, Netter P, Peyrin-Biroulet L,

Gillet P (2021) IL-17 Inhibitors and inflammatory bowel diseases: a postmarketing study in vigibase. Clin Pharmacol Ther 110(1):159–168. https://doi.org/10.1002/cpt.2155
30. Friedrich M, Pohin M, Jackson MA, Korsunsky I, Bullers SJ, Rue-Albrecht K, Christoforidou Z, Sathananthan D, Thomas T, Ravindran R, Tandon R, Peres RS, Sharpe H, Wei K, Watts GFM, Mann EH, Geremia A, Attar M, Oxford IBD Cohort Investigators, Roche Fibroblast Network Consortium, Powrie FM (2021) IL-1-driven stromal-neutrophil interactions define a subset of patients with inflammatory bowel disease that does not respond to therapies. Nat Med 27(11):1970–1981. https://doi.org/10.1038/s41591-021-01520-5
31. Alhendi A, Naser SA (2023) The dual role of interleukin-6 in Crohn's disease pathophysiology. Front Immunol 14:1295230. https://doi.org/10.3389/fimmu.2023.1295230
32. Pan Z, Lin H, Fu Y, Zeng F, Gu F, Niu G, Fang J, Gu B (2023) Identification of gene signatures associated with ulcerative colitis and the association with immune infiltrates in colon cancer. Front Immunol 14:1086898. https://doi.org/10.3389/fimmu.2023.1086898
33. West NR, Hegazy AN, Owens BMJ, Bullers SJ, Linggi B, Buonocore S, Coccia M, Görtz D, This S, Stockenhuber K, Pott J, Friedrich M, Ryzhakov G, Baribaud F, Brodmerkel C, Cieluch C, Rahman N, Müller-Newen G, Owens RJ, Kühl AA, Powrie F (2017) Oncostatin M drives intestinal inflammation and predicts response to tumor necrosis factor-neutralizing therapy in patients with inflammatory bowel disease. Nat Med 23(5):579–589. https://doi.org/10.1038/nm.4307
34. Derakhshan Nazari MH, Shahrokh S, Ghanbari-Maman L, Maleknia S, Ghorbaninejad M, Meyfour A (2023) Prediction of anti-TNF therapy failure in ulcerative colitis patients by ensemble machine learning: a prospective study. Heliyon 9(11):e21154. https://doi.org/10.1016/j.heliyon.2023.e21154
35. Jones SC, Banks RE, Haidar A, Gearing AJ, Hemingway IK, Ibbotson SH, Dixon MF, Axon AT (1995) Adhesion molecules in inflammatory bowel disease. Gut 36(5):724–730. https://doi.org/10.1136/gut.36.5.724
36. Kaiko GE, Chen F, Lai CW, Chiang IL, Perrigoue J, Stojmirović A, Li K, Muegge BD, Jain U, VanDussen KL, Goggins BJ, Keely S, Weaver J, Foster PS, Lawrence DA, Liu TC, Stappenbeck TS (2019) PAI-1 augments mucosal damage in colitis. Sci Trans Med 11(482):eaat0852. https://doi.org/10.1126/scitranslmed.aat0852
37. Barker N, van Es JH, Kuipers J, Kujala P, van den Born M, Cozijnsen M, Haegebarth A, Korving J, Begthel H, Peters PJ, Clevers H (2007) Identification of stem cells in small intestine and colon by marker gene Lgr5. Nature 449(7165):1003–1007. https://doi.org/10.1038/nature06196
38. Blache P, van de Wetering M, Duluc I, Domon C, Berta P, Freund JN, Clevers H, Jay P (2004) SOX9 is an intestine crypt transcription factor, is regulated by the Wnt pathway, and represses the CDX2 and MUC2 genes. J Cell Biol 166(1):37–47. https://doi.org/10.1083/jcb.200311021
39. Gassler N, Newrzella D, Böhm C, Lyer S, Li L, Sorgenfrei O, van Laer L, Sido B, Mollenhauer J, Poustka A, Schirmacher P, Gretz N (2006) Molecular characterisation of non-absorptive and absorptive enterocytes in human small intestine. Gut 55(8):1084–1089. https://doi.org/10.1136/gut.2005.073262
40. Gum JrJR, Hicks JW, Gillespie AM, Carlson EJ, Kömüves L, Karnik S, Hong JC, Epstein CJ, Kim YS (1999) Goblet cell-specific expression mediated by the MUC2 mucin gene promoter in the intestine of transgenic mice. Am J Physiol 276(3):G666–G676. https://doi.org/10.1152/ajpgi.1999.276.3.G666

Part II

Intestinal Stem Cells: In Vitro Models and Applications

Chapter 6

Animal-Free Intestinal Epithelial 2D "Organoid-Sheet" Cultures Using Invasin

Joost J.A.P.M. Wijnakker and Hans Clevers

Abstract

The intestinal epithelium consists of a single layer of polarized cells that serves as a crucial barrier against pathogens while mediating food digestion and nutrient absorption. The differentiated villus cells, important for the intestine's function, originate from rapidly dividing Lgr5^{+} stem cells located in the intestinal crypts. These stem cells can be expanded in vitro and form three-dimensional (3D) "mini-organs" known as intestinal organoids, which closely phenocopy the intestinal epithelium. The development and maintenance of these 3D organoids depend on a defined growth factor cocktail and on extracellular matrix (ECM) support to preserve apical-basal polarity – an essential feature for nutrient uptake, growth, and the barrier function. Matrigel or BME has been used as the standard ECM hydrogel scaffold for organoid culture. Recently, we identified the bacterial protein Invasin as a defined ECM substitute that can fully recapitulate Matrigel/BME function. Utilizing recombinant Invasin coated on transwells, we developed a two-dimensional (2D) intestinal monolayer culture system. The intestinal stem cells form a polarized, confluent epithelial sheet while maintaining stemness and differentiation capacity into functional epithelial cell types. Compared to traditional 3D cultures, this 2D system offers advantages such as improved imaging, easier access to apical and basal compartments, and compatibility with automation for high-throughput applications. Here, we present a detailed protocol for generating 2D intestinal monolayers using Invasin, starting from either fresh human intestinal tissue or from pre-established 3D organoids. These 2D Invasin-cultures recapitulate key aspects of intestinal epithelial biology and are amenable to a broad range of experimental manipulations, including studies on gene function, disease modeling, pathogen interaction, co-cultures, and cell fate determination. These defined 2D-Invasin "organoid-sheets" cultures support the realization of clinical applications using healthy and patient-derived intestinal epithelial cells.

Keywords Intestinal epithelial stem cells, Invasin, Human 2D organoid-sheet, Small intestine, Differentiation, Organoid, Transwell

1 Introduction

Over the past 15 years, organoid technology has significantly advanced our understanding of fundamental biological processes and disease mechanisms in the gut [1]. Organoids have proven particularly powerful for studying intestinal homeostasis, for modeling disease pathogenesis, and for understanding host-pathogen interactions [2].

Paloma Ordóñez-Morán (ed.), *Intestinal Stem Cells*, Methods in Molecular Biology, vol. 3059,
https://doi.org/10.1007/978-1-0716-5412-5_6,

Organoids are derived from healthy and patient-specific tissue and can be passaged indefinitely and genetically manipulated. As such, there is growing interest in using epithelial cells derived from organoids for clinical purposes, including regenerative medicine and drug screening [3, 4]. However, despite their widespread use in fundamental research, clinical translation of intestinal organoids has remained limited—partly due to the undefined and variable nature of their culture conditions.

The organoid culture platform is based on a combination of four growth factors that mimic the intestinal stem cell niche. Over time, the three key soluble components—Wntagonist/ligand, epidermal growth factor (EGF), and the bone morphogenetic protein (BMP) inhibitor Noggin—have been refined and standardized [5, 6]. However, the fourth component, the extracellular matrix (ECM), Matrigel, or Basement Membrane Extract (BME), has remained a major bottleneck. Matrigel/BME, derived from Engelbreth–Holm–Swarm (EHS) mouse sarcoma cells, contains a highly complex and poorly defined mixture of some 8000 proteins, one dominant component being laminin-111 [7]. Despite its widespread use as an ECM source in organoid culture, Matrigel presents several limitations, including batch-to-batch variability, high cost, tumor-derived origin, and poorly defined composition. These factors hinder standardization and clinical translation of human intestinal organoids. Of note, organoid research needs to be interpreted with caution, as all results are obtained using culture conditions involving a mouse tumor-derived ECM extract.

To overcome these limitations, we recently introduced a defined, standardized, and animal-free culture system based on Invasin, a recombinant protein fragment that mimics key adhesion signals of ECM proteins found in the adult intestinal epithelium (e.g., laminin-521 and laminin-511) [8]. When coated on plastic or glass surfaces, or in transwell systems, Invasin supports the attachment and long-term expansion of intestinal stem cells in defined media. These stem cells retain their proliferative capacity and differentiation potential, allowing the generation of confluent 2D epithelial monolayers.

The Invasin-based 2D transwell culture system offers several practical and experimental advantages over traditional 3DMatrigel dome cultures. It provides direct access to both the apical and basal sides of the epithelial layer, facilitating studies on host-pathogen interactions, immune cell co-cultures, and dual-sided drug testing. Moreover, differentiation protocols using air-liquid interface (ALI) culture can be easily applied, enabling robust maturation of the epithelial cells. Compared to dome-cultured 3D organoids, these 2D cultures are easier to image, more uniform, and free from the confounding effect of Matrigel/BME background proteins. As such, the Invasin system greatly simplifies organoid handling and analysis, and crucially, enables the development of defined and reproducible protocols suitable for clinical applications.

In this chapter, we describe a detailed protocol for establishing long-term 2D intestinal monolayers using Invasin-coated transwell plates. These cultures can be initiated from either fresh human intestinal biopsies or from pre-existing 3D organoid lines. We further outline the air-liquid interface method for inducing differentiation into functional intestinal epithelial cell types.

2 Materials

2.1 Coating of Transwells with Invasin

1) Transwell system (ThinCerts, 24-well, transparent 3.0 μm pore size).
2) Recombinant Invasin protein.
3) Phosphate-buffered saline (PBS).

2.2 Generation of 2D Invasin Monolayer Cultures from Freshly Isolated Human Intestinal Tissue

1) Fresh human small intestinal tissue biopsy (healthy or tumor tissue).
2) Collagenase.
3) Y-27632.
4) Advanced DMEM/F12.
5) Primocin.
6) 15 mL and 50 mL tubes.
7) 10 cm Petri dishes.
8) 5 mL and 10 mL serological pipettes.
9) Scalpels.
10) 100 μm cell strainer.
11) Trypan blue solution.
12) Hemocytometer or automatic cell counter.
13) Brightfield microscope.
14) Shaker.
15) Intestine washing buffer (ADF+++): 1:100 HEPES (1 M), 1:100 GlutaMax supplement, 100 U/mL penicillin, and 100 μg/mL streptomycin in advanced DMEM/F12.
16) Intestine expansion medium: ADF+++, 50 ng/mL EGF, 2% v/v recombinant Noggin, 3% v/v recombinant R-spondin3, 1.25 mM N-acetylcysteine (NAC), 0.5 μM A83-01, 0.5 nM recombinant WNTsurrogate, 3 μM p38i SB202190, 10 mM Nicotinamide (NIC), 1 μM PGE2, 2% B-27 (min VitaminA).

2.3 Generation of 2D Invasin Monolayer Cultures from 3D Human Intestinal Organoid Lines

In addition to materials listed in Sects. 2.1 and 2.2, the following materials are required:

1) Established 3D human intestinal organoid cultures (grown in Matrigel/BME domes).
2) Dispase.
3) TrypLE.

2.4 Air-Liquid Interphase Differentiation of Invasin-Established Intestinal Monolayers

In addition to materials listed in Sects. 2.1–2.3, the following materials are required for differentiation:

Differentiation patterning medium: ADF+++, 50 ng/mL EGF, 2% v/v recombinant Noggin, 3% v/v recombinant R-spondin3, 1% B-27 (min VitaminA), 1.25 mM N-acetylcysteine (NAC), 0.5 μM A83-01, 0.5 nM recombinantWNTsurrogate, 3 μM CHIR, 2 ng/mL IL-22.

Differentiationmaturation medium: ADF+++, 50 ng/mL EGF, 2% v/v recombinantNoggin, 3% v/v recombinant R-spondin3, 1% B-27 (min VitaminA), 1.25 mM *N*-acetylcysteine (NAC), 0.5 μM A83-01, 16.67 pM recombinant WNTsurrogate, 2 ng/mL IL-22.

3 Methods

3.1 Coating of Transwells with Invasin

1) Twelve hours prior to cell seeding, coat transwell membranes with Invasin.
2) Dilute Invasin protein to a final concentration of 5 μg/mL in PBS.
3) Add 100 μL of the diluted Invasin solution to the upper compartment (ThinCert insert) of each transwell (Fig. 1A), ensuring full coverage of the membrane surface (*see* **Note 1**).
4) Place the ThinCert insert back into the plate (Fig. 1B).
5) Incubate the transwell with the Invasin protein solution at 4°C for 12 h to allow efficient coating of Invasin to the membrane of the ThinCerts (*see* **Note 2**).
6) Immediately prior to cell seeding, aspirate the Invasin solution and proceed with cell plating.

3.2 Generation of 2D Invasin Monolayer Cultures from Freshly Isolated Human Intestinal Tissue

1) Prepare fresh digestion medium in 15 mL tubes: 5 mL ADF+++ supplemented with 1 mg/mL collagenase and 10 μM Y-27632. Pre-warm to 37°C.
2) Place freshly obtained intestinal biopsy tissue in a 10 cm Petri dish. Keep the tissue as cool as possible, for example by placing it on ice, until processing.

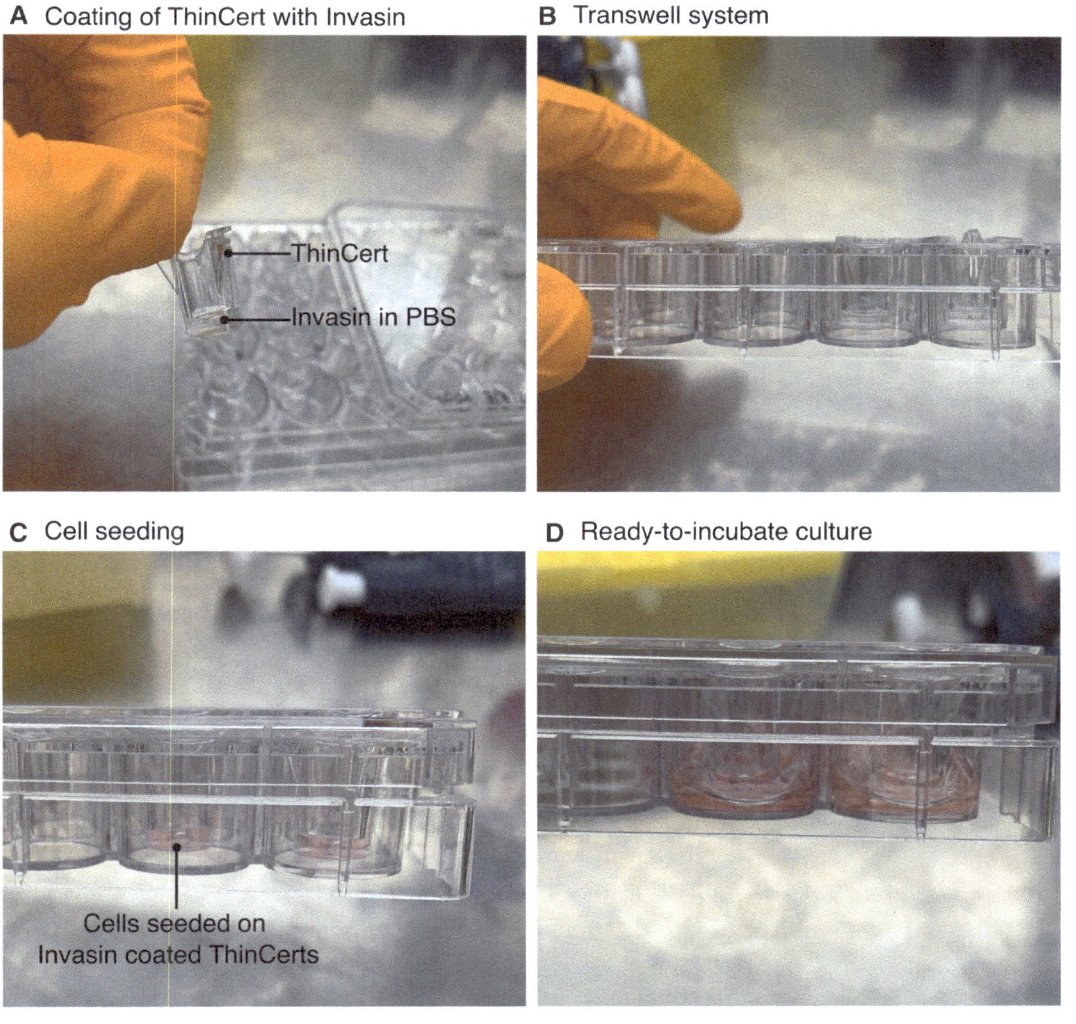

Fig. 1 Preparation of Transwell system for culturing intestinal monolayers on Invasin-coated ThinCerts. (**A**) ThinCert with 100 μL Invasin solution in PBS. (**B**) Complete transwell system with ThinCerts placed in the plate, filled with Invasin solution, ready for overnight coating at 4°C. (**C**) ThinCert immediately after seeding with 100 μL cell suspension in human intestinal expansion medium. (**D**) Complete transwell system with 100 μL cell suspension in the upper compartment and 500 μL of human intestinal expansion medium in the bottom compartment, ready to be placed into a 37°C, 5% CO_2 incubator

3) Using scalpels, mince the tissue into small fragments of ~1 mm^3.
4) Add 5 mL pre-warmed digestion medium to the 10-cm Petri dish with tissue fragments.
5) Transfer the tissue and digestion medium into a 15 mL tube.
6) Resuspend the mixture by pipetting 15 times with a 5-mL pipette to further break down the tissue.

7) Close the 15 mL tube and incubate on a shaker at 37°C for 20 min to enzymatically digest the tissue.
8) Filter the digested suspension through a 100-μm cell strainer placed on a 50 mL tube. Use a 5 mL pipette to assist passage through the strainer.
9) Rinse the cell strainer with 10 mL of cold ADF+++ supplemented with 100 μg/mL primocin and 10 μM Y-27632.
10) Centrifuge the filtrate at 400 × *g* for 5 min at 4°C. Carefully aspirate the supernatant without disturbing the cell pellet.
11) Wash the cell pellet twice with 10 mL ADF+++ supplemented with 100 μg/mL primocin and 10 μM Y-27632. Each time, centrifuge at 400 × *g* for 5 min at 4°C and aspirate the supernatant.
12) Resuspend the final cell pellet in 1 mL human intestinal expansion medium containing 100 μg/mL primocin and 10 μM Y-27632. If cell suspension is too dense for counting, dilute appropriately using the same medium.
13) Count viable cells using Trypan blue with a hemocytometer or automatic cell counter.
14) Seed 100,000–200,000 cells per Transwell insert (either as single cells, cell clumps, or crypt fragments) suspended in 100 μL human intestinal expansion medium supplemented with 100 μg/mL primocin and 10 μM Y-27632 (Fig. 1C).
15) Add 500 μL human intestinal expansion medium (with 100 μg /mL primocin and 10 μM Y-27632) to the lower compartment of the transwell system (Fig. 1D).
16) Place the transwell plate system in a 37°C incubator with 5% CO_2.
17) After 12 h, remove Y-27632 from the culture medium by replacing both the top and bottom compartments with fresh expansion medium supplemented only with 100 μg/mL primocin.
18) Refresh the medium every 2–3 days, maintaining 100 μL in the upper compartment and 500 μL intestinal expansion medium in the lower compartment.
19) Monitor cell growth using brightfield microscopy. Between days 10 and 20, distinct islands of intestinal epithelial monolayers become visible.

3.3 Generation of 2D Invasin Monolayer from Established 3D Human Intestinal Organoid Lines

1) Coat the Transwell membranes with Invasin as described in Steps 1–6 of Sect. 3.1.
2) Pre-warm dispase solution and TrypLE to 37°C.
3) Add pre-warmed dispase (1 mg/mL) directly to the medium of submerged 3D organoid domes in a cell culture plate.

4) Disrupt the domes by pipetting up and down several times using a p1000 pipette.
5) Incubate the disrupted organoids in dispase solution at 37°C for 30 min to dissolve Matrigel/BME.
6) Transfer the organoid-dispase suspension to a 15 mL tube.
7) Add 5 mL cold ADF+++, mix gently, and centrifuge at 400 × *g* for 5 min at 4°C.
8) Aspirate the supernatant and wash the cell pellet with 5 mL cold ADF+++. Centrifuge again at 400 × *g* for 5 min at 4°C.
9) Aspirate the supernatant and add 1 mL of pre-warmed TrypLE to the cell pellet. Pipette up and down ten times with a p1000 pipette to initiate dissociation.
10) Incubate the tube at 37°C for 5 min.
11) After incubation, pipette up and down another ten times to complete mechanical dissociation. Use brightfield microscopy to assess the cell state; a mix of single cells and small clusters is optimal (*see* **Note 3**) (Fig. 2B).
12) To stop the enzymatic reaction, add 5 mL of cold ADF+++.
13) Centrifuge at 400 × *g* for 5 min at 4°C and aspirate the supernatant.
14) Resuspend the final cell pellet in 1 mL human intestinal expansion medium. If cell suspension is too dense for accurate counting, dilute appropriately.
15) Count live cells using Trypan blue with a hemocytometer or automatic cell counter.
16) Seed 50,000–100,000 cells (single cells and/or cell clusters) per Invasin-coated ThinCert in the upper compartment, suspended in 100 μL human intestinal expansion medium (*see* **Note 4**) (Fig. 2C).
17) Add 500 μL of the same medium to the lower compartment of the transwell system.
18) Place the transwell plate system in a 37°C, 5% CO_2 tissue culture incubator.
19) Refresh the medium every 2–3 days, replacing 100 μL in the upper compartment and 500 μL in the lower compartment with fresh human intestinal expansion medium.
20) Within 10–15 days, a homogeneous epithelial monolayer will be visible and suitable for downstream application such as differentiation, imaging, co-culture, or functional assays (*see* **Note 5**) (Fig. 2D).

3.4 Passaging 2D Invasin Monolayer Cultures

1) Prepare Invasin-coated ThinCerts as described in Steps 1–6 of Sect. 3.1.

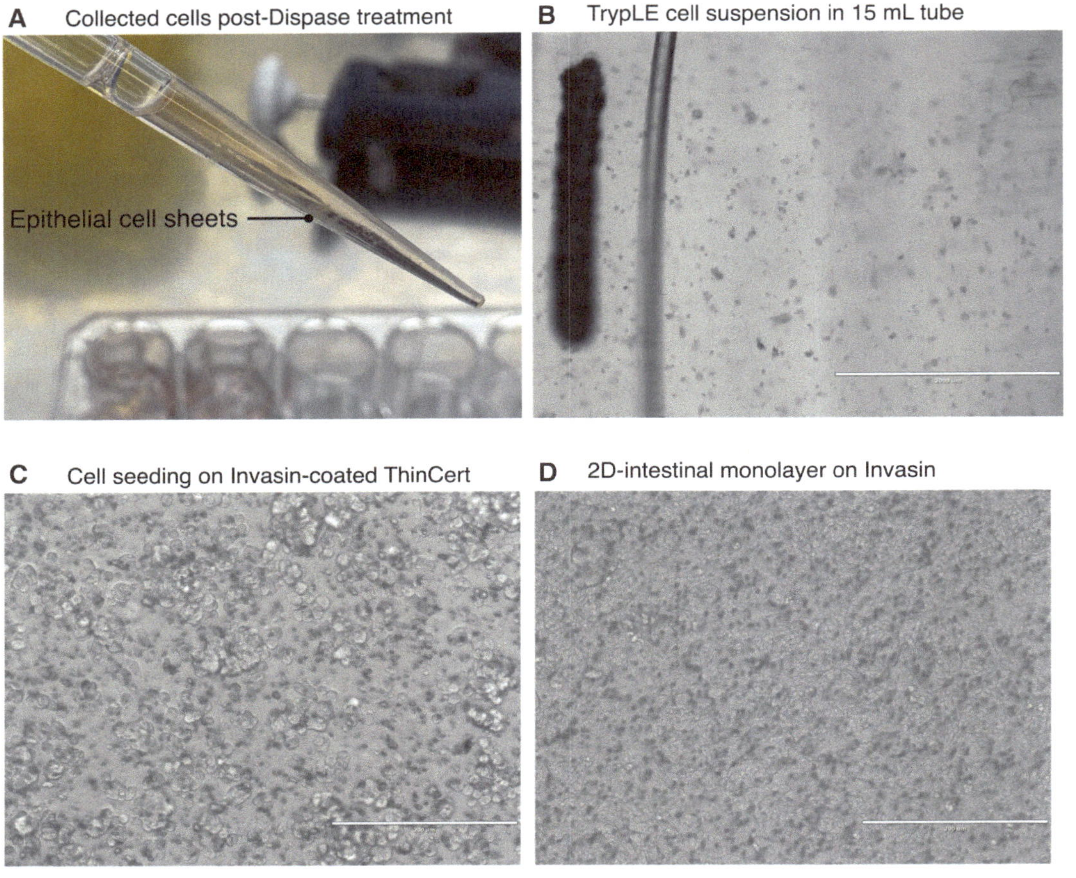

Fig. 2 Passaging of human intestinal epithelial monolayers cultured on Invasin-coated ThinCerts. (**A**) Collection of the epithelial monolayer after dispase treatment by vigorous pipetting using a p1000 pipette; the detached epithelial sheet is visible in the pipette tip. (**B**) Optimal single cell and cell fragment suspension after TrypLE treatment in a 15 mL tube, ready for re-plating. Scale bar: 2000 μm. (**C**) Representative example of seeded single cells and fragments on Invasin-coated ThinCerts shortly after plating. Scale bar: 200 μm. (**D**) Confluent human intestinal epithelial monolayer on Invasin-coated ThinCerts. Scale bar: 200 μm

2) Pre-warm dispase solution and TrypLE to 37°C.

3) When 90%–100% cell confluency is reached (*see* **Note 5**), add dispase directly into the medium present in the transwell system. Final concentration dispase: bottom compartment 4 mg/mL, upper compartment 10 mg/mL.

4) Incubate the plate at 37°C for 30 min to enzymatically loosen the epithelial monolayer from the transwell membrane (*see* **Note 6**).

5) After incubation, use a p1000 pipette to collect the epithelial sheet from the upper compartment. Pipette up and down to dislodge cells. The epithelial layer should be visible in the pipette tip (Fig. 2A).

6) If cells remain on the ThinCert, rinse with 350 μL of cold ADF+++ to collect remaining cells.
7) Transfer the collected cells to a 15 mL tube and add 10 mL of cold ADF+++.
8) Centrifuge the cells at 400 × *g* for 2 min at 4°C.
9) Aspirate the supernatant and wash the pellet with 5 mL cold ADF+++. Centrifuge 400 × *g* for 5 min at 4°C (*see* **Note 7**).
10) Remove supernatant and resuspend the cell pellet in 1 mL of pre-warmed TrypLE and pipette up and down ten times with a p1000 pipette.
11) Incubate at 37°C for 5 min.
12) Following incubation, dissociate cells again mechanically by pipetting ten times using a p1000 pipette. Assess the dissociation under a brightfield microscope; a mixture of single cells and small clusters is optimal (Fig. 2B).
13) Stop the enzymatic TrypLE reaction by adding 5 mL cold ADF+++.
14) Centrifuge at 400 × *g* for 5 min at 4°C and aspirate the supernatant.
15) Resuspend the final cell pellet in 200 μL human intestinal expansion medium (*see* **Note 8**).
16) Plate 100 μL of the cell suspension into each freshly coated Invasin ThinCert (1:2 split ratio) (Figs. 1C and 2C).
17) Add 500 μL of human intestinal expansion medium to the lower compartment of each transwell system (Fig. 1D).
18) Place the Transwell plate in a 37°C, 5% CO_2 incubator.
19) Refresh the medium every 2–3 days, replacing 100 μL in the upper compartment and 500 μL in the lower compartment with fresh human intestinal expansion medium.
20) Continue culture until 90%–100% confluency is again reached.

3.5 Air-Liquid Interphase (ALI) Differentiation

1) Ensure that the 2D intestinal epithelial monolayers cultured on Invasin-coated ThinCerts have reached 100% confluency.
2) Carefully aspirate and remove all medium from both the upper and lower compartments of the Transwell system.
3) Add 500 μL of the patterning medium to the lower compartment only. Leave the upper compartment empty, exposing the apical side of the epithelial monolayer to air.
4) Return the transwell plate to the 37°C, 5% CO_2 incubator.
5) Maintain the culture under ALI condition by refreshing the patterning medium in the lower compartment every 2–3 days for a total duration of 7 days.

6) After 7 days, aspirate the patterning medium and replace it with 500 μL of maturation medium in the lower compartment only.
7) Continue ALI culture for an additional 2–3 days, refreshing the maturation medium once during this period.
8) After a total of 10 days under ALI conditions, the epithelial monolayer should exhibit clear signs of differentiation, with the presence of enterocytes, goblet cells, Paneth cells, and Enteroendocrine cells.
9) Differentiation can be confirmed using immunostaining for lineage-specific markers by fixing the cells in the ThinCerts (Fig. 3) or gene expression analysis via qPCR.

4 Notes

1) In a 24-well ThinCert, which has a surface area comparable to that of a 96-well plate, 100 μL has been shown to sufficiently cover the entire membrane and prevent drying out. The minimum volume required for coating is 50 μL.
2) For efficient coating, incubation at 4°C for 12 h is recommended; however, a 2-h incubation at room temperature has also been shown to be sufficient.
3) The dissociation process of the epithelial cells using TrypLE can be monitored using a brightfield microscope. By placing the transparent 15 mL tube directly under the microscope, cells become visible, allowing you to assess whether single cells and cell clusters are present. If large organoids are still visible, you can extend the incubation time with TrypLE at 37°C to improve dissociation.
4) The number of cells seeded affects the growth rate in 2D culture. Although fewer than 50,000 cells can be plated, it will take longer to reach confluency.
5) For enhanced visualization of cells, the same culture method can be performed on glass or plastic 96-well plates. The plates lack the porous membrane of ThinCerts, making it easier to analyze cells and assess confluency. Although 2D intestinal cultures on glass or plastic surfaces were also successful, the use of Transwell systems proved to support more efficient intestinal growth, allowing for multiple passages. Furthermore, differentiation via the air-liquid interface in Transwell systems was found to be more effective. For visualization of cells in Transwell systems, use the highest magnification and adjust the contrast settings on a brightfield microscope to optimally

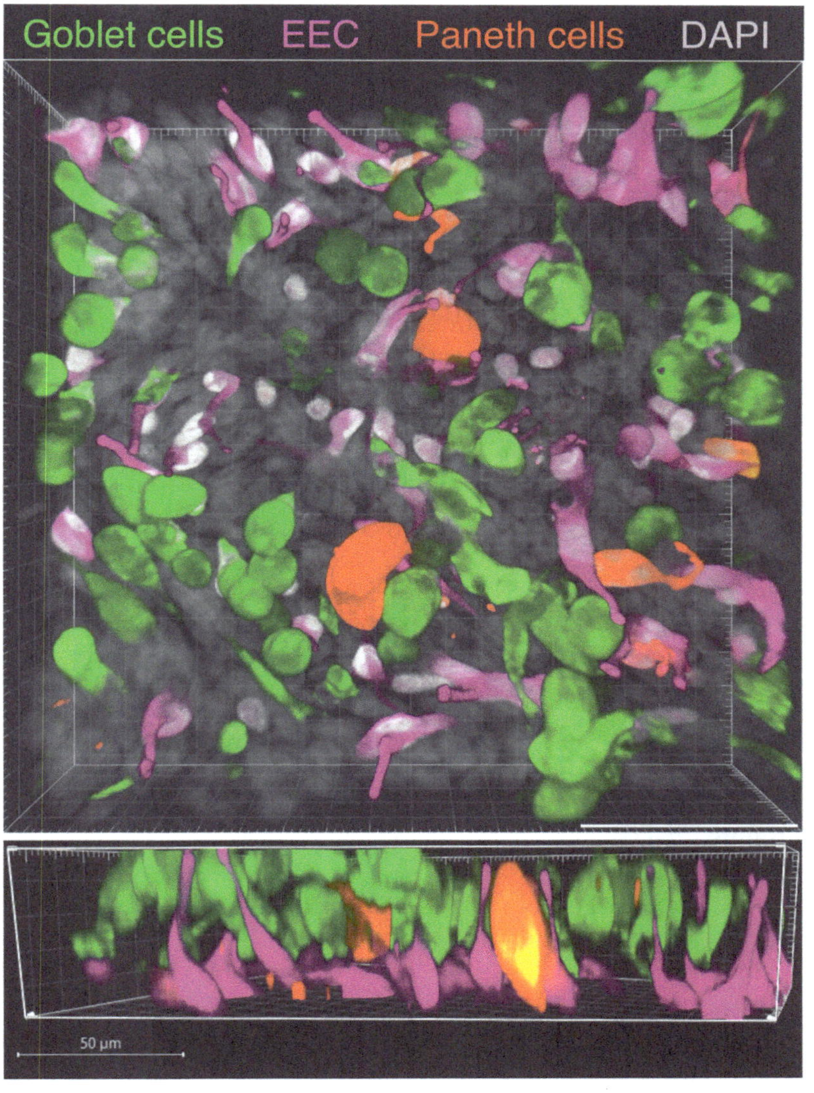

Fig. 3 Representative confocal image of a differentiated human intestinal epithelial monolayer cultured on Invasin-coated ThinCerts. Shown are top-down and orthogonal views of a differentiated monolayer derived from human ileum organoids (triple knock-in reporter line). Goblet cells are marked by MUC2 (green), enteroendocrine cells (EEC) by CHGA (purple), and Paneth cells by DEFA5 (red). Cell nuclei are stained with DAPI (white). Scale bar: 50 µm. (Adapted with permission from Wijnakker et al. [8])

visualize the cells. Other ways to improve cell visualization include adding Calcein-AM (non-toxic) to the bottom compartment or using a leftover transwell insert to stain nuclei with Hoechst (toxic). These stains can be easily visualized using a fluorescence EVOS microscope. However, Hoechst affects cell growth and should not be used for cultures intended for long-term maintenance. Confluency can also be assessed using a Transepithelial Electrical Resistance (TEER) device.

6) Longer incubation with dispase at 37°C facilitates detachment of the entire epithelial layer, making it easier to collect epithelial cells. This detachment of the epithelial sheets could be monitored using brightfield microscopy. The concentration of dispase also plays a role in the efficiency of epithelial sheet release.

7) Dispase is a proteolytic enzyme that, unlike TrypLE, cannot be inactivated by adding ADF+++. Therefore, thorough washing steps are essential to ensure no residual Dispase remains when seeding new cultures. Any leftover dispase may degrade extracellular matrix (ECM) proteins, leading to detachment of cells from the ThinCerts or plate surface, ultimately impairing cell attachment and growth.

8) Passaging of 2D Invasin monolayer culture is not dependent on the presence of Y-27632. While Y-27632 can be added during the first 12 h after seeding to improve initial cell attachment, it should be removed from the culture medium thereafter. Prolonged exposure to Y-27632 can interfere with epithelial cell organization and may prevent the formation of a homogenous intestinal epithelial monolayer.

Acknowledgments

The authors would like to thank Robin Schreurs for advice and critical reading of the manuscript.

References

1. Clevers H (2016) Modeling development and disease with organoids. Cell 165:1586–1597
2. Kretzschmar K, Clevers H (2016) Organoids: modeling development and the stem cell niche in a dish. Dev Cell 38:590–600
3. Drost J, Clevers H (2017) Translational applications of adult stem cell-derived organoids. Development 144:968–975
4. Verstegen MMA et al (2025) Clinical applications of human organoids. Nat Med 31:409–421
5. Sato T et al (2009) Single Lgr5 stem cells build crypt-villus structures in vitro without a mesenchymal niche. Nature 459:262–265
6. Yi Miao et al (2020) Next-generation surrogate wnts support organoid growth and deconvolute frizzled pleiotropy in vivo. Cell Stem Cell 27(5):840–851.e6. https://doi.org/10.1016/j.stem.2020.07.020
7. Orkin RW et al (1977) A murine tumor producing a matrix of basement membrane. J Exp Med 145:204–220
8. Wijnakker JJAPM et al (2025) Integrin-activating Yersinia protein Invasin sustains long-term expansion of primary epithelial cells as 2D organoid sheets. Proc Natl Acad Sci U S A 122:e2420595121

Chapter 7

Primary Cultures of Fetal Epithelial Progenitors

Romane Meerts, Morgane Leprovots and Marie-Isabelle Garcia

Abstract

In this chapter, we present a protocol for isolating fetal epithelial progenitors from the murine small intestine and colon, and for culturing them *ex vivo* using the three-dimensional organoid system. We also describe cellular approaches to characterize Lgr5-negative fetal progenitors and compare their identity and differentiation potential to those of Lgr5-positive adult intestinal stem cells. This system provides a valuable tool for studying the biology of fetal progenitors and exploring the developmental origins of adult intestinal stem cells.

Keywords Organoid, Embryonic, Progenitors, Stem cells, Small intestine, Colon, Mouse

1 Introduction

Intestinal development is highly conserved across mammals. In mice, intestinal development begins around embryonic day (E) 9, when the primitive gut tube forms as a stratified endodermal epithelium surrounded by mesoderm-derived mesenchyme [1]. Between E9.5 and E13.5, the intestine elongates, and both the epithelium and mesenchyme expand, resulting in lumen formation [2]. At E14, the initially stratified epithelium remodels into a simple monolayer in a proximal-to-distal wave. This epithelial reorganization marks the onset of villus morphogenesis, driven by mesenchymal invaginations. Villi appear first in the small intestine (SI) at E15, and later in the large intestine (colon) at E16, corresponding to human gestational weeks 9 and 10, respectively [3]. As villi emerge, epithelial proliferation becomes progressively restricted to the intervillus regions, which will later develop into the adult crypts. By E16.5, villi are well established and cytodifferentiation has begun, with absorptive and secretory lineages appearing within the villus domain [4]. In the adult intestine, the epithelium is structured into crypt-villus units in the SI and into crypt units in the colon. Homeostasis is maintained through continuous epithelial renewal every 4–6 days, a process sustained by an actively cycling stem cell population that specifically

Paloma Ordóñez-Morán (ed.), *Intestinal Stem Cells*, Methods in Molecular Biology, vol. 3059,
https://doi.org/10.1007/978-1-0716-5412-5_7,

expresses the marker Lgr5. These Lgr5-positive (Lgr5^{+ve}) stem cells reside at the base of the crypts and divide approximately every 24 h to generate all epithelial lineages of the adult intestinal epithelium [5]. In 2009, advances in understanding the intestinal stem cell niche led to the development of a three-dimensional *ex vivo* organoid culture system. This technique allows for the long-term expansion of Lgr5^{+ve} adult stem cells as organoids [6]. These organoids mimic the structure and cellular complexity of the intestinal epithelium, including both crypt and villus regions. Similarly, adult colon stem cells can be cultured as organoids, provided that some adjustments are made to the culture medium [7]. Building on this, Mustata et al. [8] adapted the system to fetal intestinal progenitors, whose molecular signature was poorly characterized, and demonstrated that fetal epithelial cells could also be maintained and expanded long-term in this *ex vivo* culture system. Unlike organoids derived from adult stem cells, which exhibit a crypt-like budding morphology, fetal progenitors grow as hollow spheres, termed spheroids, that are poorly differentiated and represent an earlier developmental stage, preceding the Wnt-mediated wave of cytodifferentiation. In the small intestine, the proportion of spheroids versus organoids progressively decreases from the fetal to postnatal period, reflecting the maturation of the epithelium. Notably, fetal spheroids are enriched in markers such as Trop2 (Tacstd2) and Connexin 43 (Cnx43/Gja1), both of which are expressed within the fetal intestinal epithelium at E14 but absent in adults during homeostasis. Noteworthy, fetal enterospheres with similar characteristics can be grown from human intestinal tissue isolated from gestational week 10 [9].

These distinctions between fetal progenitors and adult stem cells underscore the specific cellular and molecular features of the developing intestine. The *ex vivo* culture of fetal intestinal progenitors as spheroids represents a useful model to investigate the mechanisms guiding intestinal development, such as stem cell emergence and lineage specification. Understanding these processes may help to define new therapeutic strategies for treating intestinal diseases and may also contribute to regenerative medicine approaches by highlighting molecular programs shared between fetal intestine and adult tissue regeneration.

2 Materials

2.1 Animals

Transgenic *Lgr5*-EGFP-IRES-*Cre*ERT2 mice on a Hsd:ICR (CD-1) genetic background, aged 2–6 months, were used for this study. This mouse strain was originally generated by the laboratory of Prof. Hans Clevers [5] and is available from The Jackson Laboratory (*Lgr5*$^{tm1(cre/ERT2)Cle}$, JAX stock: #008875).

To time pregnancies, females were mated in the late afternoon and inspected the next morning for the presence of a copulatory vaginal plug, defined as embryonic day (E) 0.5. Upon detection of a plug, male and female mice were separated. Pregnancies progressions were confirmed by daily weight monitoring.

All animals were bred and maintained under specific pathogen-free (SPF) conditions, in accordance with FELASA standards. Environmental conditions were controlled: temperature was maintained at 22 ± 1 °C, relative humidity at 55% ± 10%, and animals were housed under a 12-h light/dark cycle in individually ventilated cages. Food and water were provided *ad libitum*.

All procedures involving animals were conducted in compliance with the European Union guidelines for the care and use of laboratory animals and were approved by the local ethical committee (CEBEA) under the approved protocol for organ collection (939N).

2.2 Dissection of Embryos

1. Stereomicroscope (e.g., Zeiss, Stemi 2000-C, see Fig. 1).
2. Cold light illuminator (e.g., Euromex, EK-1, see Fig. 1).
3. Surgical tools (Fine Science Tool, see Fig. 1):
   a. Dumont #5 and #5 Fine forceps (11251-20 and 11254-20).
   b. Fine and Extra-Fine Scissors (91460-11 and 14084-08).
   c. Disposable scalpels (Swan Morton, 0501).
4. Plastic Petri dish, 92 × 16 mm (Sarstedt, 82.1473.001).
5. 10 mL polypropylene tube, 100 × 16 mm (Sarstedt, 62.9924.284).
6. 1.5 mL reaction tube with caps (Greiner Bio-One, 616201).
7. Dulbecco's phosphate-buffered saline (DPBS), without $CaCl_2$ and $MgCl_2$ (Thermo Fisher Scientific, 14190-094).

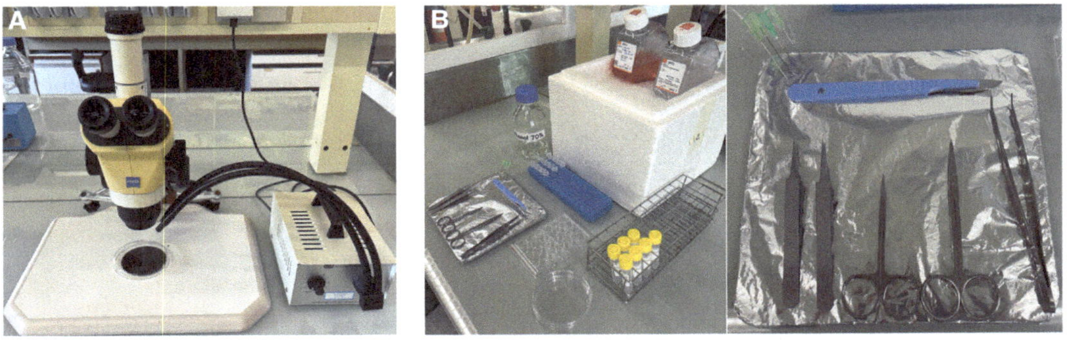

Fig. 1 Representative images of equipment and materials used for isolation of the embryonic mouse intestinal tract. (**A**) Stereomicroscope with a cold light source. (**B**) Ice-cold embryo recovery medium and DPBS, ethanol 70%, 1.5 mL and 10 mL tubes, Petri dish, 24-well plate, and dissection tools

8. Leibovitz's L-15 medium (Thermo Fisher Scientific, 11415-049).
9. D(+)-Glucose (Merck, 1.08337.1000).
10. Embryo recovery medium (*see* Table 1).
11. Ethanol, Eurodenaturated, ≥96% (VWR, 85829.360).
12. Ethanol 70% (v/v) in distilled water.
13. Ice box.

2.3 Isolation of Epithelial Progenitors

1. Class II Biohazard Safety Cabinet (e.g., Esco, AC2-6S1).
2. Refrigerated centrifuge (e.g., Jouan, GR4-11).
3. Orbital shaker (e.g., Edmund Bühler GmbH, KS-15).
4. Cell strainer, pore size 70 μm (Falcon, Corning, 352350).
5. DPBS without $CaCl_2$ and $MgCl_2$ (see Sect. 2.1.7).
6. UltraPure 0.5 M EDTA, pH 8.0 (Thermo Fisher Scientific, 15575020).
7. Intestinal fetal progenitors' isolation buffer (see Table 2).
8. Fetal bovine serum (FBS) (Thermo Fisher Scientific, 10270-106).

2.4 Primary Culture and Maintenance of Fetal Progenitors and Stem Cells

1. Cell culture incubator (37°C, humidified, 5% CO_2).
2. Inverted bright-field microscope equipped with camera (e.g., Motic, AE31E and Moticam Pro 282A).
3. Inverted fluorescent microscope (e.g., Zeiss, Axio Observer 7).

Table 1
Embryo recovery medium

Reagents	Final concentration	Volume	Preparation notes
Leibovitz's L-15 medium	–	500 mL	
Glucose (1 M)	35 mM	17.5 mL	Prepare 1 M stock by dissolving 180 g glucose in 1 L dH_2O. Sterile filter (0.22 μm)

Table 2
Intestinal fetal progenitors' isolation buffer

Intestinal progenitors' isolation buffer	DPBS—EDTA (5 mM)
DPBS (No Mg^{2+}, Ca^{2+})	50 mL
EDTA (0.5 M)	500 μl

4. TrypLE Express Enzyme (1×), phenol red (Thermo Fisher Scientific, 12605028).
5. DPBS without $CaCl_2$ and $MgCl_2$ (see Sect. 2.1.7).
6. Advanced DMEM/F-12 (Thermo Fisher Scientific, 12634010).
7. Recombinant murine EGF (Peprotech, 315-09).
8. Recombinant murine Noggin (Peprotech, AF-250-38).
9. Recombinant murine R-spondin 1 (R&D Systems, 7150-RS-250).
10. HEPES buffer solution (1 M) (Thermo Fisher Scientific, 15630-056).
11. N-acetyl cysteine (Sigma-Aldrich, A7250).
12. N-2 Supplement (100×) (Thermo Fisher Scientific, 17502048).
13. B-27 Supplement (50×), minus vitamin A (Thermo Fisher Scientific, 12587010).
14. Nicotinamide (Sigma-Aldrich, N0636).
15. GlutaMAX Supplement (100×) (Thermo Fisher Scientific, 35050-038).
16. Gentamycin, 50 mg/mL (Thermo Fisher Scientific, 15750037).
17. Amphotericin B, 250 μg/mL (Thermo Fisher Scientific, 15290026).
18. Penicillin-streptomycin (100×) (Thermo Fisher Scientific, 15140122).
19. Y-27632 dihydrochloride (Rho Kinase inhibitor) (Merck, Y0503).
20. 12-well culture plate (VWR, 734-2324).
21. 24-well culture plate (Sarstedt, 83.3922).
22. Matrigel basement membrane matrix (Corning, 354234).
23. Wnt3a-conditioned medium produced with L Wnt-3A cells (ATCC, CRL-2647).
24. Small intestine progenitors and stem cells culture medium (ENR; see Table 3).
25. Colon progenitors and stem cells culture medium (ENR-W; see Table 3).

2.5 Cryopreservation and Thawing

1. Liquid nitrogen tank (−196°C)
2. Ultra-low temperature freezer (−80°C)
3. Advanced DMEM/F-12 (see Sect. 2.3.6).
4. FBS (see Sect. 2.2.8).

Table 3
Fetal progenitor and stem cell culture media for small intestine and colon organoids (*see* Note 8)

Reagents	Final concentration	ENR (small intestine)	ENR-W (colon)
Advanced DMEM/F-12	–	+	+
Wnt3a-conditioned medium (100%)	50 % (v/v)	–	+
Recombinant murine EGF (100 µg/mL)	50 ng/mL	+	+
Recombinant murine Noggin (100 µg/mL)	100 ng/mL	+	+
Recombinant murine R-spondin 1 (100 µg/mL)	80 ng/mL (ENR), 100 ng/mL (ENR-W)	+	+
GlutaMAX Supplement (200 mM)	2 mM	+	+
HEPES (1 M)	10 mM	+	+
B-27 Supplement (without vit. A) (50×)	1×	+	+
N-2 Supplement (100×)	1×	+	+
N-acetyl-L-cysteine (500 mM)	1 mM	+	+
Nicotinamide (1 M)	10 mM	–	+
Penicillin-Streptomycin (100×)	1×	+	+
Amphotericin B (250 µg/mL)	2.5 µg/mL	+	+
Gentamycin (50 mg/mL)	40 µg/mL	+	+
Y-27632 (ROCK inhibitor) (10 mM)	10 µM	+	+

Table 4
Freezing and thawing media

Reagents	Final concentration	Freezing	Thawing
Advanced DMEM/F-12		+	+
FBS (100%)	10% (v/v)	+	–
DMSO (>99.9%)	10% (v/v)	+	–
Y-27632 (10 mM)	10 µM	–	+

5. Dimethyl sulfoxide (DMSO) (Sigma-Aldrich, D8418)
6. Y-27632 (see Sect. 2.3.19).
7. Freezing and thawing media (see Table 4)

8. Cryovial (Greiner Bio One, 123263)
9. CoolCell LX cell freezing container (Corning, 432002).

2.6 Fluorescent Activated-Cell Sorting (FACS)

1. Flow cytometer/cell sorter (e.g., BD Biosciences, FACSAria I).
2. 5 mL round-bottom polystyrene test tube (Falcon, Corning, 352052).
3. Eppendorf Protein LoBind microcentrifuge tube (Eppendorf, VWR, 525-0133).
4. Cell strainer, pore size 70 μm (see Sect. 2.2.4).
5. Cell strainer, pore size 40 μm (Falcon, Corning, 352340).
6. TrypLE Express enzyme (see Sect. 2.3.4).
7. DPBS without $CaCl_2$ and $MgCl_2$ (see Sect. 2.1.7).
8. UltraPure 0.5 M EDTA, pH 8.0 (see Sect. 2.2.6).
9. Bovine serum albumin (BSA) (Sigma-Aldrich, A3294).
10. Y-27632 (see Sect. 2.3.19).
11. ENR and ENR-W culture medium (see Table 3).

2.7 Tissue, Spheroids, Organoids Embedding in OCT Compound

1. DPBS without $CaCl_2$ and $MgCl_2$ (see Sect. 2.1.7).
2. 10 mM phosphate buffer, pH 7.5 (PBS) (see Table 5).
3. Formaldehyde 4% aqueous solution, buffered (VWR, 11699408).
4. Sucrose (Millipore, 1.07651).
5. Ultra pure water.
6. 20% and 30 % (w/v) sucrose solutions in ultra-pure water.
7. Disposable cryomolds biopsy (10 × 10 × 5 mm) (Tissue-Tek, 4565).
8. Tissue freezing medium (OCT compound) (Leica, 14020108926).

Table 5
10 mM Phosphate buffer, pH 7.5 (PBS)

Components (powders)	1 l of distilled water	Preparation notes
$Na_2HPO_4.2H_2O$	1.8 g [10 mM]	Add 800 mL of dH2O to a jar. Dissolve the salt powders in dH2O. Adjust the pH to 7.5. Add dH2O to bring the total volume to 1 l.
KH_2PO_4	0.24 g [1.8 mM]	
NaCl	8 g [137 mM]	
KCl	0.2 g [2.7 mM]	

9. SuperFrost Plus adhesion microscope slide (Epredia, J1800MNZ).
10. FluorSave reagent (antifade mounting medium) (Millipore, 345789).
11. Cover slip, rectangular, 60×24 mm (VWR, 631-1569).

2.8 Immunostainings

1. Cryostat (e.g., Leica, CM3050S).
2. Inverted fluorescent microscope (e.g., Zeiss, Axio Observer 7).
3. Microwave (for antigen retrieval).
4. PBS (see Table 5).
5. Trisodium citrate dihydrate (VWR, 27833.294).
6. Triton X-100 (Sigma-Aldrich, T8787).
7. Horse serum (Thermo Fisher Scientific, 16050122).
8. BSA (see Sect 2.5.9).
9. Primary and secondary antibodies (see Table 8 for detailed list and dilutions).
10. DAPI nucleic acid stain (Sigma-Aldrich, D9542).

3 Methods

3.1 Dissection of Embryos

Disinfect the work surface and dissection tools using 70% ethanol before starting. Although a laminar flow hood is not required, work in a clean environment to minimize contamination. Disinfect tools with 70% ethanol between each embryo to prevent cross-contamination.

1. Sacrifice the pregnant mouse at the desired embryonic stage (E14.5–18.5) using cervical dislocation or an alternative method approved by the institutional ethics committee.
2. Lay the mouse on its back and fix the limbs to a dissection board using pins. Disinfect the abdomen with 70% ethanol. Using forceps, lift the abdominal skin and perform a transverse incision with scissors to expose the uterine horns containing the embryos (see Fig. 2**A**).
3. Excise the uterine horns and transfer them to a Petri dish containing ice-cold embryo recovery medium (see Table 1). Keep the embryos on ice throughout the procedure (*see* **Note 1**).
4. Isolate each embryo with the yolk sac intact and place it on a clean Petri dish under a stereomicroscope equipped with a cold light source (see Fig. 1). Carefully open the yolk sac using fine forceps, avoiding damaging the embryo's abdomen (*see* **Note 2**).

5. Confirm the embryonic stage by external morphological criteria (Theiler [10]) (see Fig. 2**B**).
6. Euthanize each embryo by decapitation before proceeding with digestive tract dissection. If genotyping is required, collect a small tail biopsy.
7. Position the embryo with its left side facing upward. Using two pairs of forceps, make a transverse abdominal incision (perpendicular to the body axis) to access the abdominal cavity. Identify the liver-stomach-intestine complex and gently extract it using forceps (see Fig. 2**C**).
8. Using two pairs of forceps, sequentially remove the liver, pancreas, and mesentery (see Fig. 2**C**). To isolate the small intestine, excise the stomach along with approximately 5 mm of the proximal duodenum. Subsequently, remove the cecum to separate the small intestine from the colon.
   a. Proceed to Sect. 3.7 if tissue embedding in OCT compound is required.
9. Using a scalpel, cut both the small intestine and colon into 5 mm fragments. Transfer the tissue fragments into two separate tubes containing the appropriate volume of ice-cold DPBS (see Table 6).
10. Keep all tubes containing dissected tissues on ice. Proceed to the dissection of the next embryo.

3.2 Isolation of Fetal Progenitors

1. Centrifuge the tubes at 230 × *g* for 5 min at 4°C.
2. Carefully discard the supernatant.
3. Resuspend the tissue pellet in an adequate volume of ice-cold DPBS supplemented with 5 mM EDTA (see Table 6).
4. Incubate the tubes horizontally in an ice-containing box on a shaker at 75 rpm for 30 min.
5. Centrifuge at 230 × *g* for 5 min at 4°C.
6. From this point onward, all steps should be performed under a tissue culture hood.
7. Aspirate the supernatant and resuspend each pellet in an appropriate volume of sterile DPBS (see Table 6).
8. Mechanically dissociate the epithelial layer from the mesenchyme by pipetting up and down with a P1000 micropipette pre-coated with FBS (*see* **Note 3**).
   The number of pipetting cycles should be adjusted according to the developmental stage. Typically, 12 cycles for E14-15 tissues, and 15–20 cycles for E16-18 tissues.
9. Filter each suspension through a 70 μm cell strainer into a new tube (1.5 mL reaction or 10 mL conical tube depending on volume).

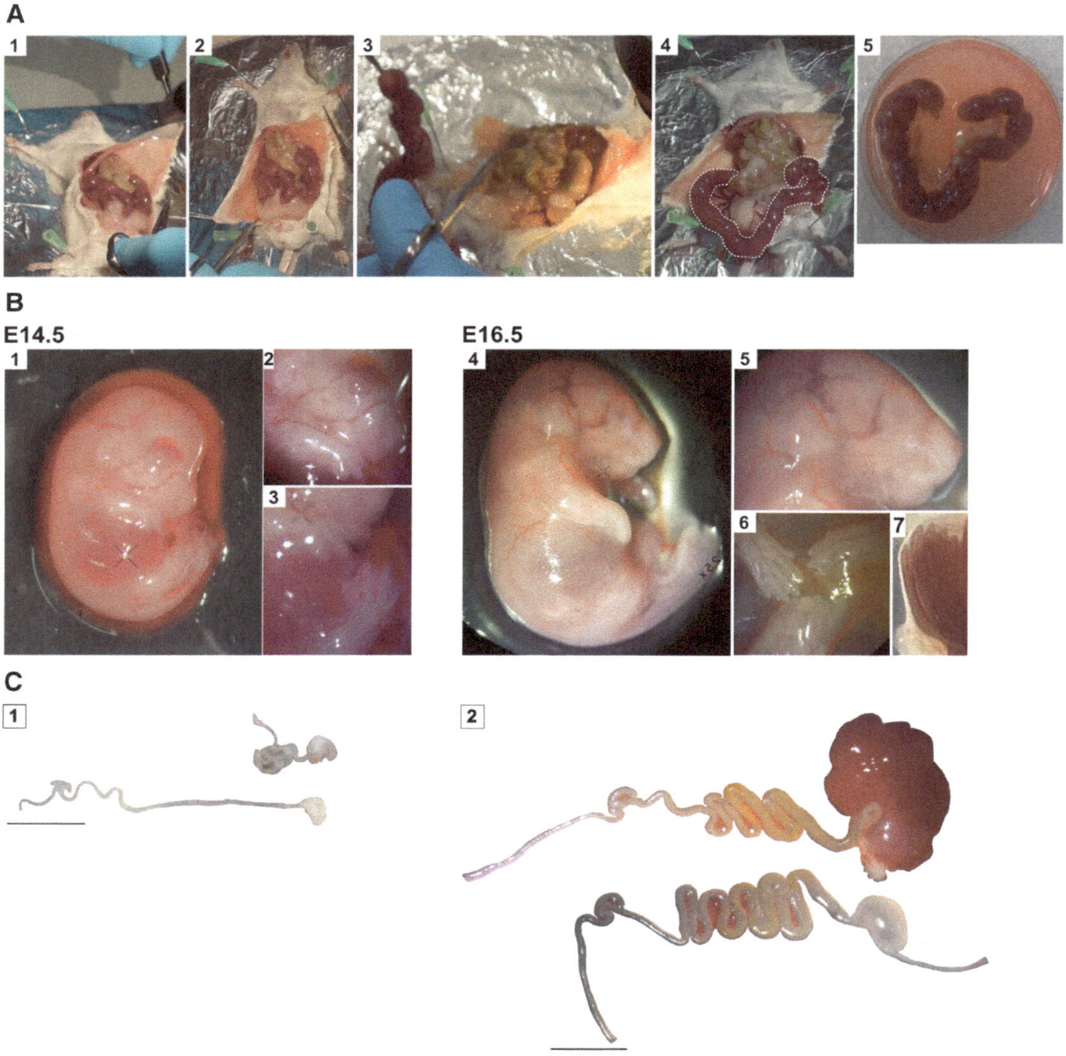

Fig. 2 Isolation of the fetal mouse gastrointestinal tract. (**A**) Representative images illustrating the step-by-step dissection of a pregnant mouse at embryonic day (E) 14.5 and the isolation of uterine horns containing embryos. Uterine horns are delimited by a dotted line. (**B**) Representative images of E14.5 (1) and E16.5 embryos (4), with magnified views of external morphological criteria used for developmental staging [10]. Staging is based on vaginal plug detection and external morphological criteria. At E14.5, digits in the forefoot are separated (2), while those in the hindfoot show deep indentations but remain fused (3). The pinna extends over approximately half of the external auditory meatus (2). Numerous hair follicles are visible across the skin, except on the head, and the eyelids remain open (2). At E16.5, digits 2–5 in the forefoot are almost parallel (6), and hindfoot digits are fully separated (7). The pinna nearly covers the external auditory meatus, eyelids are fused, and a transparent epithelial membrane covers the cornea (5). The skin appears wrinkled (6). (**C**) Dissection and cleaning of the embryonic gastrointestinal tract from E14.5 (1) to E16.5 (2) embryos, preserving the integrity of the duodenum, jejunum, ileum, cecum, and colon. These tissues are subsequently processed to isolate epithelial progenitor cells from the small intestine (duodenum, jejunum, and ileum) or the colon, which are then seeded in a 3D ex vivo culture system. At E16.5, note the presence of meconium in the small intestine, which gradually accumulates during embryonic development (2). Scale bar, 5 mm

Table 6
Recommended buffer volumes (DPBS or DPBS–EDTA 5 mM) for intestinal fetal progenitors' isolation at different developmental stages

Developmental stage	Volume/small intestine	Volume/colon
E14.5	0.5 mL	0.5 mL
E15.5	1 mL	1 mL
E16.5	1.5 mL	1 mL
E17.5	3 mL	1.5 mL
E18.5	3 mL	1.5 mL

10. Evaluate the density cell suspension using an inverted bright-field microscope to determine appropriate plating design in Sect. 3.3.1.
11. Centrifuge at 300 × *g* for 5 min at 4 °C.
12. Discard the supernatant. The resulting pellet can be used for downstream applications.
    a. Proceed to Sect. 3.3 for the primary culture of fetal intestinal progenitors and stem cells (*see* **Note 4**).
    b. Alternatively, proceed as described in Sect. 3.5 for single-cell isolation via FACS.

3.3 Primary Culture of Fetal Progenitors and Stem Cells

Before starting, thaw Matrigel aliquots on ice and always keep them on ice throughout the process. Minor temperature changes can trigger polymerization.

1. Resuspend the pellet (from Step 12, Sect. 3.2) in the appropriate volume of Matrigel (see Table 7) and homogenize the suspension thoroughly on ice (*see* **Note 5**).
2. Plate the Matrigel–cell suspension as a central drop in the well. Using a pipette tip, spread the drop from the center to cover the well bottom, avoiding contact with the walls.
3. Incubate the plate at 37°C for 15 min to allow the Matrigel to polymerize.
4. After polymerization, add the appropriate volume of ENR or ENR-W seeding medium for small intestine and colon progenitors, respectively (see Tables 3 and 7).
5. Return the plate to the incubator (37°C, 5% CO_2).
6. After 2 days, replace the seeding medium withmaintenance medium to remove the Rho-associated kinase (ROCK) inhibitor.
7. Medium should be replaced every 2–3 days using ENR or ENR-W maintenance medium (see Table 3).

Table 7

Recommended well formats, Matrigel, and medium volumes for initial seeding of fetal progenitors and stem cells isolated from small intestine or colon, based on developmental stage

Developmental stage	Well-format	Number of wells per biological sample	Matrigel volume	Medium per well
Small intestine-derived progenitors and stem cells				
E14-16	12-well	×1	100 μl	650 μl
E16-18	6-well	×1	200 μl	1.3 mL
Colon-derived progenitors and stem cells				
E14-16	48-well	×1	25 μl	200 μl
E16-18	24-well	×1	50 μl	350 μl

Note that the morphology of cultured elements isolated from the small intestine varies depending on the embryonic stage of the samples, appearing either as spheroids or organoids (see Fig. 3**A**), whereas colon-derived cells generally grow as spheroids (see Fig. 3**B**).

3.4 Passaging Fetal Progenitors and Stem Cells

Small intestine-derived spheroids/organoids should be passaged after 5–7 days in culture, while colon-derived organoids typically require 8–10 days, depending on their growth rate. Passage can be performed either by selecting specific structures to expand a defined population (Sect. 3.4.1), or by harvesting part or the entirety of the well for culture maintenance (Sect. 3.4.2).

3.4.1 Replating by Element Selection

1. Using an inverted bright-field microscope, select specific spheroids or organoids to be collected by surrounding the desired structures with a fine-tip permanent marker.
2. Prepare one 1.5 mL tube per sample, each containing 1 mL of ice-cold DPBS. Using a P200 micropipette pre-filled with 50 μl of DPBS, aspirate the selected elements from the well and transfer them to the prepared tube. Repeat as necessary.
3. Centrifuge the tube at 300 ×*g* for 5 min.
4. Remove the supernatant.
5. (a) For small intestine-derived spheroids/organoids: Resuspend the pellet in 300 μl of ice-cold DPBS. Mechanically dissociate the structures by pipetting up and down with a P200 micropipette (~40 cycles), until large fragments are no longer visible resulting in a homogeneous cell cluster suspension. (b) For colon-derived organoids: Add 300 μl of TrypLE Express and incubate at 37°C for 5–7 min for enzymatic dissociation (*see* **Note 6**). Follow with mechanical dissociation by pipetting

up and down (~20 cycles) with a P1000 micropipette to achieve fragmentation into cell clusters and single-cell suspension.

6. Add 1 mL of DPBS to wash the cell suspension and/or to dilute the TrypLE Express.
7. Centrifuge at 300 × *g* for 5 min.
8. Remove the supernatant and resuspend the pellet in an appropriate volume of Matrigel (see Table 7).
9. Proceed with Steps 1–7 outlined in Sect. 3.3 for embedding and plating.

3.4.2 Replating Without Element Selection

Based on our experience, a 1:2 split ratio is generally used for small intestine-derived spheroids and organoids, while a 1:3 ratio is used for colon-derived organoids. These ratios should be adjusted depending on culture density or culture conditions. Excessive replating density may result in the accumulation of cellular debris and reduced culture quality.

1. Aspirate the culture medium from the well.
2. Add 1 mL of ice-cold DPBS and disrupt the Matrigel by pipetting up and down with a P1000 micropipette until a homogeneous suspension is obtained. This step releases the embedded spheroids and organoids.
3. Transfer the entire suspension (or a fraction, depending on the density and experimental needs) into a new tube.
4. Add 1–3 mL of ice-cold DPBS to further solubilize the Matrigel and wash out cellular debris. The volume should be adjusted based on well format and culture density.
5. Centrifuge at 300 × *g* for 5 min.
6. Discard the supernatant.

   a. Proceed with Steps 4-8 outlined in Sect. 3.4.1 for dissociation, washing, neutralization, embedding, and plating.
   b. If single-cell isolation by FACS is required, refer to Sect. 3.5.

3.5 Fluorescence Activated Cell Sorting (FACS) Isolation

The use of transgenic mouse models, such as *Lgr5*-EGFP-IRES-*Cre*ERT2 mice, enables the efficient isolation of specific cell populations (e.g., $Lgr5^+$ adult intestinal stem cells) by FACS, based on endogenous expression of the fluorescent reporter EGFP.

3.5.1 Single-Cell Suspension Preparation

Fetal small intestine or colon epithelium (cellular clusters)

1. Resuspend the pellet (from Step 11, Sect. 3.2) in 1 mL of TrypLE Express supplemented with 10 μM Y-27632.
2. Shake at 70 rpm and 37°C for 20 min.

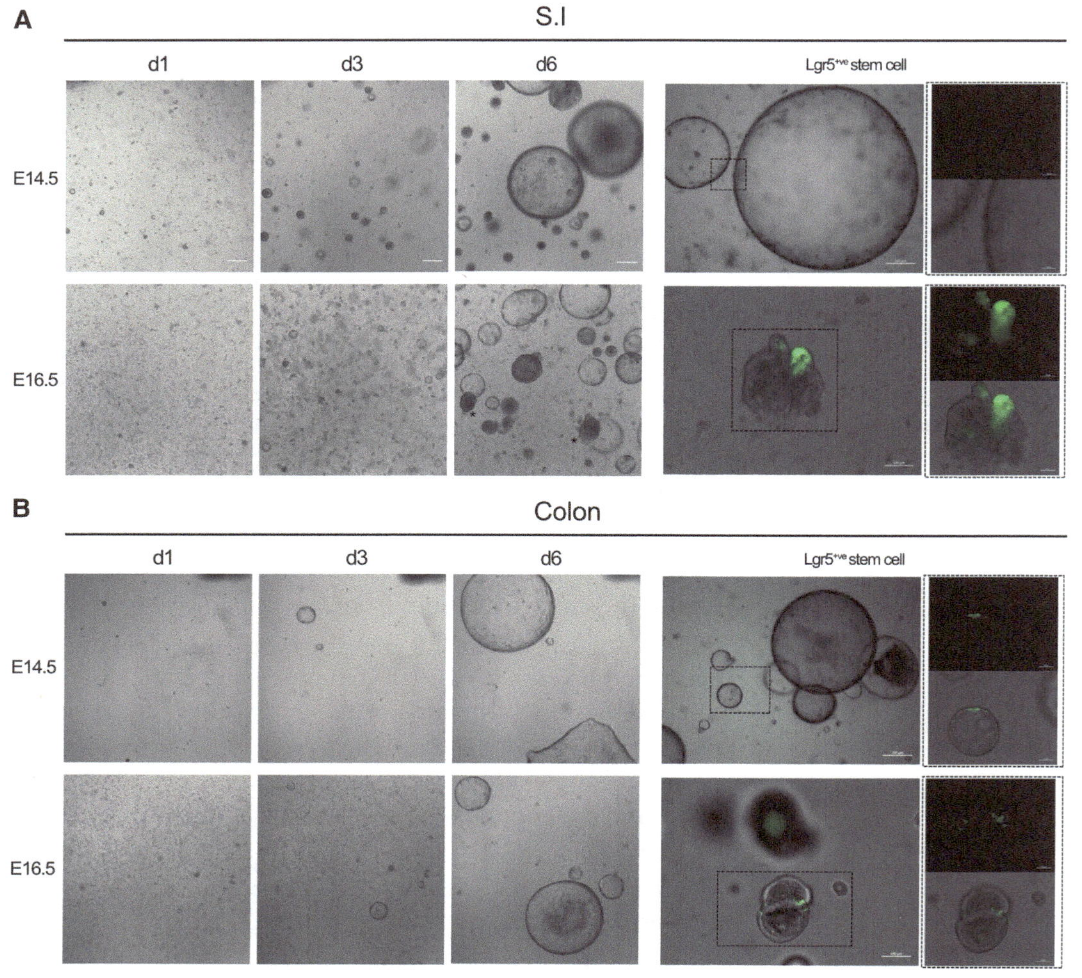

Fig. 3 Primary culture of fetal intestinal progenitors. Bright-field and fluorescent images showing the ex vivo culture of heterozygous Lgr5 EGFP-IRES-CreERT2 fetal epithelial progenitors isolated from either the small intestine (S.I) or colon at embryonic days (E) 14.5 and 16.5. For each condition, representative bright-field images were taken at days 1, 3, and 6 at initial seeding. A GFP fluorescent image acquired between days 6 and 8, depending on the condition, is included to assess Lgr5 expression. (**A**) In the S.I, E14.5-derived progenitors give rise to spherical structures (termed spheroids), whereas E16.5-derived progenitors generate a heterogeneous population of spheroids and more complex budding structures (termed organoids, marked with asterisks). Spheroids are derived from fetal progenitors, while organoids are constituted by adult-like stem cells within the ex vivo 3D culture system. Lgr5-GFP fluorescence serves as a marker for the emergence of adult intestinal stem cells. At E14.5, fetal progenitors-derived spheroids remain GFP^{-ve}, whereas at E16.5, GFP signal is observed in organoid-like structures containing adult stem cells. (**B**) In the colon, no significant morphological differences are observed between E14.5 and E16.5 cultures, both stages predominantly form spheroids. This morphological uniformity may result from the addition of Wnt3a-conditioned medium in the culture medium. Notably, in contrast to the small intestine, a subset of E14.5 colon spheroids expresses GFP in the ex vivo 3D culture system. The proportion of GFP^{+ve} cells and spheroids appears to increase with developmental age, reflecting progressive *in vivo* emergence of the colonic Lgr5^{+ve} stem cell population. Scale bars, 200 μm (brightfield images and GFP-fluorescence images at E14.5), 100 μm (GFP-fluorescence images at E16.5), 50 μm (GFP-fluoresence magnified views)

3. Monitor dissociation progress using an inverted bright-field microscope.
4. Further dissociate residual cellular clusters by pipetting up and down with a P1000 micropipette (~20 cycles). The number of cycles should be adjusted based on visual assessment using an inverted bright-field microscope.
5. Once most of the suspension is composed of single cells, neutralize the TrypLE Express by adding 3 mL of FACS buffer (2% BSA, 2 mM EDTA, and 10 μM Y-27632 in PBS).
6. Filter the suspension through a 70 μm cell strainer into a new tube.
7. Centrifuge at 300 × *g* for 10 min at 4°C.
8. Discard the supernatant.
9. Resuspend the pellet in 1 mL of FACS buffer.
10. Filter the suspension through a 40 μm cell strainer into a new tube to remove residual clumps.

3.5.2 Spheroids or Organoids

1. Resuspend the pellet (from Step 6, Sect. 3.4.2) in 0.5 mL of TrypLE Express supplemented with 10 μM Y-27632.
2. Incubate at 37°C for 10 min.
3. Mechanically dissociate residual clusters by pipetting up and down with a P1000 micropipette (20–30 cycles).
4. Assess the dissociation level using an inverted bright-field microscope.
5. Once most of the suspension is composed of single cells, neutralize the TrypLE Express by adding 2 mL of FACS buffer.
6. Centrifuge at 300 × *g* for 10 min at 4°C.
7. Discard the supernatant.
8. Resuspend the pellet in 0.5 mL of FACS buffer.

3.5.3 Cell Sorting, Collection, and Seeding in Culture

Sort cells based on endogenous GFP fluorescence (*Lgr5*-EGFP-IRES-*Cre*ERT2). If required, epithelial cells can be isolated by surface staining with anti-EpCAM antibody (BD Biosciences, 563477) prior to sorting. Alternatively, fetal progenitors can be isolated using the same procedure by staining with fluorochrome-conjugated antibody against the fetal progenitor marker Trop2 (Tacstd2) (R&D Systems, FAB1122A).

1. Collect sorted cell populations into 1.5 mL low-binding reaction tubes containing 1 mL of ENR (for small intestine) or ENR-W (for colon) culture medium, both supplemented with 10 μM Y-27632. Keep all tubes containing cells on ice.
2. Centrifuge at 300 × *g* for 5 min.

3. Discard the supernatant.
4. Proceed with embedding and plating steps as described in Sect. 3.3, Steps 1–7.

3.6 Cryopreservation of Small Intestine and Colon-Derived Spheroids/Organoids

3.6.1 Freezing Protocol

This protocol is optimized for spheroid/organoid cultures grown in a 12-well format at moderate to high density. Reagent volumes can be scaled proportionally based on the culture format and number of wells. Cryopreservation is recommended during the proliferative phase, prior to full maturation. This corresponds approximately to days 2–3 for small intestine-derived organoids and days 4–5 for colon-derived organoids. These time windows may vary depending on culture conditions.

1. Aspirate the culture medium from the well.
2. Add 1 mL of ice-cold DPBS and disrupt the Matrigel by pipetting up and down with a P1000 micropipette until a homogeneous suspension is obtained, releasing the embedded spheroids and organoids.
3. Transfer the entire suspension into a new 10 mL conical tube.
4. Add 3 mL of ice-cold DPBS to further solubilize the Matrigel and remove cellular debris.
5. Centrifuge the tube at 150 × *g* for 10 min at 4°C.
6. Discard the supernatant.
7. Resuspend the pellet in 1 mL of freezing medium (see Table 4).
8. Transfer the suspension into pre-labeled cryovial and place the vial in a controlled-rated freezing container.
9. Put the container in a −80°C freezer.
10. After 24–48 h, transfer the vial to a liquid nitrogen tank for long term storage.

3.6.2 Thawing Protocol

Before starting the thawing procedure, thaw Matrigel aliquots on ice and keep them cold throughout the process. Pre-warm a water bath to 37°C. For each thawed sample, prepare a 10 mL conical tube containing 3 mL of pre-warmed thawing medium (see Table 4).

1. Take out the cryovial from the liquid nitrogen tank and immediately place it on dry ice.
2. Thaw the vial in a 37°C water bath with continuous agitation.

3. Just before complete thawing, remove the vial from the water bath and disinfect it with 70% ethanol.
4. Transfer the content of the vial into the tube containing 3 mL of pre-warmed thawing medium. Mix gently by pipetting up and down (~5 cycles) with a P1000 micropipette.
5. Centrifuge at 150 × *g* for 10 min.
6. Aspirate the supernatant carefully.
7. Proceed with embedding and plating as described in Sect. 3.3, Steps 1–7.

3.7 Embedding of Tissue or Spheroids/Organoids in Optimal Cutting Temperature (OCT) Compound

3.7.1 Collection and Fixation Tissue

1. Following dissection (Sect. 3.1, Step 8), transfer tissue (small intestine or colon) into a 1.5 mL reaction tube.
2. Add 500 μl of formaldehyde 4% aqueous solution, buffered (PFA).
3. Fix the tissue by incubating it overnight at 4°C.
4. Proceed to tissue cryopreservation and then to embedding in OCT compound as described in Sects. 3.7.2 (Steps 1–6) and 3.7.3 (Steps 1 and 5–9), respectively.

3.7.2 Spheroids or Organoids

1. At the desired time point, aspirate the culture medium from the wells
2. Add 1 mL of ice-cold DPBS.
3. Gently disrupt the Matrigel by pipetting up and down using a P1000 micropipette. Avoid vigorous pipetting to preserve structural integrity.
4. Transfer the suspension to a 1.5 mL reaction tube pre-coated with FBS to minimize cell adhesion.
5. Centrifuge at 150 × g for 10 min at 4°C.
6. Carefully remove the supernatant.
7. Add 500 μl of PFA and fix for 20 min at room temperature. To avoid aggregation, gently tap the tube to resuspend the pellet. Do not pipette after adding PFA.
8. Centrifuge at 150 × *g* for 10 min.
9. Discard PFA into a designated toxic waste container.
10. Wash the pellet once with 1 mL of PBS.
11. Centrifuge at 150 × *g* for 10 min.
12. Discard the supernatant.

3.7.3 Cryopreservation

1. Prepare fresh 20% and 30% (w/v) sucrose solutions in ultra-pure water. For short-term use, store at 4°C; for long-term storage, keep at −20°C.
2. Add 500 μl of 20% sucrose solution to each tube.
3. Incubate overnight at 4°C.
4. If the sample is sedimented, remove the 20% sucrose solution and replace it with 500 μl of 30% (w/v) sucrose solution.
5. Incubate overnight at 4°C.
6. Once sedimented, remove the 30% sucrose solution as much as possible (*see* **Note** 7).

3.7.4 Embedding in OCT Compound

1. Add a thin layer of OCT compound to the bottom of a cryomold (10 × 10 × 5 mm) to prevent direct contact between the specimen and the cryostat blade during sectioning.
2. Add approximately 500 μl of OCT compound to each tube.
3. Cut off the first third of a P200 micropipette tip and pre-load it with ~30 μl of OCT compound.
4. Gently aspirate the sample into the pipette tip, avoiding air bubbles.
5. Transfer the sample into the cryomold in a homogeneous manner.
6. Flash-freeze the mold by placing it flat on dry ice.
7. Once fully frozen, store the OCT blocks at −20°C or −80°C (for long-term storage) until sectioning.
8. Using a cryostat, collect 6 μm sections on SuperFrost Plus slides.
9. Store the slides at −20°C prior to staining.

3.8 Immunostaining on OCT-Embedded Sections

1. Air-dry the cryosections at room temperature for 20 min.
2. Wash the slides in 10 mM phosphate buffer (PBS), pH 7.5 (see Table 5), for 2 min; repeat twice (total of three washes).
3. Antigen retrieval (if required)
   a. Prepare 10 mM citrate buffer by dissolving 1.47 g of trisodium citrate dihydrate in 500 mL of ultra-pure water. Adjust pH to 6.0 using HCl or NaOH.
   b. Place slides in a heat-resistant staining jar filled with citrate buffer.
   c. Perform heat-induced antigen retrieval as follows: heat the buffer until boiling, pause to allow a slight temperature decrease, and reheat. Repeat this boil-cool cycle for 5 min; repeat this cycle three times.

d. Allow the slides to cool down to room temperature for 15 min.

e. Wash slides in PBS for 2 min; repeat twice.

4. Permeabilize cell membranes with 0.1% (v/v) Triton X-100 in PBS (permeabilization buffer) for 10 min at room temperature.
5. Block non-specific binding sites by incubating slides for 1–2 h at room temperature in a blocking buffer composed of 1% (w/v) bovine serum albumin (BSA), 5% (v/v) horse serum, and 0.1% (v/v) Triton X-100 in PBS.
6. Apply 20–25 μl of primary antibodies diluted (see Table 8) in blocking buffer to each section and incubate overnight at 4°C in a humidified chamber to prevent drying. Antibody dilutions should be optimized empirically. A simple humidified chamber may be built using a culture dish with moistened filter paper.
7. Wash slides in PBS for 4 min; repeat twice.

Table 8
List of primary and secondary antibodies used and their recommended dilution factors

Antibody	Dilution	Host	Catalog number	Source
Trop2 (Tacstd2)	1:75	Goat	AF1122	R&D systems
Connexin 43 (Gja1)	1:100	Rabbit	3512	Cell signaling
CD44	1:100	Rat	550538	BD pharmingen
β-Catenin	1:100	Mouse	610154	BD pharmingen
GFP	1: 200	Rabbit	A-11122	Invitrogen
UEA I-biotin conjugated	1:200	/	150141	GALAB Technologies GmbH
Anti-goat AF488 secondary antibody	1:500	Donkey	705-545-003	Jackson Immuno Research
Anti-rabbit AF488 secondary antibody	1:500	Donkey	711-545-152	Jackson Immuno Research
Anti-rabbit Cy3 secondary antibody	1:500	Donkey	711-165-152	Jackson Immuno Research
Anti-rat Cy3 secondary antibody	1:500	Donkey	712-165-153	Jackson Immuno Research
Anti-mouse AF647 secondary antibody	1:500	Donkey	715-605-150	Jackson Immuno Research
APC Streptavidin	1:100	/	016-130-084	Jackson Immuno Research

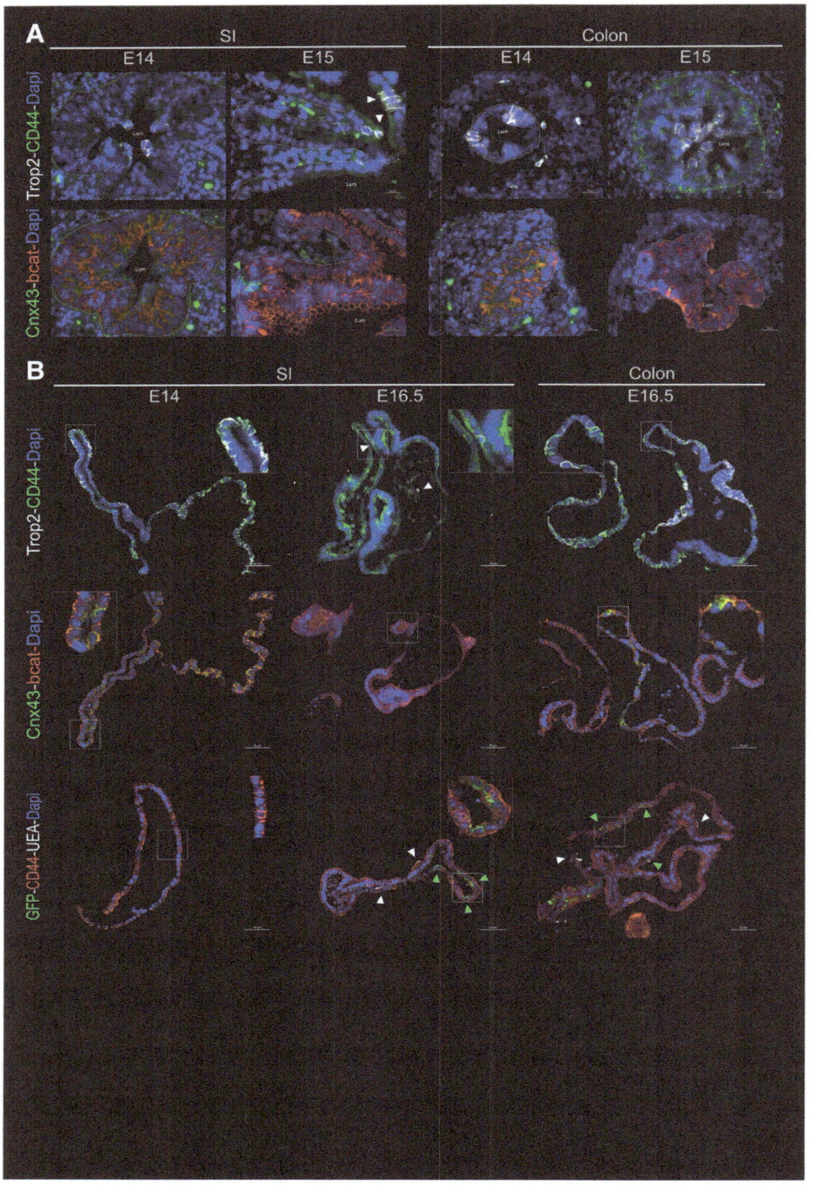

Fig. 4 Immunofluorescence staining of fetal intestine and primary culture of fetal intestinal progenitors. Fluorescent images show immunostaining for the epithelial marker β-catenin (β-cat), fetal markers Trop2 and Connexin 43 (Cnx43), and the adult crypt compartment marker CD44 on (**A**) frozen sections of mouse small intestine (SI) and colon at embryonic days (E) 14 and 15, and (**B**) ex vivo cultures derived from fetal progenitors isolated from SI at E14 and E16.5, and from colon at E16.5. White and green arrowheads indicate Trop2^{+ve} and Cnx43^{+ve} epithelial cells, respectively. In panel (**B**), additional stainings were performed for the adult intestinal stem cell marker Lgr5 (reported via EGFP signal) and the differentiated Goblet cell marker UEA-lectin, in combination with CD44. White and green arrowheads indicate UEA^{+ve} (Mucin-producing) and Lgr5^{+ve} cells, respectively. DAPI was used to counterstain nuclei. Lum: lumen. Scale bars: 50 μm (ex vivo cultures), 20 μm (tissue sections). (**A**) At E14, Trop2 and Cnx43 are expressed by fetal epithelial progenitors in both the small intestine (SI) and colon. By E15, Trop2 expression in the SI becomes restricted to a few cells at the villus tips,

8. Incubate with fluorochrome-conjugated secondary antibodies and nucleic acid stain (e.g., DAPI) diluted in blocking buffer, for 1 h at room temperature, protected from light.
9. Wash slides in PBS for 2 min; repeat twice.
10. Gently blot away excess PBS using filter paper. Ensure no liquid remains, but do not allow sections to dry.
11. Apply approximately three drops of antifade mounting medium (e.g., FluorSave) to each slide.
12. Mount coverslips (24 × 60 mm), avoiding introducing air bubbles.
13. Allow slides to dry in the dark for half a day at room temperature. For short-term storage, keep at 4°C, protected from light. For long-term storage, transfer them to −20°C.

As shown in Fig. 4, the *ex vivo* culture system accurately reflects the temporal and regional differences observed *in vivo* in the expression of fetal progenitor markers (Trop2/Tacstd2, Cnx43/Gja1), adult stem cell markers (Lgr5/EGFP), and differentiated Goblet cell markers (UEA-lectin).

4 Notes

1. Embryo recovery medium is used to maintain isolated embryos at physiological osmolarity and prevent desiccation during dissection prior to tissue isolation. Dissection of all embryos should be completed within 90 min after incubation in the recovery medium to minimize tissue degradation.
2. Be careful when removing the umbilical cord at E14.5–15.5 stages; the tissue is very soft, and this may pull out the internal digestive tract and compromise the rest of the dissection.
3. Pre-coat micropipette tips with FBS to prevent tissue fragments or cells from adhering to the tip. This precaution minimizes sample loss during mechanical dissociation.

Fig. 4 (Continued) while Cnx43 is detected only in rare cells within the intervillus region. In the colon at E15, Trop2 remains widely expressed, whereas Cnx43 expression is reduced, appearing in only a few cells in the intervillus region. CD44 expression in the SI remains lowly expressed between E14 and E15, whereas in the colon, an upregulation of CD44 is observed at E15 compared to E14. (**B**) Ex vivo spheroids derived from E14.5 SI progenitors recapitulate the expression patterns of Trop2 and Cnx43 observed in tissue sections at E14. In organoids derived from E16.5 SI progenitors, expression of both Trop2 and Cnx43 is reduced, indicating progression toward differentiation. Organoids derived from E16.5 colon progenitors display heterogeneous expression of Trop2 and Cnx43. Staining for Lgr5-EGFP and UEA-lectin reveals that organoids from both SI and colon at E16.5 exhibit differentiation features compared to those derived from E14 progenitors

4. This procedure enables the isolation of small fetal intestine or colon epithelium as cellular clusters or aggregates. In culture, fetal progenitors and stem cells from the small intestine give rise to hollow spheres (termed spheroids) and organoid structures, respectively [8]. In contrast, for fetal and adult colon epithelium, the addition of Wnt-3A to the culture medium does not allow phenotypic distinction between progenitor- and stem cell-derived structures. Note that mouse fetal progenitors from the pre-glandular stomach can be isolated and grown *ex vivo* using the same experimental protocol and culture conditions as those used for small intestinal fetal progenitors [11].
5. The recommended well format and Matrigel volume for initial seeding (see Table 7) may need to be adjusted depending on the size of the cell pellet obtained after tissue dissociation and isolation,
6. If passaging colon-derived organoids cultured in a 12-well plate format (or larger), the volume of TrypLE Express may be increased to 1 mL. In this case, TrypLE Express should be neutralized by the addition of 3 mL of DPBS prior to centrifugation.
7. Residual 30% sucrose will lead to gaps between the tissue and OCT compound. Proper removal is essential to obtain high-quality cryosections.
8. Growth factors concentration may require adjustment based on their source, as biological activity can vary.

Acknowledgments

This work was supported by the Université Libre de Bruxelles (ULB) and the Fonds de la Recherche Scientifique (FNRS). R.M. is a recipient of an Aspirant (ASP) research fellowship from the FNRS. The authors would like to thank Jérôme Jeandriens for his assistance with cryo-sectioning and Nathalie Tubez for her feedback and critical reading of the manuscript.

References

1. Wells JM, Melton DA (1999) Vertebrate endoderm development. Annu Rev Cell Dev Biol 15(1):393–410. https://doi.org/10.1146/annurev.cellbio.15.1.393
2. Spence JR, Lauf R, Shroyer NF (2011) Vertebrate intestinal endoderm development. Dev Dyn 240(3):501–520. https://doi.org/10.1002/dvdy.22540
3. Montgomery RK, Mulberg AE, Grand RJ (1999) Development of the human gastrointestinal tract: twenty years of progress. Gastroenterology 116(3):702–731. https://doi.org/10.1016/S0016-5085(99)70193-9
4. Chin AM, Hill DR, Aurora M, Spence JR (2017) Morphogenesis and maturation of the embryonic and postnatal intestine. Semin Cell Dev Biol 66:81–93. https://doi.org/10.1016/j.semcdb.2017.01.011
5. Barker N, van Es JH, Kuipers J, Kujala P, van den Born M, Cozijnsen M, Haegebarth A, Korving J, Begthel H, Peters PJ, Clevers H (2007) Identification of stem cells in small

intestine and colon by marker gene Lgr5. Nature 449(7165):1003–1007. https://doi.org/10.1038/nature06196
6. Sato T, Vries RG, Snippert HJ, van de Wetering M, Barker N, Stange DE, van Es JH, Abo A, Kujala P, Peters PJ, Clevers H (2009) Single Lgr5 stem cells build crypt-villus structures *in vitro* without a mesenchymal niche. Nature 459(7244):262–265. https://doi.org/10.1038/nature07935
7. Sato T, Stange DE, Ferrante M, Vries RGJ, van Es JH, van den Brink S, van Houdt WJ, Pronk A, van Gorp J, Siersema PD, Clevers H (2011) Long-term expansion of epithelial organoids from human colon, adenoma, adenocarcinoma, and Barrett's epithelium. Gastroenterology 141(5):1762–1772. https://doi.org/10.1053/j.gastro.2011.07.050
8. Mustata RC, Vasile G, Fernandez-Vallone V, Strollo S, Lefort A, Libert F, Monteyne D, Pérez-Morga D, Vassart G, Garcia M-I (2013) Identification of Lgr5-independent spheroid-generating progenitors of the mouse fetal intestinal epithelium. Cell Rep 5(2):421–432. https://doi.org/10.1016/j.celrep.2013.09.005
9. Fordham RP, Yui S, Hannan NRF, Soendergaard C, Madgwick A, Schweiger PJ, Nielsen OH, Vallier L, Pedersen RA, Nakamura T, Watanabe M, Jensen KB (2013) Transplantation of expanded fetal intestinal progenitors contributes to colon regeneration after injury. Cell Stem Cell 13(6):734–744. https://doi.org/10.1016/j.stem.2013.09.015
10. Theiler Karl (1989). The house mouse: atlas of embryonic development. Springer, Berlin Heidelberg: Imprint: Springer
11. Fernandez-Vallone V, Leprovots M, Strollo S, Vasile G, Lefort A, Libert F, Vassart G, Garcia M-I (2016) Trop2 marks transient gastric fetal epithelium and adult regenerating cells after epithelial damage. Development 143(9):1452. https://doi.org/10.1242/dev.131490

Check for updates

Chapter 8

An Integrated Workflow for Culturing and AI-Based Profiling of Mouse Intestinal Epithelium

Farah Qaderi, U Kaicheng, Yeganeh Farsi, Junhan Zhao and Nima Saeidi

Abstract

Two-dimensional (2D) culture systems are powerful in vitro tools that have revolutionized the study of intestinal stem cell biology. These culture systems derived from mouse intestinal and colonic epithelia enable the modeling of tissue architecture, barrier function, and disease mechanisms under physiologically relevant conditions. This protocol details methods for isolating and maintaining mouse intestinal crypts as 2D monolayers. Monolayers are established via enzymatic dissociation and seeding onto ECM-functionalized hydrogels. This protocol also describes a deep learning pipeline that performs virtual fluorescent staining of label-free phase-contrast images of live organoids. This protocol enables the reproducible generation and computational analysis of murine intestinal epithelial cultures for studies in epithelial biology, drug testing, and disease modeling. This integrated system is designed for facilitating high-throughput imaging, immunolabeling, and functional analysis, expanding its use in regenerative medicine, tissue engineering, and gastrointestinal research.

Keywords Intestine, AI, Bioinformatics, Imaging, Monolayers, Pipeline

1 Introduction

The intestinal epithelium plays a fundamental role in nutrient absorption and maintains a physical barrier between the body and the contents of the gastrointestinal tract. This highly dynamic tissue undergoes rapid and continuous renewal, a process sustained by stem cells located at the base of the crypts in both the small intestine and colon [1, 2]. These intestinal stem cells generate a hierarchy of proliferative and differentiated cell types that preserve tissue homeostasis and function throughout life.

The ability to culture intestinal epithelial cells ex vivo has greatly expanded our capacity to model tissue biology and disease. A landmark advancement in 2009 introduced a method to grow intestinal stem cells in vitro in a three-dimensional (3D) matrix that supports

---

Supplementary Information The online version contains supplementary material available at https://doi.org/10.1007/978-1-0716-5412-5_8

Paloma Ordóñez-Morán (ed.), *Intestinal Stem Cells*, Methods in Molecular Biology, vol. 3059,
https://doi.org/10.1007/978-1-0716-5412-5_8,

the formation of organoids [3, 4]. While this 3D approach has become a widely used model for studying intestinal development, regeneration, and disease, it presents technical challenges such as limited accessibility to the epithelial surface [5, 6].

To address these limitations, organoid-derived two-dimensional (2D) monolayer cultures have been developed as a complementary platform that provides direct access to both the apical and basal surfaces of epithelial cells. This accessibility makes 2D cultures particularly amenable to imaging, molecular analysis, and high-throughput drug screening [1, 5, 7].

Despite the power of these culture systems, traditional phenotypic analysis relies on immunofluorescence, which can be expensive, slow, and often requires fixation of the cells. This has led to a reliance on fluorescent labeling for understanding cell structure and function, which has some serious limitations: high reagent cost, batch-to-batch variation, and phototoxicity that often changes cell behavior and prevents longitudinal tracking of the same live sample [8]. To address these challenges, a new paradigm of computational microscopy has recently been developed: virtual staining. In this approach, deep learning models are trained to predict fluorescent labels from label-free images, such as those acquired using phase-contrast or differential interference contrast (DIC) microscopy [9]. In other words, by learning the morphological features that correspond to specific molecular markers, AI-based methods can produce high-fidelity fluorescent images in the absence of any external dye. This protocol integrates an artificial intelligence (AI) based approach, PhaseFIT, for virtual staining. By transforming simple phase-contrast images into rich, multi-channel fluorescent outputs, this method allows for rapid, scalable, and non-destructive quantification of organoid phenotypes, such as crypt-villus architecture and cell population dynamics.

Here, we present detailed protocols for generating and maintaining 2D monolayer cultures from mouse small intestinal and colonic epithelium. These methodologies enable precise control over the cellular microenvironment, serving as powerful tools for in vitro modeling of intestinal biology. Intestinal stem cells are responsible for life-long tissue maintenance and replenishment of the epithelium. In the colon, stem cells reside at the bottom of deep crypts, protected from the harsh environment in the lumen [4]. This protocol outlines procedures adapted from established methods and optimized for consistency, yield, and experimental reproducibility. These 2D culture models have been used extensively in studies of intestinal homeostasis, host-pathogen interactions, regenerative medicine, and disease modeling [10, 11].

2 Materials

Below is a comprehensive list of laboratory equipment, consumables, reagents, and media used for the culture of 2D monolayers from mouse small intestinal and colonic epithelium. All reagents should be of cell-culture grade. Sterile technique is essential throughout.

Basic Lab Supplies

- 6-well or 24-well, or 48-well plates (tissue culture treated for 2D).
- 15 mL and 50 mL conical tubes.
- 70 μm cell strainers.
- Pipettes and tips (low retention recommended, 1000 μL, 200 μL, 20 μL, and 10 μL)
- Centrifuge.
- Biosafety cabinet.
- Water bath (37°C).
- CO_2 incubator (5% CO_2, 37°C).
- Ice bucket.
- Serological pipettes, 5 mL (Falcon, cat no. 357543), 10 mL (Falcon, cat no. 357551).
- Aluminum cooling block.
- Glass dish, 150×75 mm (VWR, cat no. 10754-778).
- Countess automated cell counter and chamber slides (Invitrogen, cat no AMQAX1000).
- Microscope glass slide (VWR, cat no. 16004-368).
- Microscope glass coverslip (VWR, cat. no. 48382-042).
- Scalpel (Exel International, cat no. 29550).
- Tweezer (Excelta, cat. no KST-5).
- Ethanol, 70% (vol/vol)! CAUTION: Ethanol is highly flammable and causes serious eye irritation. Use with caution and wear appropriate personal protective equipment.

Reagents

- Advanced DMEM/F12, 500 mL (Gibco, cat. no. 12634-010).
- HEPES buffer, 100× (Gibco, cat no. 15630-080).
- Penicillin–streptomycin, 100× (Gibco, cat no. 15140-148).
- Glutamax, 100× (Gibco, cat. no. 35050-061).
- Fetal bovine serum, certified, heat-inactivated, US origin (Gibco, cat no. 10082147).
- Bovine serum albumin (BSA, Millipore-Sigma, cat no A9418).
- Dimethyl sulfoxide (Sigma, cat no D2650-100ML).
- R-Spondin-1-conditioned medium.
- B27 supplement, 50× (Gibco, cat no. 17504-001).
- N2 supplement, 100× (Gibco, cat no. 17502-001).
- N-acetyl L-cysteine (Sigma-Aldrich, cat no A9165).
- Y-27632 dihydrochloride (Biogems, cat no. 1293823).

- Murine epidermal growth factor (EGF; Peprotech, cat no AF-315-09).
- Human Noggin (Peprotech, cat no. 120-10C).
- Dulbecco's phosphate-buffered saline (DPBS) without calcium and magnesium, pH 7.4 (DPBS$^{-/-}$; Gibco, cat no. 10010023).
- TypLE Express (Gibco, cat no. 1260413).
- UltraPure 0.5 M EDTA, pH 8.0 (Invitrogen, cat no. 15575020).
- PREemt disinfectant solution (VWR, cat no. 10822-51).
- Trypan Blue (Invitrogen, cat no T10282).
- Phalloidin-iFluor 488 reagent (ab176753-300TEST).
- DAPI solution (1 mg/mL) (Invitrogen, cat no. 62248).
- BlockAid blocking solution (Invitrogen, cat no B10710).
- Formaldehyde, 4% (vol/vol) (Sigma-Aldrich, cat no. 100496).

2.1 Isolation of Colon/Intestinal Crypts

- PREempt disinfectant solution.
- 70% ethanol (vol/vol).
- 1× PBS.
- 0.5 M EDTA stock solution (to prepare 0.5 mM EDTA/PBS).
- Advanced DMEM/F12 medium.
- ENR culture medium supplemented with Y-27632 and CHIR99021 (seeding medium).
- 50 mL conical tubes.
- 50 mm Petri dishes.
- 70 μm cell strainers.

Equipment

- Biosafety cabinet.
- CO_2 euthanasia system.
- Surgical scissors and dissection instruments.

Imaging

- EVOS M5000 cell imaging system (Thermo Fisher Scientific, cat no AMF5000).

2.2 2D Monolayer Culture of Murine Small Intestinal and Colon Crypts

- Bind silane components:
  - Ultrapure water.
  - 3-(Trimethoxysilyl)propyl methacrylate.
  - Acetic acid.

- SANPAH (light-sensitive crosslinker; 100 mg/vial, 50 mg/mL in DMSO, 60 μL aliquots).
- HEPES buffer (0.1 M; dilute 1 M HEPES 1:10 in PBS).
- Collagen I (3 mg/mL stock, diluted to 0.1 mg/mL in cold PBS).
- Advanced DMEM/F12.
- PBS.
- Glass dish, 150 × 75 mm (VWR, cat no. 10754-778).
- Coverslips (18 mm).
- Curved and pointed forceps.
- UV light source.
- Aluminum foil.
- Organoid culture medium (ENR, see Table 1) supplemented with Y-27632 and CHIR99021 (Y + CHIR).
- Gel composition (Table 2).

2.3 Immuno-fluorescence for 2D Monolayer Staining

- Phosphate-Buffered Saline (PBS).
- Paraformaldehyde (PFA), 4% in PBS (freshly prepared).
- Triton X-100 (0.3% in PBS).
- Bovine Serum Albumin (BSA), 2% in PBS.
- Tween 20 (0.05%).

Table 1
ENR medium composition (100 mL total)

Component	Volume	Final Concentration
R-spondin1-conditioned Media	10 mL	10%
Advanced DMEM/F12	85.16 mL	—
GlutaMAX (100×)	1 mL	1×
HEPES 1 M	1 mL	10 mM
Primocin	200 μL	100 μg/mL
Normocin	200 μL	100 μg/mL
B27	1 mL	1×
N2	500 μL	1×
N-Acetylcysteine (500 mM)	250 μL	1 mM
EGF (100 μg/mL)	10 μL	50 ng/mL
Noggin (100 μg/mL)	100 μL	100 ng/mL

Table 2
Gel composition (1 mL total)

Gel components						
Stiffness (kPa)	0.15 kPa	0.60 kPa	2.4 kPa	9.6 kPa	38.4 kPa	153.6 kPa
Ultrapure water	895 μL	885 μL	785 μL	740 μL	630 μL	395 μL
40% Acrylamide	75 μL	75 μL	188 μL	188 μl	300 μL	300 μL
2% Bis	20 μL	29 μL	17 μL	59 μl	59 μL	295 μL
Polymerizing agents						
10% ammonia persulfate (APS)	10 μL in 1 mL					
TEMED	0.5 μL in 1 mL (Add at last just before use)					

- Primary antibodies (as required).
- Secondary antibodies (fluorescently labeled, as required).
- DAPI (1:1000 in PBS).

Equipment
Imaging

- Zeiss LSM 880 confocal microscope (Carl Zeiss Microscopy, cat no ZEISS LSM 880).

2.4 Virtual Staining and Quantification

This section lists the required software and third-party libraries to run the PhaseFIT workflow.

Programming Language: Python

Deep Learning Framework: PyTorch, Torchvision

Image Processing and Analysis: OpenCV-Python, Scikit-image, Scikit-learn

2.4.1 Equipment

This section details the computational hardware for training the PhaseFIT model.

Computer Workstation: A high-performance computing workstation with 32 GB of RAM.

Graphics Processing Unit (GPU): Model development and training were performed on a workstation with four NVIDIA Tesla V100 GPUs with 32 GB VRAM.

3 Methods

3.1 Isolation of Colon/Intestinal Crypts (Fig. 1)

Timing: ~2.5 h
Harvest Tissue

1. Sterilize the biosafety cabinet using PREempt disinfectant solution and 70% ethanol (vol/vol).

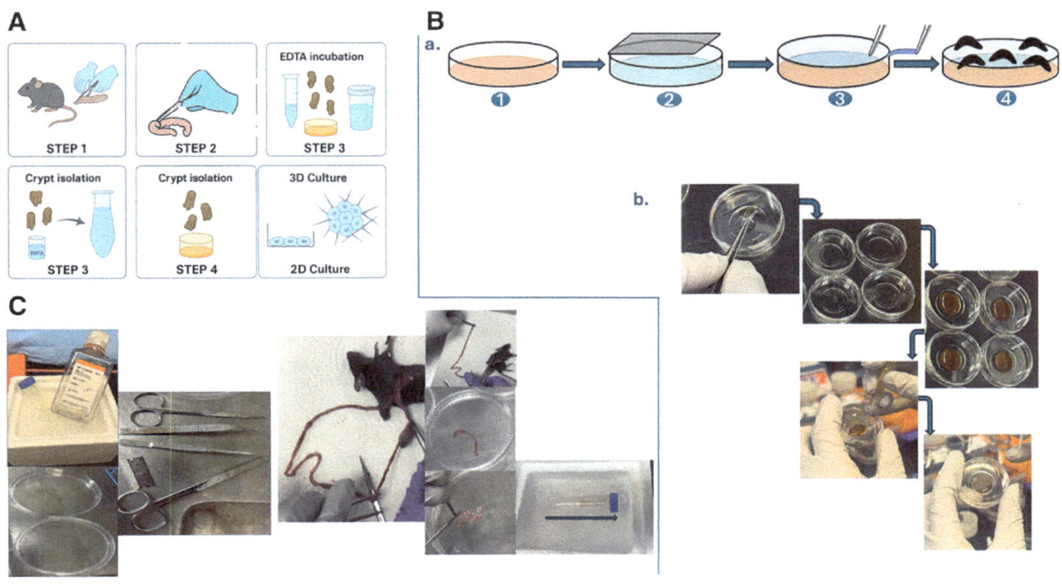

Fig. 1 Workflow for isolation and culture of murine intestinal crypts in 2D system. (**A**) Schematic representation of crypt isolation. Mice are first anesthetized and prepared for dissection. The colon or small intestine is excised (step 1), and intestinal segments are washed and cut into 2–5 mm pieces (step 2). Segments are incubated in PBS containing 2–10 mM EDTA to chelate divalent cations and loosen epithelial junctions (step 3). Following incubation, tissue fragments are shaken to release crypts (step 4), which are then filtered, pelleted, and resuspended in PBS. Isolated crypts are used for establishing 2D monolayers (step 5). (**B**) (**a**) Schematic representation of the 2D monolayer-coated wells. Culture wells are coated with bind silane (1), followed by polymerization of a polyacrylamide gel beneath a coverslip (2). Gels are activated with Sulfo-SANPAH and coated with ECM proteins (e.g., collagen I) (3). After equilibration, crypts are seeded to form a confluent epithelial monolayer (4). (**b**) Representative images show plating of hydrogel domes. (**C**) Representative images of mouse dissection and tissue preparation. Essential supplies include cold PBS, PBS/EDTA, sterile dissection tools, and 50 mL collection tubes. Following midline incision, the small intestine or colon is exposed and dissected, flushed with PBS to remove fecal content, and transferred to chelation buffer (PBS/EDTA). Small intestine segments are stored on ice; colon tissues are incubated at room temperature for up to 30 min for the small intestine and 40 min to 1 h for the colon on a shaker

2. Obtain a bucket of ice.
3. Prepare 1× PBS and 50 mL 10 mM EDTA/PBS (1 mL 0.5 M EDTA in 50 mL PBS) on ice and transport to the animal facility along with surgical tools (scissors and 50 mm petri dish, and 50 mL tube in ice).
4. Euthanize the mouse with CO_2 and ensure death by performing cervical dislocation, under approved protocol, and dissect the colon or small intestine.
5. Open longitudinally and wash with cold PBS to remove feces (*see* **Note 1**).
6. Cut the tissue into 2–4 mm pieces and put the pieces in a 50 mL tube with 0.5 mM EDTA/PBS and label.

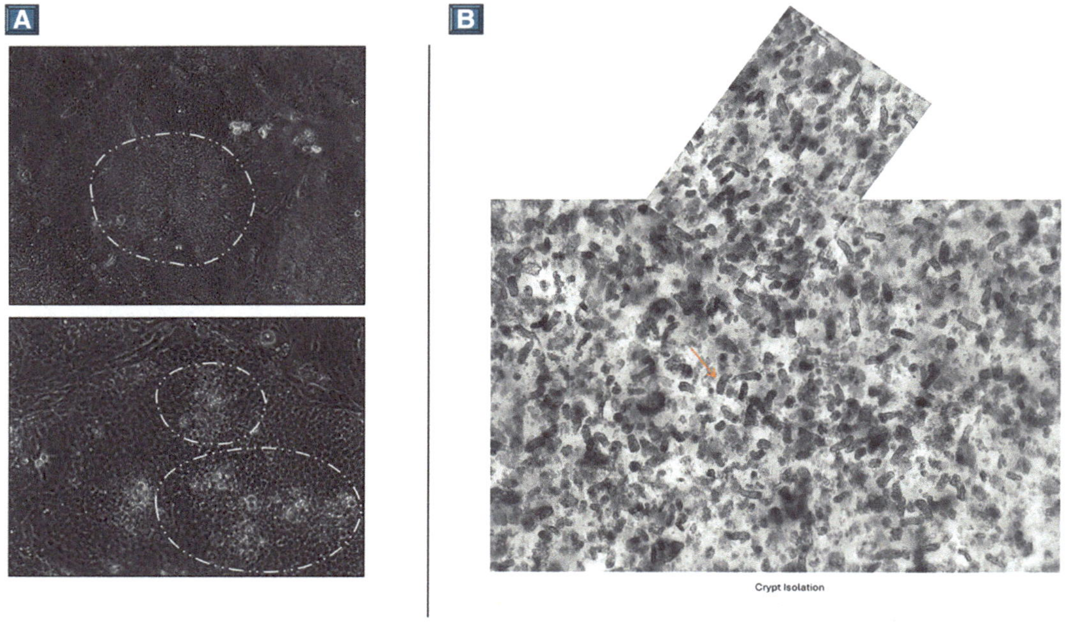

Fig. 2 Morphological and immunofluorescent characterization of colonic epithelial cells in culture. (**A**) Phase contrast images of murine colonic epithelial cells grown as 2D monolayers. Compared to small intestinal cultures, colonic monolayers exhibit distinct epithelial morphology with reduced crypt density, particularly when derived from distal colon or rectal regions. Crypt-like domains appear as localized cellular clusters with tightly packed cells. (**B**) Crypt isolation efficiency and culture performance are influenced by the anatomical origin of the tissue and the condition of luminal contents at dissection. Dashed ovals highlight representative crypt domains. Scale bars: 150 μm

7. Incubate tissue in 0.5 mM EDTA/PBS at 4°C for 30–45 min on a rocker (colon should be for an hour, no need to put it in 4°C) (*see* **Note 2**).
8. After 30 min, decant the EDTA solution, and add ~25 mL of PBS. Shake vigorously for 2 min to release crypts (Solution becomes cloudy) (*see* **Note 3**).
9. Pass the solution through 70 μm filter into a new 50 mL conical tube. Retain tissue in the origin tube, add another ~25 mL, and repeat the process.
10. Centrifuge crypt suspension at 100 × *g* for 5 min at 4°C. The centrifuge must be set to 4°C.
11. Decant fluid and Resuspend pellet in 10 mL Advanced DMEM/F12. (Mix thoroughly with pipette gently to avoid breaking crypts.)
12. Spin at 100 × *g* for 5 min. The centrifuge must be set to 4°C.
13. Resuspend crypts in seeding media (ENR Culture media + Y and Chir) (If cell/crypt counting is needed, add 100 μL initially and collect 10 μL for cell/crypt counting) (*see* **Note 4**).

3.2 2D Monolayer Culture of Murine Small Intestinal and Colon Crypts (Fig. 2)

Estimated Timing:

- **Gel preparation**: ~2.5 h.
- **Crypt seeding**: ~4–6 h.

A. *Gel Coating and Activation*

1. **Bind Silane Coating**
   a. Prepare a bind silane solution:
      - 200 mL ultrapure water.
      - 160 μL 3-(Trimethoxysilyl)propyl methacrylate.
      - 160 μL acetic acid.
      - Mix thoroughly and use at room temperature

   b. Add ~800 μL of bind silane to the center of each well and incubate for 30 min.
   c. (Optional) Wash 3× with ultrapure water and ensure wells are completely dry.

2. **Gel Preparation**
   Depending on the experimental goal, antibody application, and desired matrix stiffness, different hydrogel formulations can be combined and tuned to optimize 2D shell culture conditions (*see* **Note 5**).

   a. Prepare 1000 μL of gel solution. Mix thoroughly after adding each component.
   b. Add 70 μL of gel solution to each well.
   c. Place an 18 mm coverslip gently onto the droplet using curved tweezers.
   d. Allow the gel to polymerize for 40 min to 1 h (*see* **Note 6**).
   e. Add ~2 mL ultrapure water per dish and wrap in foil. Store at 4°C if not using immediately.

3. **Coverslip Removal**

   a. Using pointed forceps, gently remove coverslips along the edge to avoid damaging the gel.
   b. Discard coverslips into a sharp container.
   c. Ensure SANPAH is prepared before aspirating water; **do not let the gel dry**.
   d. Carefully aspirate water around the gel without touching it.

4. **SANPAH Activation**

   a. Prepare SANPAH solution in the dark: dilute 60 μL SANPAH aliquot with 8 mL of 0.1 M HEPES (*see* **Note 7**).
   b. Add 250 μL of SANPAH solution as a dome on the gel surface, ensuring full coverage.

c. Expose uncovered dishes to UV light for 10 min.

d. Aspirate SANPAH solution carefully.

5. **Washing**

a. Wash 3× with 0.1 M HEPES. Incubate for 5 min during the secondnd and third washes.

b. Wash 3× with PBS, with 5-min incubations for all washes.

6. **Collagen Coating**

a. Prepare 0.1 mg/mL Collagen I in cold PBS (1:30 dilution) (*see* **Note 8**).

b. Add 200 μL of the collagen solution as a dome on each gel.

c. Store overnight at 4°C, protected from drying or collapse of the dome (*see* **Note 9**).

7. **Pre-culture Wash**

a. Wash 2–3× with PBS before seeding.

b. Add ~2 mL Advanced DMEM/F12 per well and equilibrate in the incubator.

B. Crypt Isolation and Seeding
Refer to the crypt isolation protocol from earlier steps. Seeding depends on crypt density and quality.

1. **Crypt Preparation**

a. Based on isolated crypt concentration, prepare 3–6 mL of ENR medium supplemented with Y-27632 and CHIR99021.

b. Gently mix crypts and media by slowly pipetting.

2. **Crypt Quality Check**

a. Before seeding, observe the suspension under a microscope to confirm healthy crypt morphology and density.

3. **Seeding**

a. Add ~150 μL of crypt-containing ENR + Y + CHIR medium directly onto each gel-coated well (*see* **Note 10**).

b. Avoid shaking; maintain dome structure.

4. **Initial Attachment**

a. Transfer plates carefully into the incubator (37°C, 5% CO_2).

b. Incubate undisturbed for 4 h to allow crypts to attach (*see* **Note 11**).

5. **Post-seeding Wash**
   a. After 4 h, gently remove the media.
   b. Wash 1–2× with PBS, gently rocking the plate to remove debris and dead cells.
6. **Media Addition and Maintenance**
   a. Add 1.5 mL of fresh ENR medium per well (Table 1).
   b. Examine wells under a microscope to verify attachment and confluency (should resemble the density shown in Fig. 3).
   c. Culture plates in the incubator and avoid frequent disturbance.
   d. Replace medium every 2–3 days depending on growth density (*see* **Notes 12–16**).

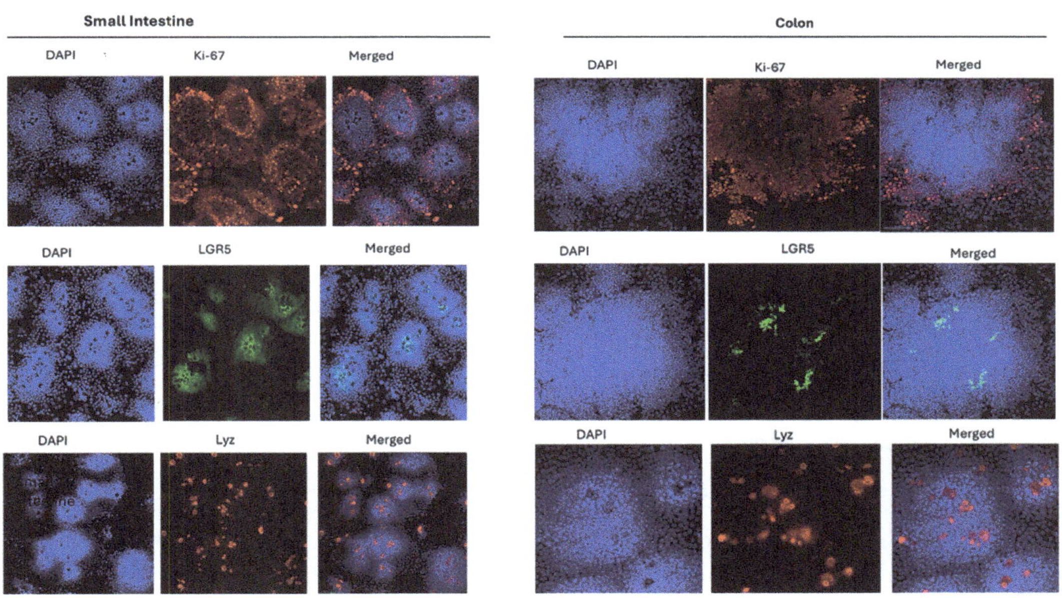

Fig. 3 Immunofluorescent characterization of murine small intestinal and colonic 2D monolayer cultures. Representative immunofluorescence (IF) staining of epithelial monolayers derived from murine small intestine (left) and colon (right), showing spatial expression of key epithelial cell markers. Top row: proliferative cells were identified using Ki-67 (red), a nuclear marker for actively cycling cells. Ki-67+ cells are distributed in clustered zones throughout both small intestinal and colonic cultures, consistent with compartmentalized epithelial proliferation. Middle row: LGR5 (green), a well-established marker of intestinal stem cells, shows localized expression within crypt-like domains in both tissue types, highlighting maintenance of stem cell identity in the monolayer model. Bottom row: Lysozyme (Lyz, red), a Paneth cell marker, labels a subset of differentiated secretory cells distributed among epithelial clusters. DAPI (blue) was used for nuclear counterstaining. Merged images demonstrate co-localization and cellular organization within each culture. All images acquired at equivalent magnification; scale bars: 100 μm

3.3 Immunofluorescence for 2D Monolayer Staining (Fig. 3)

Day 1—Fixation and Permeabilization

1. Wash: Remove culture media and gently wash cells once with PBS at room temperature (RT).
2. Fixation: Add 4% PFA in PBS and incubate for 10 min at RT (*see* **Note 17**).
3. Permeabilization: Aspirate the fixative and incubate cells with 0.3% Triton X-100 in PBS for 30 min at RT. *Do not shake the plate* (*see* **Note 18**).
4. Blocking: Remove Triton solution and add 2% BSA + 0.05% Tween 20 in PBS for 30 min at RT to block non-specific binding. *Do not shake the plate.*
5. Primary Antibody Incubation: Dilute the primary antibody in blocking buffer (2% BSA + 0.05% Tween 20 in PBS) and add 300 μL per well. Incubate overnight at 4°C.

Day 2—Detection and Imaging

1. Wash: Remove primary antibody and wash cells three times with PBS for a total of 5 min at RT.
2. Secondary Antibody Incubation: Dilute the fluorescent secondary antibody and DAPI (1:1000) in blocking buffer. Add to each well and incubate for 1–2 h at RT, protected from light (*see* **Note 19**).
3. Final Wash and Storage: Wash cells three times with PBS. Add 1.5 mL PBS per well, cover with foil, and store at 4°C. Monitor wells every other day to ensure they remain hydrated. Replenish PBS as needed. Samples may be stored for up to 1 month at 4°C before imaging.

3.4 Virtual Staining and Quantification (Fig. 4)

This protocol documents the procedure for setting up the software environment, data preparation, and running the PhaseFIT model.

A: Code Download and Environment Setup

1. Download the Code Repository (https://github.com/MGH-Gut/PhaseFIT). Obtain the project files with the model architecture and evaluation scripts.
2. Set up a Python Environment. Create and activate a new Python environment for this project.
3. Install Required Libraries. Install the required Python libraries using a package manager like pip:

   - pip install torch torchvision opencv-python scikit-image scikit-learn

4 Verify the Installation. Run get_model.py from the root project directory to confirm the environment is set up correctly and the PhaseFIT model can be loaded.

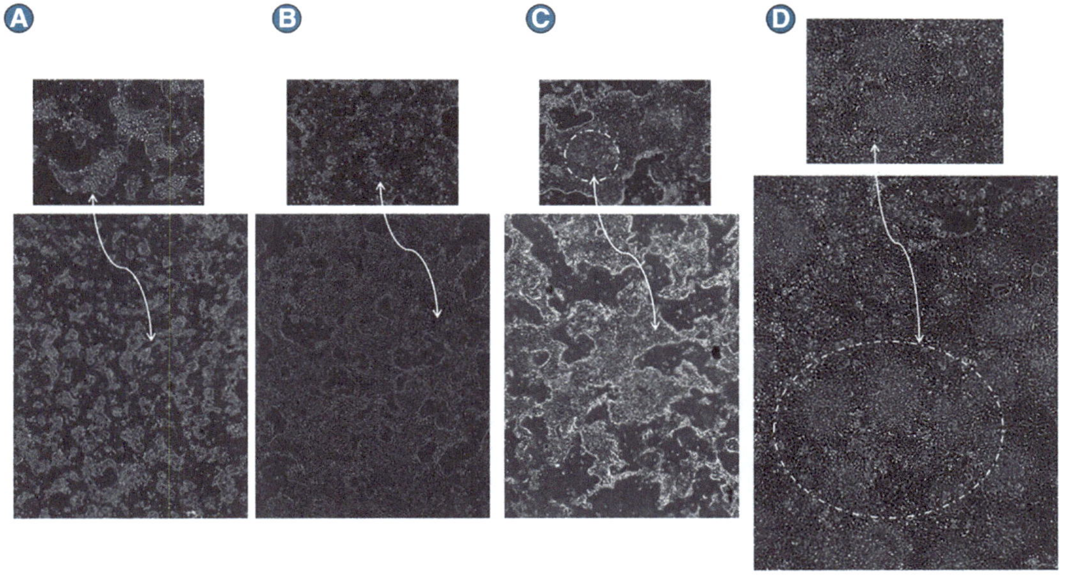

Fig. 4 Temporal progression of murine small intestinal epithelial cells in 2D monolayer culture. (**A**). Day 0 of 2D monolayer culture. Isolated intestinal crypts or dissociated epithelial cells are seeded onto pre-coated polyacrylamide hydrogels functionalized with ECM proteins. Within 4 h of seeding, cells begin to adhere and spread across the substrate. A uniform seeding density is critical to support optimal attachment and growth. Representative image shows well-dispersed crypts and early epithelial cell adherence. (**B**). Day 3. Cultures display dense, cohesive monolayers. Cells have proliferated to fill the available surface, and crypt-derived epithelial clusters exhibit healthy morphology and uniform organization. (**C**). Day 5. As the culture progresses, regions of active cell turnover emerge. Consistent with the in vivo epithelial renewal cycle, proliferative cells may undergo apoptosis and sloughing, while new crypt-like clusters begin to repopulate the monolayer. This dynamic transition reflects tissue self-renewal and remodeling. (**D**). Day 7. Mature epithelial monolayers exhibit high confluency and distinct crypt-like domains. This time point represents a stable and functionally mature 2D culture system suitable for downstream assays, including imaging, barrier function analysis, and pharmacological testing. Insets highlight cellular architecture at higher magnification. Scale bars: 150 μm (all panels)

```
python get_model.py
```

B: Data Pre-processing

1. Organize Raw Data. Place raw phase-contrast images and corresponding ground-truth fluorescent images in separate folders. The training dataset consists of 889 image pairs (*see* **Notes 20 and 21**).
2. Prepare Input Images. The model's encoder expects a 3-channel input. Convert the single-channel phase-contrast images into 3-channel tensors.
3. Normalize Input Images. Normalize the 3-channel input tensors using the standard ImageNet mean and standard deviation (*see* **Note 22**).
   - Mean: [0.485, 0.456, 0.406]
   - Standard deviation: [0.229, 0.224, 0.225]

4. Crop Images into Patches. Crop each pair of original 1536×1024 px images (source and target) into smaller 256×256 px patches with a 50% overlap between adjacent patches.
5. Binarize Target Masks. Convert the target fluorescent patches into binary images using a thresholding technique. The thresholds are determined by the background pixel values of the three types of color images.

C: Model Training

This section describes the procedure for fine-tuning the Phase-FIT model using transfer learning.

1. Set Random Seed. For reproducibility, set a fixed random seed for all relevant libraries (random, numpy, torch) to ensure deterministic behavior (*see* **Note 23**).
2. Data Partitioning. Partition the dataset using a fivefold cross-validation strategy. For each of the five repetitions, use four parts for training and the held-out part for validation.
3. Implement Data Augmentation. To increase sample diversity, apply the following data-augmentation techniques to the training patches as listed in the publication:
   - Random cropping.
   - Horizontal and vertical flips.
   - Random scaling.
   - Brightness and contrast changes.
   - Addition of Gaussian noise.
4. Define the Training Loss Function. The network is trained using a combined Focal-Loss + Dice-Loss formulation. For a given pixel i with ground-truth label y_i and predicted probability p_i:
   - Focal Loss:

$$L_{\mathrm{FL}} = -\sum_i \left[\left(1 - p_i\right)^{\gamma} y_i \log p_i + (p_i)^{\gamma}\left(1 - y_i\right)\log\left(1 - p_i\right)\right]$$

The focusing parameter is set to $\gamma = 2$.

   - Dice Loss:

$$L_{\mathrm{DICE}} = -\frac{2\sum_i^N y_i p_i}{\sum_i^N y_i^2 + \sum_i^N p_i^2}$$

5. Set Training Hyperparameters. The training process should be configured with the following parameters:
   - Optimizer: Adam
   - Initial Learning Rate: 0.0003
   - Learning Rate Schedule: The learning rate is reduced by a factor of 10 at epoch 10, and again by a factor of 10 at epoch 20.
   - Mini-batch Size: 24
   - Number of Epochs: 30
   - Dropout: A dropout layer with $p = 0.5$ is incorporated into the model architecture.
6. Initiate Training. Initiate model training using the parameters defined above. This will fine-tune the SE-ResNeXt-50 encoder initialized with ImageNet weights and train the decoder portion of the network. The decoder utilizes a spatial and channel squeeze & excitation (scSE) module to refine features.

D: Model Inference and Evaluation

1. Run Inference. Apply the trained model to a new phase-contrast image to generate a prediction. The model operates on 256×256 px tiles and outputs a probability mask for each tile.
2. Merge and Smooth Tiles. In a post-processing step, merge the predicted tiles to form a complete image. An Aggregated Contextual Transformation (AOT) block is used to smooth the tile concatenation and remove edge artifacts.
3. Generate Final Color Image. Convert the final probability mask into an 8-bit image by multiplying each pixel's probability by 255.
4. Evaluate Performance (Optional). Run the supplied evaluation_metrics.py to compute performance against ground-truth images. The PhaseFIT pipeline processes images at a rate of $<1\ \text{s}\cdot\text{mm}^{-1}$ on a V100 GPU, which is sufficient to complete virtual painting for a 96-well assay. Common metrics include:
   - Dice Coefficient (F1-Score).
   - Intersection over Union (IoU).
   - Pixel Accuracy.
   - Precision & Recall (Sensitivity).
   - Area Under the ROC Curve (AUC).
   - Structural Similarity Index (SSIM).
   - Mean Squared Error (MSE) & Peak Signal-to-Noise Ratio (PSNR).

E: Quantitative Analysis of Organoid Phenotypes

The virtual fluorescent images can be used for downstream quantitative analysis as demonstrated in the original study (*see* **Note 24**).

1. Quantify Crypt and Villus Area. Using the virtual nuclear stain, measure crypt regions characterized by dense nuclei and villus regions to compute the crypt-to-villus area ratio.
2. Quantify Cell Populations. Using the three virtual stains, count total nuclei, LGR5$^+$ cells, and UEA-I$^+$ cells to determine their proportions within the total cell population.

3.4.1 Anticipated Results

Following this computational protocol should produce virtual fluorescent images that are qualitatively and quantitatively comparable to conventional immunofluorescence and support robust phenotypic analysis.

- Virtual Nuclear Staining (Hoechst): Shows clear nuclei and allows for the identification of dense crypt zones.
- Virtual Stem-Cell Staining (LGR5-EGFP): Generates signals confined to the crypt regions, identifying the stem cell population.
- Virtual Secretory-Cell Staining (UEA-I): Produces dot-like signals corresponding to Paneth cells (in crypts) and other secretory cells (in villus regions).

4 Notes

1. Always perform dissections and tissue handling on ice whenever possible to minimize epithelial degradation. Colon tissue is generally more fragile than the small intestine and benefits from shorter handling times.
2. EDTA concentration and incubation time are critical for efficient crypt release. Over-chelation can lead to excessive single-cell dissociation and reduced viability. Begin with the lower time range and visually assess crypt integrity under a microscope.
3. When shaking to release crypts, avoid vigorous motion; short, sharp inversions are sufficient. Over-shaking leads to mucus contamination and mechanical damage.
4. The number of crypts obtained varies considerably with animal age and tissue region. Proximal segments of the small intestine typically yield more intact crypts than the distal colon.

5. For best results, use freshly prepared binding and gel reagents. Acrylamide polymerization efficiency and SANPAH activity both decline upon prolonged storage or repeated freeze-thaw cycles.
6. During gel polymerization, avoid introducing air bubbles—these create uneven surfaces and interfere with uniform cell attachment. If bubbles form, gently move them aside with a sterile pipette tip before polymerization completes.
7. SANPAH is light-sensitive; prepare and handle it in dim light, and discard any solution that turns brown or opaque.
8. Collagen coating density strongly influences cell spreading and morphology. If cells appear rounded or fail to attach, verify the collagen concentration and coating uniformity before troubleshooting other factors.
9. After collagen coating, keep gels hydrated at all times. Even brief drying can cause cracking and delamination, leading to uneven growth.
10. For single-cell cloning, supplement cultures with 10 μM Y-27632 for 48 h post-plating to enhance survival and reduce anoikis during early attachment. Removing it too early can result in poor confluency; keeping it beyond 48 h may, however, affect differentiation.
11. The first few hours after seeding are critical. Do not disturb the plate during the initial attachment period—movement can shear loosely bound crypts.
12. Inspect cultures daily under phase contrast. A healthy culture should show evenly distributed, bright, refractive crypt clusters that flatten into cohesive sheets within 48–72 h.
13. Medium changes should be done slowly along the well wall to avoid detaching the monolayer. Use a pipette rather than vacuum aspiration.
14. For antibiotic-free cultures, use Primocin or MycoZap during the first passages only; routine long-term exposure is unnecessary and may affect epithelial health.
15. Confirm mycoplasma-free status monthly for long-term cultures. Mycoplasma contamination alters proliferation and compromises imaging quality.
16. If crypts fail to attach, check the gel stiffness. Softer matrices (<0.6 kPa) support the spreading of epithelial cells but can detach easily during washes; stiffer matrices (>10 kPa) restrict cell spreading and favor cluster formation.
17. Fixation with freshly prepared 4% PFA ensures good preservation of cell morphology. Old or unfiltered PFA causes uneven staining and background fluorescence.

18. During immunostaining, gentle pipetting and minimal agitation help preserve monolayer integrity. Excessive washing is a common cause of partial detachment.
19. Always protect fluorescent dyes and stained samples from light. Cover plates with foil and minimize microscope exposure to avoid photobleaching.
20. For virtual staining, make sure raw phase-contrast images are evenly illuminated and free of debris. Uneven lighting introduces artifacts in AI prediction.
21. Maintain consistent magnification and imaging settings across all datasets used for model training and validation. Variations in brightness or contrast can significantly affect network performance.
22. When setting up the computational environment, note that the model will attempt to download pre-trained ImageNet weights for the SE-ResNeXt-50 encoder during the first run. An active internet connection is required, and a successful setup will print the tensor size (e.g., torch.Size ([6, 1, 256, 256])).
23. For reproducible AI results, keep a copy of the software environment specifications and random seed used for training. These details ensure future users can replicate your model outputs.
24. Store stained or virtual-stained samples at 4°C in PBS, sealed to prevent evaporation. Re-image within one month for the best fluorescence intensity.

Acknowledgments

J.Z. and N.S. are co-founders and shareholders of Alvus Health Inc. F.Q. and K.U. were employed at Alvus Health Inc. F.Q. K. U and J.Z. are co-inventors on a patent application related to AI and human organoids that has been assigned to Alvus Health Inc. At the time of this publication's completion, there is no overlap between the company's primary area of expertise and the content presented herein. In addition, J. Z reports an advisory role and receipt of consulting fees and equity from Encapsulate Inc. and serves on the research advisory board of Shriners Children's. None of these activities are related to the subject matter of this manuscript.

References

1. Sato T, Vries RG, Snippert HJ, Van De Wetering M, Barker N, Stange DE et al (2009) Single Lgr5 stem cells build crypt-villus structures in vitro without a mesenchymal niche. Nature 459(7244):262–265
2. Yin X, Farin HF, Van Es JH, Clevers H, Langer R, Karp JM (2014) Niche-independent high-purity cultures of Lgr5+ intestinal stem cells and their progeny. Nat Methods 11(1):106–112

3. Miyoshi H, Stappenbeck TS (2013) In vitro expansion and genetic modification of gastrointestinal stem cells in spheroid culture. Nat Protoc 8(12):2471–2482
4. Soares MP, Weiss G (2015) The Iron age of host–microbe interactions. EMBO Rep 16(11):1482–1500
5. Clevers H (2016) Modeling development and disease with organoids. Cell 165(7):1586–97
6. Snippert HJ, Van Der Flier LG, Sato T, Van Es JH, Van Den Born M, Kroon-Veenboer C et al (2010) Intestinal crypt homeostasis results from neutral competition between symmetrically dividing Lgr5 stem cells. Cell 143(1):134–144
7. He S, Lei P, Kang W, Cheung P, Xu T, Mana M et al (2023) Stiffness restricts the stemness of the intestinal stem cells and skews their differentiation toward goblet cells. Gastroenterology 164(7):1137–1151.e15
8. Nitta N, Sugimura T, Isozaki A, Mikami H, Hiraki K, Sakuma S et al (2018) Intelligent image-activated cell sorting. Cell 175(1):266–276.e13
9. Christiansen EM, Yang SJ, Ando DM, Javaherian A, Skibinski G, Lipnick S et al (2018) In silico labeling: predicting fluorescent labels in unlabeled images. Cell 173(3):792–803.e19
10. Merenda A, Fenderico N, Maurice MM (2020) Wnt signaling in 3D: recent advances in the applications of intestinal organoids. Trends Cell Biol 30(1):60–73
11. Watanabe S, Kobayashi S, Ogasawara N, Okamoto R, Nakamura T, Watanabe M et al. (2022) Transplantation of intestinal organoids into a mouse model of colitis. Nat Protoc 17(3):649–671

Check for updates

Chapter 9

Updated Protocols for 3D and 2D Culture and Histological Analysis of Intestinal Organoids

Hikaru Hanyu, Shinya Sugimoto and Toshiro Sato

Abstract

The intestinal epithelium maintains continuous self-renewal and multi-differentiation capacities through the integrated regulation of multiple signaling pathways, including Wnt, bone morphogenetic protein, epidermal growth factor (EGF), and Notch signaling pathways. Based on this principle, a defined combination of EGF, Noggin, and the Wnt agonist R-spondin has been shown to support long-term expansion of mouse intestinal stem cells and the formation of organoids with indefinite self-renewal and full differentiation capacity. In human intestinal organoid cultures, the addition of small molecule inhibitors targeting p38 and TGF-β pathways facilitated epithelial propagation, while replacement of EGF and a p38 inhibitor with insulin-like growth factor-1 and fibroblast growth factor-2 enabled efficient multilineage differentiation. Recently, the monolayer culture of intestinal organoid has been increasingly used due to its advantages of direct access to the apical membrane and its ability to co-culture with gut microbiota. Here, we summarize our recent technological advances in human intestinal organoid culture and histological analysis, providing updated protocols that will deepen the understanding of intestinal homeostasis and the pathogenesis of intestinal diseases.

Keywords Intestinal stem cells, Organoids, Organoid monolayer, Small intestine, Colon, Human, Immunostaining, Formalin fixed paraffin embedded

1 Introduction

Intestinal homeostasis is maintained by several niche (microenvironment) signals. Wnt, bone morphogenetic protein (BMP), epidermal growth factor (EGF), and Notch are well-established regulators of the intestinal stem cell state, which maintains cellular diversity by balancing self-renewalproliferation and differentiation [1]. Wntsignaling preserves stem cell identity and promotes the proliferation of stem and transit-amplifying (TA) cells [2], while paradoxically driving terminal differentiation of Paneth cells [3]. In contrast, BMP signaling is active in the villus compartment, where it suppresses stemness. Noggin-mediated inhibition of BMP signaling induces the formation of crypt-like structures along the villi [4]. EGF signaling strongly

Paloma Ordóñez-Morán (ed.), *Intestinal Stem Cells*, Methods in Molecular Biology, vol. 3059, https://doi.org/10.1007/978-1-0716-5412-5_9,

promotes the proliferation of stem and TA cells and is indispensable for the maintenance of stemness. Notch signaling also plays a critical role in preserving the undifferentiated state, as its inhibition induces the differentiation of secretory lineage cells from proliferative stem and TA cells [5].

On the basis of these research findings on stem cell niches, Sato et al. first demonstrated that a combination of niche factors, EGF, Noggin, and the Wnt agonist R-spondin is essential for culturing mouse intestinal stem cells in vitro, giving rise to structures termed organoids [6]. Subsequently, the culture conditions were adapted to long-term culture of human small intestinal and colonic cells through the addition of Wnt3a, an inhibitor of activin-like kinase 4/5/7 (A83-01), an inhibitor of the p38 mitogen-activated protein kinase (SB202190), and gastrin [7]. Furthermore, we showed that removal of the p38 inhibitor and supplementation with insulin-like growth factor-1 (IGF-1) and fibroblast growth factor-2 (FGF-2)—referred to as the IF condition—enabled the generation of organoids with high culture efficiency and multilineage differentiation capacity [8]. Under the IF condition, we observed most of the cell types found in vivo, including stem cells, TA cells, early enterocytes, goblet cells, enteroendocrine cells, M cells, Paneth cells, tuft cells, and BEST4^{+} cells [8, 9]. Furthermore, the addition of differentiation-promoting factors can drive specific lineage commitments [10–12].

Recent advances in organoid culture extend beyond medium optimization to new culture technologies. Following the establishment of a primary mouse intestinal monolayer culture system [13], human intestinal organoid monolayer cultures have been widely applied. In this approach, dissociated organoids are seeded onto cell culture inserts coated with Matrigel™ and supplemented with optimized medium in both the apical and the basal compartments. This system provides direct access to the apical surface, rendering it suitable for various applications, including drug absorption assays [14] and co-culture experiments with intestinal bacteria [15]. Moreover, monolayer culture has been shown to facilitate organoid maturation more effectively than conventional 3D culture. For instance, some studies have utilized air–liquid interface culture [16] or external flow systems [17, 18].

In this protocol, we describe updated methods for 3D and 2D human intestinal organoids, as well as approaches for analyzing histological features in both culture methods.

2 Materials

2.1 Intestinal Organoid Culture (3D)

1. Matrigel™, basement membrane matrix, growth factor reduced (GFR), phenol red-free.

2. Basal culture medium: 10 mM HEPES, 100 U/mL penicillin, and 100 μg/mL streptomycin in advanced DMEM/F12 medium.
3. B27 supplement (50×).
4. GlutaMAX (100×).
5. *N*-Acetyl-L-cysteine [500× stock; 81.5 mg/mL in distilled water (500 mM)].
6. [Leu^{15}]-Gastrin I (10,000× stock; 100 μM in 0.1% BSA/PBS).
7. Afamin-Wnt3a serum-free conditioned medium prepared from a cell line [19] (*see* **Note 1**).
8. Recombinant human R-spondin (100× stock; 100 μg/mL in 0.1% BSA/PBS) (*see* **Note 2**).
9. Recombinant mouse Noggin (1000× stock; 100 μg/mL in 0.1% BSA/PBS).
10. Recombinant mouse EGF (10,000× stock; 500 μg/mL in 0.1% BSA/PBS).
11. Recombinant human IGF-1 (1000× stock; 100 μg/mL in 0.1% BSA/PBS).
12. Recombinant human FGF-2 (FGF-basic) (1000× stock; 50 μg/mL in 0.1% BSA/PBS).
13. A83-01 (1000× stock; 500 μM in DMSO).
14. Y-27632 (1000× stock; 10 mM in PBS).
15. Human WENRAIF medium: basal culture medium supplemented with 5% (vol/vol) Afamin-Wnt3a serum-free conditioned medium, 1 μg/mL recombinant human R-spondin, 100 ng/mL recombinant mouse Noggin, 50 ng/mL EGF, 100 ng/mL recombinant human IGF-1, 50 ng/mL recombinant human FGF-2, 500 nM A83-01, 1× B27 supplement, 1× GlutaMAX, 1 mM *N*-acetyl-L-cysteine, and 10 nM [Leu^{15}]-Gastrin I.
16. TrypLE Select Enzyme (10×), no phenol red (*see* **Note 3**).
17. Dulbecco's phosphate-buffered saline without Ca^{2+} and Mg^{2+} (DPBS).
18. 15 mL centrifuge tube.
19. Flat-bottom 48-well cell culture plate.
20. Cryopreserved organoids.

2.2 Cryopreservation of Organoid

1. Recovery™ Cell Culture Freezing Medium.
2. Cool Cell® Cell Freezing Container.
3. 1 mL cryotube.

2.3 Intestinal Organoid Culture (2D)

1. Matrigel™.
2. Basal culture medium.
3. TrypLE Select Enzyme (10×), no phenol red.
4. DPBS.
5. Cell strainer (20 μm).
6. Human WENRAIF medium.
7. Y-27632 (1000× stock; 10 mM in PBS).
8. Cell culture inserts [24-well, 0.4 μm pores ($2 \times 10^6/cm^2$, transparent)].
9. Flat-bottom 24-well cell culture plate.
10. 15 mL centrifuge tube.

2.4 Formalin Fixed Paraffin Embedded (FFPE) Section Preparation (3D)

1. DPBS.
2. Paraformaldehyde (PFA; 4%).
3. iPGell®.
4. Ethanol.
5. Xylene.
6. Paraffin.
7. 15 mL low-protein-binding centrifuge tube.
8. 1.5 mL centrifuge tube.
9. Tissue cassette.
10. Base mold.
11. Forceps.
12. Glass slide.

2.5 FFPE Section Preparation (2D)

1. Basal culture medium.
2. DPBS.
3. PFA (4%).
4. Ethanol.
5. Xylene.
6. Paraffin.
7. Fine forceps.
8. Scalpel.
9. Tissue cassette.
10. Kimwipe.
11. Base mold.
12. Glass slide.

2.6 HE Staining of FFPE Section

1. Xylene.
2. Ethanol.

3. Distilled water.
4. Hematoxylin.
5. Eosin.
6. Mounting medium.
7. Cover glass.

2.7 Immunostaining of FFPE Section

1. Xylene.
2. Ethanol.
3. Distilled water.
4. Antigen retrieval buffer (Tris/EDTA buffer, pH 9.0 or Citrate buffer, pH 6.0).
5. Blocking buffer.
6. DPBS.
7. Antibodies.
8. Hoechst 33342 (1000×; 10 mg/mL).
9. Microwave.
10. Food steamer.
11. Antifade mounting medium.
12. Immedge™ hydrophobic barrier pen.
13. Cover glass.

3 Methods

3.1 Intestinal Organoid Culture (3D)

1. Thaw Matrigel™ on ice (*see* **Note 4**). Prewarm 48-well plates at 37°C.
2. Prepare cryopreserved organoids (*see* **Note 5**).
3. Quickly thaw cryopreserved organoids at 37°C in a water bath and transfer them into a 15 mL centrifuge tube.
4. Quickly add 10 mL of basal culture medium.
5. Centrifuge at 400 × *g* for 3 min. Remove the supernatant.
6. Add Matrigel™ (20 μL/wells), and carefully resuspend the organoids, avoiding making bubbles.
7. Drop 20 μL of the crypt suspension in Matrigel™ (containing 1–2 × 10^3 cells) into the center of each well of a prewarmed 48-well plate (*see* **Note 6**).
8. Place the plate in a CO_2 incubator for 10 min at 37°C under 5% CO_2.

9. After the Matrigel™ completely polymerizes, add 300 μL of WENRAIF medium to each well. To prevent anoikis, add 10 μM Y-27632 to the medium for the first 2–3 days.
10. Culture in a CO_2 incubator at 37°C under 5% CO_2. Replace the optimized medium every 2–4 days (*see* **Note 7**).

3.2 Passage of Intestinal Organoids (3D)

1. Thaw Matrigel™ on ice. Prewarm 48-well plates at 37°C.
2. Remove the culture medium from the wells of the cell culture plates.
3. Collect the 3D intestinal organoids in a 15 mL centrifuge tube by pipetting with a 1:3 dilution of TrypLE Select (10×) (500 μL/well).
4. Incubate the tubes at 37°C in a water bath for 10–20 min until the organoids are dissociated into single cells. During dissociation, gently pipette the solution every 5 min.
5. Add 10 mL of basal medium to the tube.
6. Centrifuge the crypts at 400 × *g* for 3 min. Carefully remove the supernatant.
7. Count the cells and transfer the required volume of suspension (for $1–5 \times 10^3$ cells/well) into a new tube.
8. Repeat Steps 5–10, Sect. 3.1.

3.3 Cryopreservation of Organoid (3D)

1. Passage organoids and seed them at 1×10^4 cells/well density in 48-well plates.
2. Culture in a CO_2 incubator at 37°C under 5% CO_2 for 2–4 days.
3. Remove the culture medium and add 500 μL of Recovery™ Cell Culture Freezing Medium to each well with a 1000 μL pipette.
4. Scrape and suspend the Matrigel™ containing the organoid culture in the freezing solution and transfer the suspension into a 1 mL cryotube.
5. Place the cryotubes in a CoolCell® cell freezing container and store at −80°C.
6. After overnight freezing, transfer the tubes into liquid N_2 or −150°C freezer. Samples can be stored in liquid N_2 or −150°C freezer for at least several years.

3.4 Intestinal Organoid Culture (2D)

1. Before coating the cell culture inserts with Matrigel™, cool the inserts at −20°C for 5 min or 4°C for 30 min.
2. Prepare a 20% Matrigel™ solution in basal culture medium.
3. Set the cooled cell culture inserts into the 24-well plates.

4. Add 100 μL of the 20% Matrigel™ solution into the apical side of each cell culture insert, allowing it to coat the whole surface of the membrane, and then remove the solution (*see* **Note 8**).
5. Place the plate in an incubator for 5 min at 37°C.
6. Wash the Matrigel™-coated cell culture inserts with basal culture medium.
7. Add 200 μL of basal culture medium into the apical side of each cell culture insert, and place the plates in a 37°C incubator until the cells are seeded.
8. Collect the 3D intestinal organoids in a 15 mL centrifuge tube by pipetting with a 1:3 dilution of TrypLE Select (10×) solution.
9. Incubate the tubes at 37°C in a water bath for 10–20 min until the organoids are completely dissociated into single cells. During dissociation, pipette the solution every 5 min.
10. Add basal medium and centrifuge at 400 ×*g* for 3 min. Remove the supernatant.
11. Resuspend the pellets in basal medium and filter them through a cell strainer.
12. Transfer the suspension (1.5–2 × 10^5 cells/well) in a new centrifuge tube and centrifuge at 400 ×*g* for 3 min. Remove the supernatant.
13. Resuspend the pellets in WENRAIF medium supplemented with 10 μM Y-27632 (200 μL/well).
14. Remove the basal culture medium from the cell culture inserts. Add 200 μL of the cell suspension into the apical side of each cell culture insert.
15. Leave the plates at room temperature for 10 min (*see* **Note 9**).
16. Add 700 μL of WENRAIF medium supplemented with 10 μM Y-27632 into the basal side of each cell culture insert (*see* **Note 10**).
17. Incubate in a CO_2 incubator (5% CO_2, 37°C).
18. (optional) For further differentiation, incubate the plate on a rotary shaker at 150 rpm for 4 days to generate luminal flow at confluence (3–4 days post-seeding).

3.5 FFPE Section Preparation (3D)

Before making paraffin block, culture 3D intestinal organoids in 48-well cell culture plates at a density of 3 × 10^3 cells/well for 5–7 days.

1. Remove the culture medium from the wells of the cell culture plates.

2. Collect the 3D intestinal organoids (6–9 wells) in a 15 mL low-protein-binding centrifuge tube by pipetting with 4% PFA (500 μL/well).
3. Place the tubes on ice for 15 min and tap the tubes every 5 min to prevent the organoids from aggregating.
4. Centrifuge the crypts at 200–400 × *g* for 1 min. Carefully remove the supernatant.
5. Wash with ice-cold DPBS for 3 min.
6. Centrifuge the crypts at 200–400 × *g* for 1 min. Carefully remove the supernatant.
7. Repeat Steps 5–6, Sect. 3.5.
8. Add 50 μL of DPBS and suspend the organoids. Transfer the suspension into the 1.5 mL tube (Fig. 1A).
9. Solidify the cell suspension using iPGell® kit. Add 10 μL of A-solution and pipette adequately.
10. Add 50 μL of B-solution and immediately pipette three times (*see* **Note 11**).
11. Place the tubes for 1 min at room temperature. Place the tube upside down to check that the sample is completely solidified (Fig. 1B).
12. Add 1.3 mL of 4% PFA and gently tap the tube to allow the gelled sample to be removed from the inner wall (Fig. 1C).
13. Place the tubes for overnight on a rocking shaker at room temperature.
14. Wash the gelled sample with DPBS.
15. Place the gelled sample in the tissue cassette (*see* **Note 12**) (Fig. 1D).

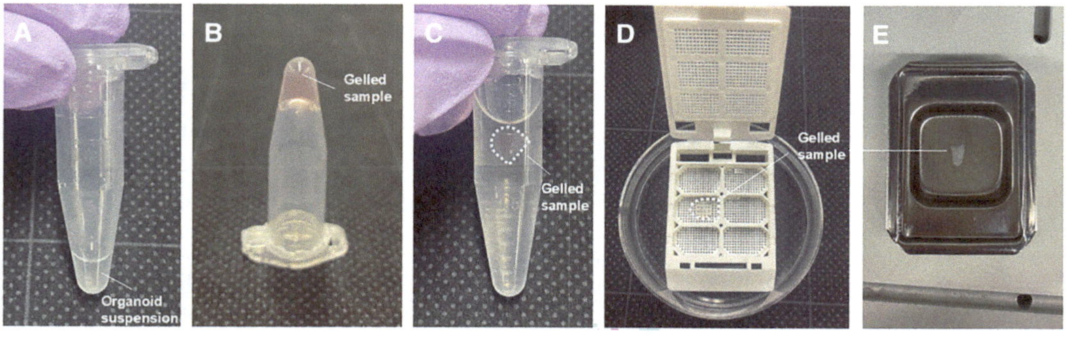

Fig. 1 Overviews of FFPE section preparation (3D organoids). (**A**) Organoids floating in DPBS after the suspension was pipetted. (**B**) Organoid suspension was solidified in the tube. (**C**) The gelled sample was floating in 4% PFA. (**D**) The gelled sample was placed in the tissue cassette. (**E**) The gelled sample was placed in melted paraffin

16. Soak the cassette in 70% ethanol overnight at room temperature (*see* **Note 13** and **Note 14**).
17. Soak the cassette in 80% ethanol for 1 h at room temperature.
18. Soak the cassette in 90% ethanol 1 h at room temperature.
19. Soak the cassette in 100% ethanol for 1 h at room temperature (×3).
20. Soak the cassette in a 1:1 solution of 100% ethanol and 100% xylene for 1 h at room temperature.
21. Soak the cassette in 100% xylene for 1 h at room temperature (×3).
22. Soak the cassette in melted paraffin overnight at 60°C.
23. Pour the melted paraffin into the base mold, and soak the sample (Fig. 1E).
24. Cool the mold for 20–30 min, and remove the paraffin block from the mold.
25. Section the sample at 4 μm on a glass slide.
26. Dry the slides at room temperature or 37°C overnight.

3.6 FFPE Section Preparation (2D)

Before staining, culture 2D intestinal organoids in 24-well cell culture inserts for ≥4 days.

1. Remove the culture medium from the wells of the cell culture inserts and plates.
2. Wash the apical side of each cell culture insert with basal culture medium.
3. Add 4% PFA to both sides of the cell culture inserts (e.g., 200 μL/well into the apical side and 500 μL/well into the basal side) and incubate for 20 min at room temperature.
4. Remove the 4% PFA and wash both sides of each cell culture insert with DPBS (×3).
5. Remove the membrane with the organoids from the culture inserts by peeling or cutting off the edge of the membrane (Fig. 2A).
6. Cut the membrane in half using a scalpel (Fig. 2B).
7. Place a Kimwipe soaked in 25% ethanol (Fig. 2C) in the tissue cassette, and place the sample on top (Fig. 2D).
8. Place another Kimwipe soaked in 25% ethanol on top and close the cassette.
9. Soak the cassette in 25% ethanol for 2 h at room temperature (*see* **Note 13**).
10. Soak the cassette in 50% ethanol for 2 h at room temperature.
11. Soak the cassette in 75% ethanol overnight at 4°C.

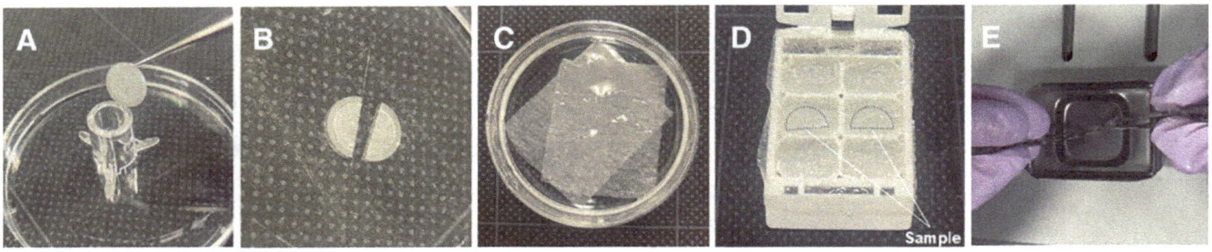

Fig. 2 Overviews of FFPE Section Preparation (2D monolayer organoids) (**A**) The organoid monolayer attached membrane was removed from the cell culture insert frame. (**B**) The membrane was cut in half using a scalpel. (**C**) The Kimwipes were soaked in 25% ethanol before sandwiching the membrane. (**D**) The membrane was placed on the Kimwipe. (**E**) The sample was placed flat side down and held vertically with forceps in melted paraffin

12. Soak the cassette in 100% ethanol for 2 h at room temperature.
13. Soak the cassette in 100% xylene for 1 h at room temperature.
14. Soak the cassette in fresh 100% xylene for 1 h at room temperature.
15. Soak the cassette in paraffin overnight at 60°C.
16. Pour the melted paraffin into the base mold.
17. Place the sample flat side down and hold it vertically with forceps until it cools and solidifies (Fig. 2E).
18. Keep the mold cold for 20–30 min, and remove the paraffin block from the mold.
19. Section the sample at 4 µm on glass slide.
20. Dry the slides at room temperature or 37°C overnight.

3.7 H&E Staining of FFPE Section

1. For deparaffination, soak the slides in 100% xylene for 5 min (×2).
2. Soak the slides in 100% ethanol for 5 min (×2).
3. Soak the slides in 90% ethanol for 3 min.
4. Soak the slides in 80% ethanol for 3 min.
5. Soak the slides in distilled water for 3 min.
6. For nuclear staining, soak the slides in Hematoxylin solution for 1–3 min.
7. Wash the slides with distilled water for 1–2 min.
8. For cytoplasmic staining, soak the slides in Eosin solution for 1–10 s.
9. Wash the slides with distilled water by moving the rack up and down three to five times.
10. Wash the slides with 80% ethanol by moving the rack up and down three to five times.

11. Soak the slides in 100% ethanol for 2 min
12. Remove excess liquid from the slides, and add one to two drops of mounting medium on the slide. Carefully place a cover glass over the section.
13. Dry the slides at room temperature overnight.
14. Acquire the images by microscope (Fig. 3A, B).

3.8 Immunostaining of FFPE Section

1. For deparaffination, repeat Steps 1–5, Sect. 3.7.
2. During deparaffination, prepare the antigen retrieval buffer. Preheat the buffer in a microwave oven for 4 min, and place it in the warmed food steamer. Check that the buffer temperature is around 90°C (*see* **Note 15**).
3. Soak the slides in preheated antigen retrieval buffer and boil them for 15 min.
4. Wash the slides with distilled water.
5. Draw a border two to four times around the section using the Immedge™ hydrophobic barrier pen.
6. Add two to three drops of blocking buffer on the slide, incubate at room temperature for 10 min.
7. Wash the slides with DPBS.
8. Add primary antibody in DPBS on the slide, incubate at room temperature for 1 h (*see* **Note 15**).
9. Wash the slides with DPBS (×3).

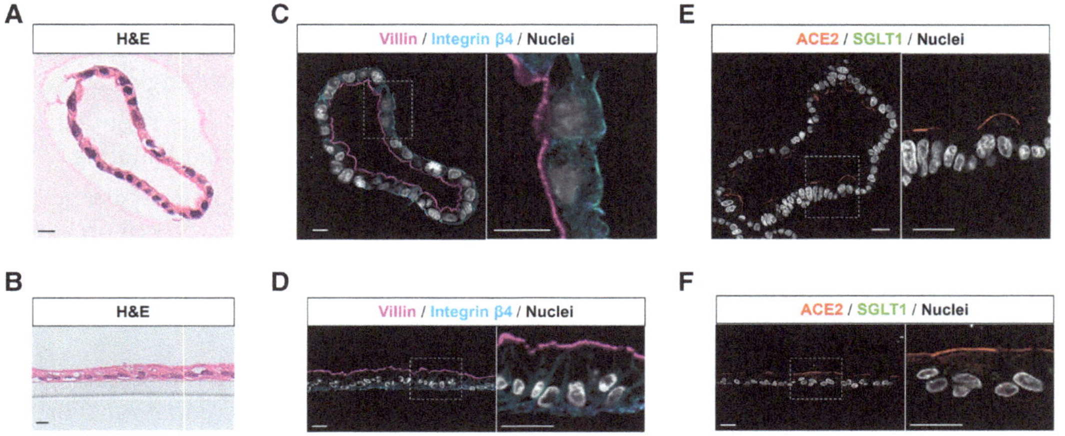

Fig. 3 Representative images of FFPE section staining (**A**) H&E staining of 3D human duodenal organoids. (**B**) H&E staining of a human duodenal organoid monolayer, with a vertical cross-sectional view. (**C**, **E**) Immunofluorescence of 3D human duodenal organoids: Villin (magenta), Integrin-β4 (cyan), ACE2 (red), SGLT1 (green), and nuclei (white). (**D**, **F**) Immunofluorescence of human duodenal organoid monolayers: Villin (magenta), Integrin-β4 (cyan), ACE2 (red), SGLT1 (green), and nuclei (white). Further optimization of the culture medium may induce maturation of organoids expressing transport proteins, including SGLT1. Scale bars, 10 μm

10. Add secondary antibody in DPBS on the slide, incubate at room temperature for 30 min. Keep it in a dark place.
11. Wash the slides with DPBS (×2).
12. Wash the slides with distilled water (×2).
13. Remove excess liquid from the slides, and add one to two drops of antifade mounting medium on the slide. Carefully place a cover glass over the section.
14. Dry the slides at room temperature overnight. Keep it in a dark place.
15. Acquire the images by fluorescence microscope (Fig. 3C–F).

4 Notes

1. We used in-house-prepared Afamin-Wnt3a serum-free conditioned medium (CM) at a final concentration of 5% as human WENRAIF medium. We confirmed that commercially available Afamin/Wnt3a CM (J2-001, MBL) can be used as an alternative at a final concentration of 10%. In addition, 0.5 nM Wnt3a mimetic peptide (PG-008, PeptiGrowth) can be used in place of in-house-prepared Afamin-Wnt3a.
2. Recombinant R-spondin can be substituted with 10% (vol/vol) in-house-prepared conditioned medium produced by HEK293T cells [20].
3. Dilute TrypLE Select (10×) threefold with DPBS (1:2). We used TrypLE Select instead of TrypLE Express because the organoids were more easily dissociated into single cells.
4. Matrigel™ begins to polymerize and solidify at room temperature; therefore, keep it chilled on ice during handling.
5. Organoids can be cryopreserved 2–3 days after passage. For the generation of organoids, refer to the protocol previously published by our group [21].
6. Prewarming the 48-well plate allowed us to easily drop the Matrigel™ in a dome shape.
7. Removal of EGF from WENRAIF medium (WNRAIF) for 2 days allows a moderate level of cell differentiation. For further differentiation, specific culture conditions can drive lineage-specific cell fates. For example, removal of Noggin and addition of BMP-4 induce enterocyte differentiation; removal of EGF and addition of an EGFR inhibitor induce differentiation of secretory lineage cells, including enteroendocrine cells

[10]; addition of RANKL induces M cell differentiation [11], and removal of Wnt, Noggin, and R-spondin together with the addition of IFN-γ promotes differentiation of BEST4$^+$ cells [9]. Furthermore, the addition of the γ-secretase inhibitor DAPT has been reported to induce differentiation into enterocytes, enteroendocrine cells, and goblet cells in human jejunal organoids [12]. For Paneth celldifferentiation, several mechanical passages are required rather than a single dissociation step with TrypLE Select. However, reconstruction of villus structures remains unattainable in 3D organoids, even under differentiation-promoting culture conditions.

8. Optionally, the same 20% Matrigel™ solution can be reused to coat multiple cell culture inserts. The 20% Matrigel™ solution can be stored at −20°C or 4°C and reused for a few weeks.
9. This step helps the cells evenly attach to the membrane surface by preventing the cells from clumping.
10. EGF is required for the initial few days to allow the spreading of 2D cells. Once the 2D cells reach confluence, medium optimization allows cell differentiation similar to that observed in 3D culture. Air–liquid interface culture has been reported to promote cell differentiation in 2D systems [16]. A recent study indicated that a 2D culture system can be applied to co-culture with gut microbes [15].
11. After mixing solutions A and B, the suspension begins to solidify immediately. Optionally, a wide-bore pipette tip can be used to minimize bubble formation during pipetting.
12. For small samples, such as gelled suspensions, tissue cassettes with separate compartments facilitate dehydration and paraffin immersion (Fig. 1D).
13. A rotary automatic paraffin embedding device can be used as an alternative to manual preparation.
14. Deparaffinization is conducted according to the manufacturer's instructions for iPGel®. To preserve the integrity of the gelled sample, a soaking step in a 1:1 mixture of 100% ethanol and 100% xylene is required.
15. For optimization of antibodies, factors such as the types of antigen retrieval buffer (Tris/EDTA buffer pH 9.0, or Citrate buffer, pH 6.0) and the primary antibody incubation time (1 h at room temperature or overnight at 4°C) should be considered. Multiple conditions should be compared when introducing a new antibody.

Acknowledgments

This work was in part supported by the Japan Agency for Medical Research and Development (AMED) (grant number JP24bm1123058), JST FOREST Program (grant number JPMJFR2458), Exploratory Research for Advanced Technology (ERATO) (grant number JPMJER2303), JSPS KAKENHI (grant numbers JP26K02474, JP23K27665 and JP22H04995), and Fukuzawa Memorial Fund for the Advancement of Education and Research.

References

1. Sato T, Clevers H (2013) Growing self-organizing mini-guts from a single intestinal stem cell: mechanism and applications. Science 340(6137):1190–1194
2. Korinek V et al. (1998) Depletion of epithelial stem-cell compartments in the small intestine of mice lacking Tcf-4. Nat Genet 19(4):379–83
3. Farin HF, Van Es JH, Clevers H (2012) Redundant sources of Wnt regulate intestinal stem cells and promote formation of Paneth cells. Gastroenterology 143(6):1518–1529.e7
4. Haramis A-PG et al. (2004) De novo crypt formation and juvenile polyposis on BMP inhibition in mouse intestine. Science 303(5664):1684–1686
5. van Es JH et al. (2005) Notch/γ-secretase inhibition turns proliferative cells in intestinal crypts and adenomas into goblet cells. Nature 435(7044):959–963
6. Sato T et al. (2009) Single Lgr5 stem cells build crypt-villus structures in vitro without a mesenchymal niche. Nature 459(7244):262–265
7. Sato T et al. (2011) Long-term expansion of epithelial organoids from human colon, adenoma, adenocarcinoma, and Barrett's epithelium. Gastroenterology 141(5):1762–1772
8. Fujii M et al. (2018) Human intestinal organoids maintain self-renewal capacity and cellular diversity in niche-inspired culture condition. Cell Stem Cell 23(6):787–793.e6
9. Wakisaka Y et al. (2025) Understanding the differentiation pathway to bestrophin 4-positive cells in human colonic epithelium. Gastroenterology 169(1):154–157.e5
10. Basak O et al. (2017) Induced quiescence of Lgr5+ stem cells in intestinal organoids enables differentiation of hormone-producing enteroendocrine cells. Cell Stem Cell 20(2):177–190.e4
11. de Lau W et al. (2012) Peyer's patch M cells derived from Lgr5+ stem cells require SpiB and are induced by RankL in cultured "miniguts". Mol Cell Biol 32(18):3639–3647
12. Meran L et al. (2020) Engineering transplantable jejunal mucosal grafts using patient-derived organoids from children with intestinal failure. Nat Med 26(10):1593–1601
13. Moon C et al. (2014) Development of a primary mouse intestinal epithelial cell monolayer culture system to evaluate factors that modulate IgA transcytosis. Mucosal Immunol 7(4):818–828
14. Ayehunie S et al. (2018) Human primary cell-based organotypic microtissues for modeling small intestinal drug absorption. Pharm. Res 35(4):72
15. Sasaki N et al. (2020) Development of a scalable coculture system for gut anaerobes and human colon epithelium. Gastroenterology 159(1):388–390.e5
16. Wang Y et al. (2019) Long-term culture captures injury-repair cycles of colonic stem cells. Cell 179(5):1144–1159.e15
17. Nikolaev M et al. (2020) Homeostatic mini-intestines through scaffold-guided organoid morphogenesis. Nature 585(7826):574–578
18. Sugimoto S et al. (2021) An organoid-based organ-repurposing approach to treat short bowel syndrome. Nature 592(7852):99–104
19. Mihara E et al. (2016) Active and water-soluble form of lipidated Wnt protein is maintained by a serum glycoprotein afamin/α-albumin. Elife 5:e11621
20. Ootani A et al. (2009) Sustained in vitro intestinal epithelial culture within a Wnt-dependent stem cell niche. Nat Med 15(6):701–706
21. Sugimoto S, Fujii M, Sato T (2020) Organoid derivation and orthotopic xenotransplantation for studying human intestinal stem cell dynamics, in: Intestinal stem cells. Springer, pp 303–320

Chapter 10

Rapid and Efficient Genetic Manipulation of Intestinal Organoids Under Transient Two-Dimensional Culture Conditions

Qiaochu Fu, Chao Gao, Cewen Chen, Shinya Tanaka and Masamichi Imajo

Abstract

Genetic manipulation of cells represents one of the fundamental techniques in current biological research, essential for elucidating the regulatory mechanisms underlying various biological phenomena, modeling many diseases, and investigating their therapeutic interventions. However, in the context of organoid biology, the manipulation of target genes and the selection of resultant genetically modified clones often encounter significant technical challenges due to the typically low gene introduction efficiency. This difficulty is largely attributed to the three-dimensional culture of organoids within the basement membrane matrix, which can be substantially mitigated by employing transient two-dimensional culture conditions during the introduction of genes. This section describes a method involving the temporary use of two-dimensional culture of intestinal organoids on collagen gel, followed by gene introduction using lipofection or lentiviral vectors. This method allows rapid and efficient genetic manipulation of intestinal organoids and can be applied to organoids from other organs.

Keywords Intestinal epithelial cells, Organoid culture, Genetic manipulation, Lipofection, Lentivirus

1 Introduction

Organoids are simplified and miniaturized three-dimensional structures that recapitulate key aspects of complex organs in vitro. Since its development, organoid technology has achieved significant advancements, facilitating a more accurate replication of various biological processes, including development and disease, and enabling the exploration of the underlying molecular mechanisms [1–3]. A critical foundational technique in organoid research is the manipulation of specific genes within organoids. Typically, when introducing genes into three-dimensionally cultured organoids, it is necessary to temporarily extract the organoids from the matrix, introduce genes using viral vectors or electroporation methods for a short period, and subsequently re-embed the organoids into the matrix to continue

Paloma Ordóñez-Morán (ed.), *Intestinal Stem Cells*, Methods in Molecular Biology, vol. 3059,
https://doi.org/10.1007/978-1-0716-5412-5_10,

their cultivation. Despite the successful execution of these complex processes, the efficiency of gene introduction into organoids remains significantly lower than that observed in conventional cultured cells. Consequently, the establishment of clones with successfully introduced genes often presents considerable technical challenges [4, 5].

The technical difficulties in introducing genes into organoids mainly arise from their cultivation within a three-dimensional matrix. Therefore, temporarily culturing cells in a two-dimensional environment during the gene introduction phase can significantly streamline the process, facilitating more convenient and efficient gene delivery. Following gene introduction, by re-embedding the cells in the matrix, resuming three-dimensional culture, and proper selection of clones, it is feasible to establish organoids composed of gene-introduced clones. This enables more straightforward and efficient genetic manipulation of organoids, thereby expediting the functional analysis of genes and modeling of diseases, such as cancer, offering a valuable technical foundation for organoid research. In this chapter, we describe a detailed protocol for the rapid and efficient gene transduction into intestinal organoids under transient two-dimensional culture conditions.

2 Materials

1. Cellmatrix Type I -A (Collagen, Type I, 3 mg/mL, pH 3.0) (Nitta Gelatin).
2. Dulbecco's Modified Eagle Medium (DMEM) (SHIMADZU).
3. 5× DMEM solution: DMEM dissolved at a concentration five times greater than the final concentration (without $NaHCO_3$).
4. Reconstitution buffer: 0.05 N NaOH, 2.2% $NaHCO_3$, and 200 mM HEPES.
5. Plasticware: sterile 15 mL conical centrifuge tubes, 1.5 mL microcentrifuge tubes, cut-down P200 pipette tips, uncut P20, P200, and P1000 pipette tips, 100 mm collagen-coated dishes, and 24-well cell culture plates.
6. Inverted microscope (for cell culture).
7. Phosphate-buffered saline (PBS):137 mM NaCl, 2.7 mM KCl, 10 mM Na_2HPO_4, 1.8 mM KH_2PO_4, pH 7.4.
8. CO_2 incubator (for cell culture).
9. Matrigel, a basement membrane matrix, growth factor reduced, phenol red-free (thawed at 4°C overnight) (Corning).

10. 1 mM ethylenediaminetetraacetic acid (EDTA) in PBS.
11. 0.1% bovine serum albumin (BSA) in PBS.
12. Refrigerated microcentrifuge (for microcentrifuge tubes).
13. Advanced DMEM/F-12 (Gibco).
14. GlutaMAX (100× concentrate) (Gibco).
15. B-27 supplement (50× concentrate) (Gibco).
16. 30 mM ROCK inhibitor Y-27632 stock solution in dimethyl sulfoxide (DMSO) (3000× concentrate).
17. 50 μg/mL recombinant mouse EGF protein dissolved in 0.1% (w/v) BSA in PBS (1000× concentrate).
18. 100 μg/mL recombinant mousenoggin protein dissolved in 0.1% (w/v) BSA in PBS (1000× concentrate).
19. 100 μg/mL recombinant mouse R-spondin1 protein dissolved in 0.1% (w/v) BSA in PBS (1000× concentrate).
20. 10,000 U/mL penicillin–streptomycin (100× concentrate) (Fujifilm).
21. 10 mM GSK3 inhibitor CHIR99021 stock solution in dimethyl sulfoxide (DMSO) (1000× concentrate).
22. 100 mM valproic acid stock solution in water (100× concentrate).
23. 2D-culture medium: Advanced DMEM/F-12 containing B-27 supplement (1×), GlutaMAX (1×), penicillin–streptomycin (1×), recombinant mouse EGF (50 ng/mL), recombinant mouse R-spondin-1 (100 ng/mL), recombinant mouse noggin (100 ng/mL), CHIR99021 (10 μM), valproic acid (1 mM), and Y-27632 (10 μM).
24. HEK293T cells (ATCC).
25. 293T culture medium: DMEM supplemented with 10% fetal bovine serum (FBS) and penicillin–streptomycin (1×).
26. 0.25% Trypsin-EDTA (Gibco)
27. Packaging, envelope, and transfer plasmids for lentivirus production.
28. PEI MAX (Polysciences).
29. Opti-MEM (Gibco)
30. 0.45 μm sterile syringe filter (Millipore).
31. 4× PEG solution: 32% polyethylene glycol (PEG) 6000, 0.4 M NaCl, 40 mM HEPES, pH 7.4.
32. Centrifuge for 15 mL tubes.
33. Polybrene (Sigma-Aldrich).
34. Lipofectamine LTX (Invitrogen).

35. Culture medium: Advanced DMEM/F-12 containing B-27 supplement (1×), GlutaMAX (1×), penicillin–streptomycin (1x), recombinant mouse EGF (50 ng/mL), recombinant mouse R-spondin-1 (100 ng/mL), recombinant mouse noggin (100 ng/mL), and Y-27632 (10 μM).
36. Puromycin (Sigma-Aldrich).

3 Methods

The following protocol is for the establishment of a two-dimensional culture of intestinal epithelial cells from mouse intestinal organoids cultured under standard three-dimensional culture conditions (3.1). We then provide detailed procedures to transfer genes using lentiviral vectors and lipofection reagents (3.2), followed by resuming three-dimensional culture and antibiotic selection of manipulated cells (3.3). These methods would help readers genetically manipulate intestinal organoids more easily and efficiently, providing useful techniques for the study of basic intestinal biology and intestinal disorders.

3.1 Two-Dimensional Culture of Mouse Intestinal Organoids on Type I Collagen Gel

3.1.1 Preparation of a Thin Layer of Collagen Gel

1. Transfer 70 μl of Cellmatrix type I -A into a microcentrifuge tube using a cut-down P200 pipette tip. It is imperative to maintain the tube on ice until the collagen solution is dispensed into the cell culture plate.
2. Add 20 μl of 5× DMEM to the tube and mix thoroughly by pipetting up and down with a cut-down P200 pipette tip.
3. Add 10 μl of Reconstitution buffer to the tube and mix thoroughly by pipetting up and down with a cut-down P200 pipette tip.
4. To facilitate the uniform distribution of the collagen solution, moisten the surface of the wells in the 24-well plate by adding ice-cold PBS and aspirating it.
5. Add the collagen solution prepared in Step 3 to the wells of the 24-well plate, ensuring even distribution by gently tilting the plate.
6. Incubate the 24-well plate for 30 min in a 37°C, 5% CO_2 incubator.

3.1.2 Seeding of Intestinal Epithelial Cells Derived from Intestinal Organoids onto the Collagen Gel Layer

The following procedure is for seeding intestinal epithelial cells onto a collagen gel layer from intestinal organoids cultivated in 30 μL of Matrigel in 24-well plates under standard culture conditions (Fig. 1a). We routinely passage cells from one to three wells in 24-well plates.

1. Remove the culture medium from the wells to be passaged.
2. Using a P1000 pipette tip, add 1 mL of ice-cold 1 mM EDTA in PBS. Pipette the total solution up and down vigorously to solubilize Matrigel. Transfer the solution into a 1.5 mL microcentrifuge tube and place on ice (*see* **Note 1**).
3. Using P200 pipette tips, pipetting the organoid suspension up and down vigorously for 20–30 times to mechanically disrupt the organoids into smaller fragments.
4. Centrifuge the organoid fragment suspension at 200 × *g* for 5 min at 4°C. Discard the supernatant.
5. Resuspend the cell pellet in 1.5 mL of the 2D-culture medium.
6. Plate 500 μL of the solution onto the surface of the collagen gel layer formed in a 24-well cell culture plate.
7. Culture cells in a 37°C, 5% CO_2 incubator. The culture medium should be replaced every other day. One day after plating, most of the intestinal organoid fragments would be converted into two-dimensional epithelial cell sheets (Fig. 1b), which can be readily infected with viral vectors and transfected using conventional transfection methodologies.

3.2 Genetic Manipulation of Intestinal Epithelial Cells Cultured on Collagen Gel

Upon seeding onto a collagen gel, intestinal organoid fragments are rapidly converted into a two-dimensional epithelial monolayer without losing their active proliferative ability. This facilitates efficient gene introduction through various methods employed for conventional cultured cells, including viral vectors and lipofection methods (Fig. 2a).

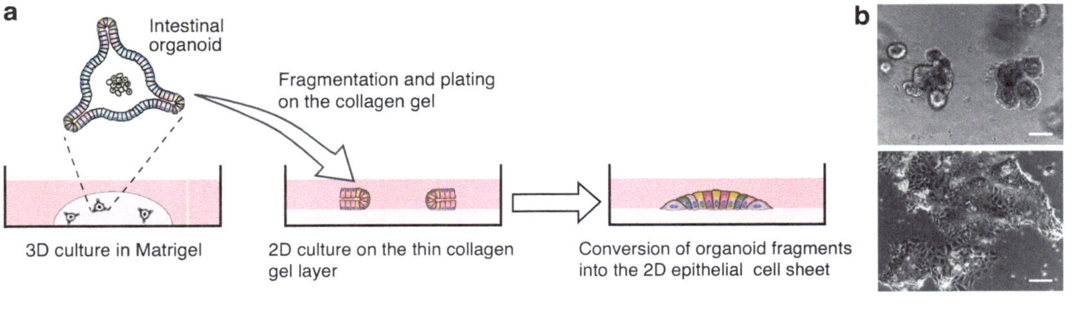

Fig. 1 Transient two-dimensional culture of intestinal organoids on collagen gel. (**a**) Schematic representation of the 2D intestinal organoid culture on collagen gel. (**b**) Images showing mouse small intestinal organoids cultured in Matrigel under standard organoid culture conditions (upper) and those cultured on collagen gel (lower). Scale bars, 100 μm

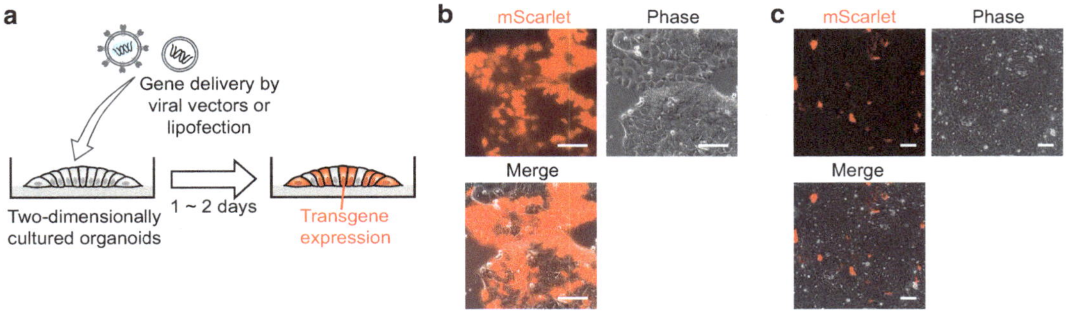

Fig. 2 Genetic manipulation of intestinal organoids cultured on collagen gel. (**a**) Schematic representation of gene delivery to intestinal organoids cultured on collagen gel. The next day after seeding on collagen gel, viral vectors or the DNA/transfection reagent complex are added to the culture dishes, leading to transgene expression within 1–2 days. (**b**) Images of the intestinal organoids infected with the lentivirus expressing red fluorescent protein, mScarlet. (**c**) Images of the intestinal organoids transfected with an expression plasmid for mScarlet. Scale bars, 100 μm

Here, we introduce protocols that can be routinely used for gene delivery into intestinal epithelial cells in a standard laboratory.

3.2.1 Production of Lentivirus

1. HEK293T cells should be cultured in DMEM supplemented with 10% FBS and antibiotics. For cell seeding, aspirate the culture medium from the 100 mm dish culturing HEK293T cells. Briefly rinse the cell layer by adding and then aspirating PBS.
2. Add 1.0 mL of 0.25% trypsin-EDTA solution to the dish, and incubate at 37°C until cells are dispersed (usually within 1 min).
3. Add 9 mL of the culture medium to the dish. Dissociate cells into a single cell suspension by pipetting up and down. Count the concentration of cells.
4. Add 10 mL of the culture medium to a new 100 mm collagen-coated dish and seed 3.0×10^6 cells. Culture cells in a 37°C, 5% CO_2 incubator for 1 day.
5. In a 15 mL conical tube, prepare a mixture of the packaging (e.g., psPAX2, 6 μg), envelop (e.g., CMV-VSV-G-RSV-Rev, 6 μg), and transfer plasmids (12 μg) in 1.5 mL Opti-MEM.
6. In another 15 mL conical tube, dilute 60 μL of PEI MAX transfection reagent in 1.5 mL of Opti-MEM.
7. Mix the plasmid solution and PEI MAX solution and incubate for 20 min at room temperature.
8. Gently add the mixed solution to the dishes culturing HEK293T cells. Slowly swirl the dish for mixing. Incubate the cells for 3 h in a 37°C, 5% CO_2 incubator.
9. After the incubation, carefully replace the culture media and culture cells in a 37°C, 5% CO_2 incubator.

10. 48 h after transfection, harvest the media containing lentivirus and filter it through a 0.45 μm sterile syringe filter. Transfer the flowthrough into a new 15 mL tube.
11. Add one-third volume of 4× PEG solution, and mix well by inverting the tube. Incubate the tube at 4°C for 24 h.
12. Centrifuge the tube at 2600 ×*g* for 45 min to precipitate the virus. Discard the supernatant. Centrifuge again at 2600 ×*g* for 5 min and discard the remaining solution.
13. Resuspend the viral pellet in 100–200 μL of Opti-MEM solution (yielding 50× to 100× concentrated viral solution). Store the concentrated viral solution at −80°C until use (*see* **Note 2**).

3.2.2 Lentivirus-Mediated Gene Transfer

1. Mix 20 μL of the 50× concentrated lentiviral solution and 500 μL of 2D-culture medium containing 10 μg/ml polybrene.
2. Remove the culture medium from the wells, culturing intestinal epithelial cells on a collagen gel layer. Add the culture medium containing lentivirus.
3. Culture cells in a 37°C, 5% CO_2 incubator for 2–4 days. The culture medium should be replaced every other day. Transgene expression can be observed from 24–36 h after infection. Typically, the infection efficiency would be 50%–80% (Fig. 2b).

3.2.3 Lipofection-Mediated Gene Transfer

Two-dimensionally cultured intestinal epithelial cells can be transfected using conventional transfection methods with several commercially available reagents. The following protocol is an example of transfection procedures using a representative transfection reagent, Lipofectamine LTX (Invitrogen) at the 24-well plate scale. We also obtained similar results with another transfection reagent, FuGENE HD (Promega).

1. Prepare a mixture of the plasmids (1 μg) and PLUS reagent in 50 μL of Opti MEM.
2. Dilute 4 μL of Lipofectamine LTX reagent in 50 μL of Opti MEM.
3. Mix the above diluted plasmid/PLUS reagent solution and Lipofectamine LTX solution and incubate for 5 min at room temperature.
4. Gently add the mixed transfection solution to the 24-well plates culturing intestinal epithelial cells. Slowly swirl the dish for mixing. Culture cells in a 37°C, 5% CO_2 incubator for 2–4 days. The culture medium should be replaced every other day. Transgene expression can be observed from 24–36 h after transfection (Fig. 2c).

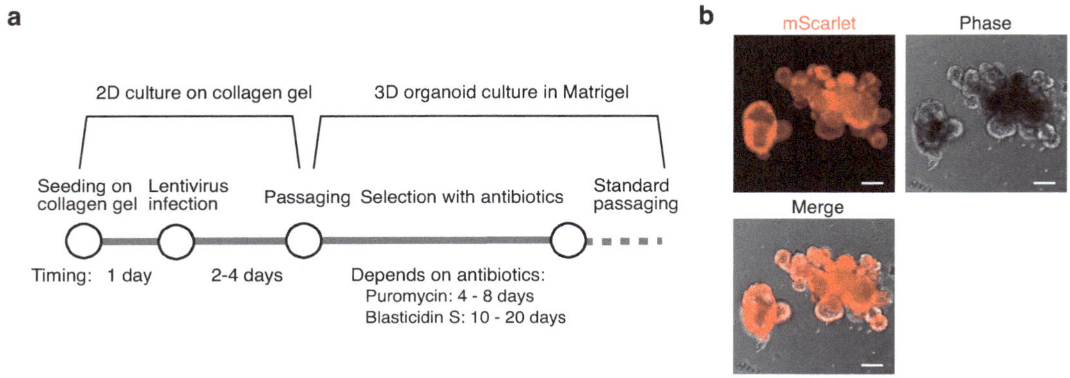

Fig. 3 Selection of genetically manipulated organoids by antibiotics. (**a**) Schematic illustration of the time course for establishing genetically modified intestinal organoids. (**b**) Images of the intestinal organoids stably expressing mScarlet. Organoids were infected with the lentivirus expressing mScarlet on the collagen gel, followed by passaging into Matrigel and selection with antibiotics (puromycin, 1 μg/mL). After selection for 7 days, most cells were mScarlet-positive. Scale bars, 100 μm

3.3 Selection and Passaging of Genetically Modified Intestinal Epithelial Cells

Two to four days after infection (or transfection), organoids can be detached from the collagen gel by conventional trypsin treatment and re-cultured under standard organoid culture conditions. Simultaneously, genetically modified intestinal epithelial cells can be selected using antibiotics (Fig. 3a). The optimal conditions for selection may vary depending on the experiment and should be optimized for each experiment. For vectors expressing a puromycin resistance gene, 1–2 μg/mL of puromycin usually works well in eliminating non-infected (or non-transfected) cells while maintaining the viability of genetically modified cells.

1. Remove the culture medium in the wells to be passaged.
2. Briefly wash cells by adding and then aspirating 0.5 mL PBS.
3. Incubate cells in 0.5 mL of the 0.25% trypsin/EDTA solution at 37°C until the cells are detached from the collagen gel surface (usually 3–5 min). Cell detachment can be accelerated by vigorously pipetting up and down the trypsin solution.
4. Transfer the cell suspension into a 1.5 mL microcentrifuge tube and add 0.5 mL DMEM supplemented with 10% FBS to inactivate trypsin.
5. Centrifuge the cell suspension at 1500×*g* for 3 min at 4°C. Discard the supernatant.
6. Resuspend the cell pellet in 30–90 μL (depending on the cell growth status (*see* **Note 3**)) of the cold Matrigel solution.
7. Plate 30 μL of the solution into the center of each well of a 24-well cell culture plate. Incubate the culture plate for 10 min in a 37°C, 5% CO_2 incubator.

8. Gently add 500 µL of culture medium containing antibiotics. For constructs expressing a puromycin resistance gene, we routinely use 1–2 µg/mL puromycin.
9. Culture cells in a 37°C, 5% CO_2 incubator for 5–7 days. After successful puromycin selection, nearly 100% of the cells should express transgenes (Fig. 3b). Genetically modified cells generated using this protocol can be maintained under standard organoid culture conditions in the presence of appropriate antibiotics.

4 Notes

1. Before starting the cell culture procedures, plasticwares that come in contact with intestinal cells should be pre-wetted with 0.1% BSA in PBS, as the adhesion of intestinal cells to their surface can significantly reduce cell yield. Pre-wetting should be performed immediately before the experiment.
2. Concentrated viral solution can be maintained at −80°C for more than half a year without noticeable changes in titer. However, repeated freeze-thaw cycles should be avoided, as they cause a significant decrease in viral titer. It is recommended to prepare single-use aliquots of the viral solution before freezing.
3. The split ratio should be adjusted based on the cell number and viability, as transgene expression can significantly affect the cell status. As a reference, we routinely passage at one-second to one-fourth dilutions at this step.

Acknowledgments

This work was supported by JSPS KAKENHI (Grant Numbers 22K06874 and 25K10197). This work was also supported by a Research Grant from HFSP (Ref.-No: RGP004/2026) (https://doi.org/10.52044/HFSP.RGP0042026.pc.gr.243776). We thank all the members of the Bio Mix Laboratory at ICReDD and the Department of Cancer Pathology at the Faculty of Medicine for their active discussion and helpful suggestions.

References

1. Sato T, Vries RG, Snippert HJ, van de Wetering M, Barker N, Stange DE, van Es JH, Abo A, Kujala P, Peters PJ, Clevers H (2009) Single Lgr5 stem cells build crypt-villus structures in vitro without a mesenchymal niche. Nature 459:262–265

2. Date S, Sato T (2015) Mini-gut organoids: reconstitution of the stem cell niche. Annu Rev Cell Dev Biol 31:269–289
3. Zhou L, Huang J, Li C, Gu Q, Li G, Li ZA, Xu J, Zhou J, Tuan RS (2025) Organoids and organs-on-chips: recent advances, applications in drug development, and regulatory challenges. Med 6:100667
4. Menche C, Farin HF (2021) Strategies for genetic manipulation of adult stem cell-derived organoids. Exp Mol Med 53:1483–1494
5. Teriyapirom I, Batista-Rocha AS, Koo BK (2021) Genetic engineering in organoids. J Mol Med (Berl) 99:555–568

Check for updates

Chapter 11

Human Small Intestinal Organoids Differentiation Protocols Towards All Mature Epithelial Lineages

Adriana Martinez-Silgado, Lulu Huang, Joep Beumer and Hans Clevers

Abstract

The small intestine plays a crucial role in nutrient uptake and metabolism regulation while maintaining a barrier that protects against the harsh environment of the lumen. These different tasks are carried out by a multitude of cell types, continuously generated by the resident stem cell population, and whose specification is under tight regulatory control. Organoids are self-organizing 3D structures derived from tissue-resident stem cells, which retain the ability to differentiate into all other epithelial lineages. Protocols to enrich for the various cell subtypes in the intestine have greatly advanced in recent years. Here, we provide an updated overview of these protocols, which direct human small intestinal organoids differentiation into various mature cell lineages, recapitulating the in vivo gut.

Keywords Adult stem cell-derived organoids, Human small intestine, Differentiation, Epithelial lineages

1 Introduction

1.1 Structure of the Epithelium and Cell Types

The intestinal epithelium consists of a single layer of cells arranged into finger-like protrusions (villi) and invaginations (crypts). This structure greatly increases the absorptive surface area, optimizing nutrient uptake; while the crypts, which are enriched in mucus and antimicrobial products, provide a protective niche for resident stem cells from the harsh luminal environment. Intestinal stem cells (ISCs), marked by LGR5, reside at the crypt base, interspaced between niche factor-producing Paneth cells [1, 2]. As stem cells divide, some stochastically lose access to niche signals, giving rise to progenitors that differentiate into all epithelial cell types of the intestinal lining. These cell types are broadly classified into two major lineages: the absorptive lineage, comprising nutrient-absorbing enterocytes, electrolyte-regulating BEST4/CA7$^+$ cell, and immune surveillance M cells; and the

These authors Adriana Martinez-Silgado and Lulu Huang contributed equally to this work

Paloma Ordóñez-Morán (ed.), *Intestinal Stem Cells*, Methods in Molecular Biology, vol. 3059,
https://doi.org/10.1007/978-1-0716-5412-5_11,

secretory lineage, including antimicrobial peptide-producing Paneth cells, mucus-secreting goblet cells, hormone-producing enteroendocrine cells, and chemosensory tuft cells (reviewed by Bonis et al. [3]).

The self-renewal of intestinal stem cells and their differentiation into various epithelial cell types are tightly regulated by signaling cues and niche factors derived from both the epithelium and the underlying mesenchyme. Among these, the NOTCH and WNT signaling pathways are crucial for ISCs maintenance and self-renewal. When stem cells lose access to Paneth cell-derived NOTCH ligands DLL1 and DLL4, they initiate differentiation toward the secretory lineage. These secretory progenitors activate the expression of NOTCH ligand, which in turn stimulates Notch signaling in adjacent progenitors. Notch-activated progenitor cells remain proliferative and fated to the absorptive lineage. This lateral inhibition process ensures a fixed ratio of secretory and more abundant absorptive cells, as a single NOTCH ligand-expressing cell can influence multiple neighboring cells [4, 5]. Thus, loss of WNT signaling in combination with NOTCH activation results in enterocytes speciation. Within the secretory lineage, differential activities of MAPK and WNT signaling further direct the subspecification. Paneth cells migrate downward to the crypt bottom, where they mature under the influence of high WNT activity [5, 6]. Goblet cells and enteroendocrine cells both specify independently of WNT [7], with high MAPK activity biasing fate towards goblet cells over enteroendocrine cells [8]. Tuft cells require WNT signaling for their development and play a role in initiating the type 2 immune response, in turn, the immune cell-derived interleukin-4 (IL-4) and IL-13 drive tuft cell hyperplasia [9, 10]. M-cells, another immune-lineage located in the Peyer's patches-large lymphoid follicles in the intestinal mucosa- are specified in response to RANKL, a member of the TNF superfamily [11, 12] (see Fig. 1).

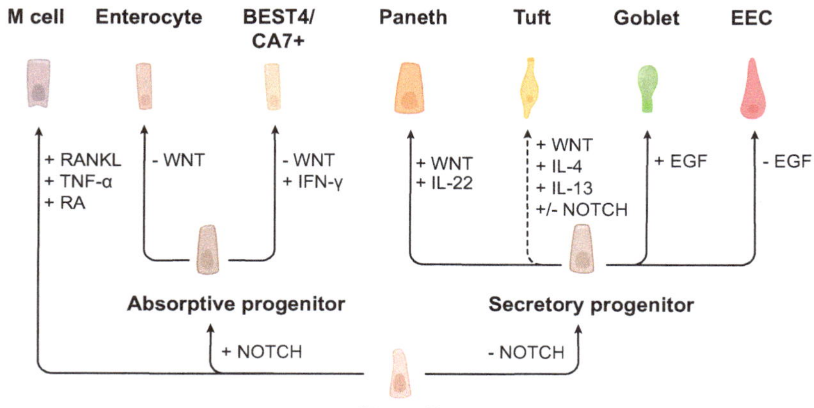

Fig. 1 Illustration of the two lineages and mature cell types in the human small intestine epithelium, and the signaling pathways involved in their specification. RA: retinoic acid.

1.2 Organoids and Organoid Differentiation

In 2009, Toshiro Sato and colleagues established a protocol to expand mouse intestinal stem cells in vitro within an extracellular matrix-like hydrogel, supplemented with Egf, Noggin to suppress Bmp, and the Wnt potentiator R-spondin [13]. This approach was later adapted to human small intestinal organoids (hSIOs), which required WNT ligands to activate WNT signaling and the inhibition of p38 MAPK signaling. Under these conditions, termed expansion medium, the organoids form spheric structures composed of a relatively homogenous population of stem and progenitor cells [14]. Alterations in the medium composition allow differentiation into multiple mature lineages, including enterocytes, goblet cells, and enteroendocrine cells. We published a compilation of these protocols in 2023 [15].

Since then, advances in our understanding of human small intestinal architecture and epithelial function have fueled the development of new protocols aimed at increasing the cellular diversity of hSIOs while preserving their proliferative potential. Organoids grown in these so-called patterning media more closely mimic the crypt–villus structure. In 2018, the development of IF medium, containing IGF-1 and FGF-2, allowed for multi-lineage differentiation alongside expansion, though it showed low levels of Paneth cells and mature enterocytes [16]. In 2022, the IL medium was introduced, which requires a two-step differentiation process with a patterning medium followed by a maturation medium in the presence of IL-22 [17]. This approach drives Paneth cell specification but is limited by IL-22-induced cell death, which prevents most organoid lines from being passaged under these conditions. In 2025, the TpC media—comprised of Trichostatin A (an HDAC inhibitor), 2-phospho-L-ascorbic acid (Vitamin C), and CP673451 (a PDGFR inhibitor)—has been shown to increase the proportion of LGR5$^+$ stem cells, supporting continuous self-renewal while enabling differentiation into all major epithelial lineages [18].

Moreover, novel protocols have been developed to enrich for M cells and rare cell types that were not previously recapitulated in human small intestinal organoids, e.g., Tuft cells and BEST4$^+$ cells [10, 19] (see Fig. 1).

1.3 Applications of Human Intestinal Organoid Differentiation

These protocols enable the enrichment and characterization of otherwise rare cell types and are especially important for studying human-specific cell types and functions. They can also be combined with co-culture protocols to investigate the crosstalk between pathogen-epithelial and immune cells [20, 21] (*see* **Notes 1 and 2** for more information on complementary protocols) (Figs. 2 and 3).

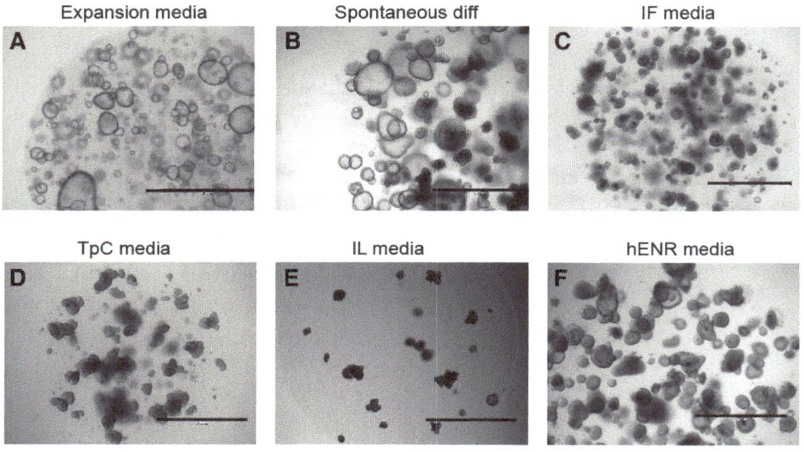

Fig. 2 Representative brightfield images of organoids grown under different medium conditions. (**A**) Organoids are ready to be passaged or differentiated. (**B**) Organoids are starting to die in the center of the hydrogel droplet and are showing signs of spontaneous differentiation. (**C–F**) Organoids cultured in the indicated differentiation media conditions, illustrating optimal growth density for each protocol. Diff: differentiation. Scale bar = 1000 μm

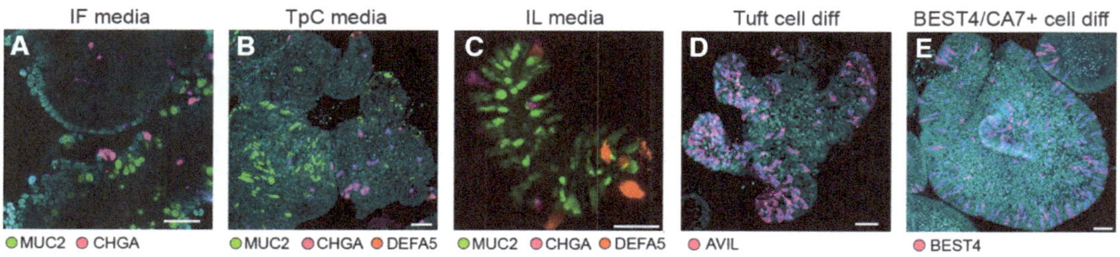

Fig. 3 Representative image of cellular composition of organoids in the described differentiation media. Diff: differentiation. Scale bar = 50 μm

2 Materials

2.1 hSIOs Maintenance and Differentiation

1. hSIOs (*see* **Note 3**).
2. Hydrogel such as Matrigel Basement Membrane Matrix (Corning) or Basement membrane extract (BME), Type II (R&D Systems).
3. Base medium: Advanced DMEM/F12 +++ (AdDF, Thermo Fisher Scientific). Advanced DMEM F12 supplemented with 10 mM HEPES, 2 mM GlutaMAX, 100 U/mL penicillin, and 100 μg/mL streptomycin.
4. TrypLE Express Enzyme (1×), phenol red (Thermo Fisher Scientific).
5. Suspension cell culture plates (24-well plates and 12-well plates).

6. Human small intestinal expansion medium (see Table 1).
7. Human small intestinal differentiation media (see Tables 2–6).

2.2 RNA Isolation and Quantitative Real-Time PCR Assay (qPCR)

1) RNA isolation kit, such as RNeasy Mini Kit (Qiagen).
2) cDNA synthesis kit such as GoScript Reverse Transcription System (Promega).
3) qPCR reagents such as SYBR green (Thermofisher).
4) Oligos for qPCR (see Table 7).
5) Appropriate qPCR instrument.

2.3 Immunofluorescence Staining

1) Cell recovery solution (Corning).
2) 4% formaldehyde.
3) Triton X-100 (Sigma).
4) Donkey serum (Jackson ImmunoResearch).
5) Primary and secondary antibodies (see Table 8).
6) Nuclear dye.
7) Mounting medium (preferably non-hardening), for example, Vectashield (Vector).

Table 1
hSIOs stem cell medium or expansion medium

Reagent	Final conc	Provider
AdDF +++	NA	Thermo Fisher Scientific
R-spondin3 or R-spondin1	R-spondin3: 2% final volume R-spondin1: 20% final volume	R-spondin3: U-Protein Express (UPE) R-spondin1: conditioned medium made in-house (*see* **note 4**)
WNT surrogate or WNT conditioned medium (CM)	WNT surrogate: 0.5 nM WNT CM: 50% final volume	WNT surrogate: U-Protein Express WNT CM: in-house conditioned medium made in-house (*see* **note 4**)
Noggin	2% final volume	U-Protein Express (UPE)
B27 supplement without Vit A	1×	Thermo Fisher Scientific
Nicotinamide	10 mM	Sigma-Aldrich
N-Acetyl-L-cysteine	1.25 mM	Sigma-Aldrich
SB202190 (P38 inhibitor)	3 μM	Sigma Aldrich
Prostaglandin E2	1 μM	Tocris
A83-01 (TGF-B inhibitor)	500 nM	Tocris
hEGF	50 ng/mL	Peprotech
Primocin (facultative)	0.1 mg/mL	Invivogen

Table 2
IF medium [16] (*see Note 5*)

Reagent	Final conc	Provider
AdDF +++	NA	Thermo Scientific
WNT surrogate or WNT conditioned medium (CM)	WNT surrogate: 0.5 nM WNT CM: 50% final volume	WNT surrogate: U-Protein Express WNT CM: in-house conditioned medium made in-house (*see* **note 4**)
R-spondin3 or R-spondin1	R-spondin3: 2% final volume R-spondin1: 20% final volume	R-spondin3: U-Protein Express (UPE) R-spondin1: conditioned medium made in-house (*see* **note 4**)
Noggin	2% final volume	U-Protein Express (UPE)
B27 supplement	1×	Thermo Fisher Scientific
N-Acetyl-L-cysteine	1.25 mM	Sigma-Aldrich
IGF-1	100 ng/mL	BioLegend
FGF2	50 ng/mL	Peprotech
EGF (optional) (*see* **Note 5**)	50 ng/mL	Peprotech
Primocin (facultative)	0.1 mg/mL	Invivogen

8) Glass microscopy slides and coverslips (18 × 18 mm).
9) Syringe filled with vaseline (optional).
10) Clear nail polish.
11) Fluorescent microscope.

3 Methods

Human intestinal organoids do not exhibit definitive differentiation in standard culture conditions (defined as expansion medium, see Table 1), as WNT-signals are an obligate part of the culture medium. As a result, these cultures are mostly composed of stem and progenitor cells. To achieve near in vivo ratios of various intestinal lineages, these progenitor cultures are first dissociated into single cells (see Sect. 3.1), then culture in IF and TpC medium with Rhoki (see Sect. 3.2), which supports both proliferation and concurrent differentiation into multiple lineages.

For cell types that are either rare or absent in IF and TpC medium -such as Paneth cells, BEST4/CA7+ cells, Tuft cells, and M cells-specific directed differentiation protocols are required. In these cases, organoids are initially grown in a conventional expansion medium

Table 3
TpC medium [18]

Reagent	Final conc	Provider
AdDF +++	NA	Thermo Fisher Scientific
WNT surrogate or WNT conditioned medium (CM)	WNT surrogate: 0.5 nM WNT CM: 50% final volume	WNT surrogate: U-Protein Express WNT CM: in-house conditioned medium made in-house (*see* **Note 4**)
R-spondin3 or R-spondin1	R-spondin3: 1% final volume R-spondin1: 10% final volume	R-spondin3: U-Protein Express (UPE) R-spondin1: conditioned medium made in-house (*see* **Note 4**)
Noggin or DMH1	Noggin: 2% final volume or 2 μM	U-Protein Express (UPE) ApexBio
B27 supplement	1×	Thermo Fisher Scientific
N-Acetyl-L-cysteine	1.25 mM	Sigma-Aldrich
IGF-1	100 ng/mL	BioLegend
FGF2	50 ng/mL	Peprotech
EGF	50 ng/mL	Peprotech
A83-01	0.5 μM	Tocris
Gastrin	10 nM	MCE
CHIR	4 μM	ApexBio
Trichostatin A (TSA)	20 nM	ApexBio
Vitamin C (pVc)	100 μg/mL	Sigma
CP673451	0.5 μM	TargetMol
Primocin (facultative)	0.1 mg/mL	Invivogen

(Table 1) for 5–7 days after passaging and then transferred to a specialized medium that promotes the differentiation of the desired cell type (see Sect. 3.3).

3.1 Dissociating Organoids to Single Cells or Small Clumps (two to four Cells) Using TrypLE

1. Prewarm TrypLE solution supplemented with 10 μM Y-27632 at 37°C.
2. Aspirate hSIOs culture medium.
3. Harvest hSIOs: use 1000 μL pippette to add 1 mL cold AdDF +++ to organoids. Disrupt the hydrogel by repeatedly aspirating and dispensing the gel droplets until the gel is dissolved. Collect in a 15 mL conical tube. Pipette the mixture up and down ~20 times in the tube to further disrupt the gel and resuspend up to 10 ml cold AdDF +++.

Table 4
IL medium [17]

Reagent	Final conc Patterning	Final conc Maturation	Provider
AdDF +++	NA		Thermo Fisher Scientific
R-spondin3 or R-spondin1	R-spondin3: 1% final volume R-spondin1 conditioned media:10% final volume	R-spondin3: 1% final volume R-spondin1:10% final volume	R-spondin3: U-Protein Express (UPE) R-spondin1: conditioned medium made in-house (*see* **note 4**)
WNT surrogate	0.5 nM	0.02 nM	WNT surrogate: U-Protein Express
CHIR	3 μM	None	Stemgent
Noggin	2% final volume	2% final volume	U-Protein Express (UPE)
B27 supplement without Vit A	1×	1×	Thermo Fisher Scientific
N-Acetyl-L-cysteine	1.25 mM	1.25 mM	Sigma-Aldrich
A83-01 (TGF-B inhibitor)	500 nM	500 nM	Tocris
hEGF	50 ng/mL	50 ng/mL	Peprotech
IL-22	None	2 ng/mL	Peprotech
Primocin (facultative)	0.1 mg/mL	0.1 mg/mL	Invivogen

Table 5
Human ENR medium (hENR)

Reagent	Final conc	Provider
AdDF +++	NA	Thermo Fisher Scientific
R-spondin3 or R-spondin1	R-spondin3: 1% final volume R-spondin1: 10% final volume	R-spondin3: U-Protein Express (UPE) R-spondin1: conditioned medium made in-house
Noggin (optional)	1% final volume	U-Protein Express (UPE)
B27 supplement	1×	Thermo Fisher Scientific
N-Acetyl-L-cysteine	1.25 mM	Sigma-Aldrich
Primocin (facultative)	0.1 mg/mL	Invivogen
EGF (optional)	50 ng/mL	Peprotech

Table 6
Additional components

Reagent	Final conc	Provider	Protocol
Rho kinase inhibitor Y-27632	10 μM	AbMole	Splitting organoids
DAPT	10 μM	Sigma-Aldrich	Tuft cell differentiation
human IL-4	5 ng/mL	Peprotech	Tuft cell differentiation
human IL-13	5 ng/mL	Peprotech	Tuft cell differentiation
RANKL	100 ng/mL	Peprotech	M cell differentiation
TNFα	20 ng/mL	Peprotech	M cell differentiation
Retinoic acid (RA)	1 μM	Sigma-Aldrich	M cell differentiation
BMP2	50 ng/mL	Peprotech	BMP stimulation
BMP4	50 ng/mL	Peprotech	BMP stimulation

4. Centrifuge organoids at 500 × *g* for 5 min at 4°C. Remove supernatant and hydrogel. If a clear separation hSIOs and hydrogel cannot be observed, remove the supernatant, resuspend the pellet in 10 mL ice-cold ADF+++, and repeat centrifugation and aspiration steps.
5. Add 1 ml TrypLE solution supplemented with 10 μM Y-27632 to the organoid pellet, resuspend and incubate 3–5 min at 37°C.
6. Using a 1000 μL pippette or a narrowed down glass pipette solution until hSIOs have dissociated to single cells or small clumps. Carefully monitor the dissociation process under the microscope.
7. Inactivate TrypLE by adding 10 mL cold ADF+++ supplemented with 10 μM Y-27632 and centrifuge at 500 × *g* for 5 min at 4°C. Remove supernatant.
8. Resuspend the cell pellet in hydrogel and plate in 10-20 μL drops (*see* **Note 7** about cell density). Incubate the plate at 37°C for 15–30 min.
9. Once the hydrogel has solidified, add pre-warmed medium (*see* **Note 8**). IF and TpC medium can be added directly. For the other differentiation protocols (Sect. 3.3), the organoid should be grown for 5–7 days in expansion medium first.

3.2 Achieving Near-Vivo Ratio of Organoids: IF and TpC Media Protocols

1. Primary tissue-derived cells can be grown directly in IF and TpC media. Alternatively, single-cell dissociated organoids that have been previously grown in conventional expansion medium can be dissociated into single cells before being transferred to

Table 7
Primers for qPCR

Cell type	Marker	Sequence fw	Sequence rev
Stem cells	*LGR5*	CTCCCAGGTCTGGTGTGTTG	GAGGTCTAGGTAGGAGGTGAAG
Enterocytes	*ALPI*	TGAGGGTGTGGCTTACCAG	GATGGACGTGTAGGCTTTGCT
Enterocytes	*APOA1*	GTGGATGTGCTCAAAGACAGCG	GCTTGCTGAAGGTGGAGGTCAC
Paneth cells	*DEFA5*	CTCCAGGAAAGAGCTGATGAGG	TCGGCAATAGCAGGTGGCTCTT
Paneth cells	*DEFA6*	ATGACCAGGACTTTGCCGTCTC	CATGACAGTGCAGGTCCCATAG
Goblet cells	*MUC2*	GAGGGCAGAACCCGAAACC	GGCGAAGTTGTAGTCGCAGAG
Goblet cells	*SPDEF*	CGAAGTGCTCAAGGACATCGAG	CGGTATTGGTGCTCTGTCCACA
Tuft cells	*HPGDS*	TTGCTGCCCTAGTGAACCAG	GCACGATGTCCTCCCAAGAG
Tuft cells	*BMX*	ACCTGAGGAGTCACGGAAAAGG	CTGTCCACCAAGCAGTTACGAG
BEST4 /CA7+ cells	BEST4	CTGGAGGGGAAGCATCTACAA	CCTGAGCATACACGTACCTCT
BEST4 /CA7+ cells	*NOTCH2*	CCTTCCACTGTGAGTGTCTGA	AGGTAGCATCATTCTGGCAGG
M cells	*GP2*	AATCAAACCCATGCCATCTACAA	CACACTGACGTTCAGGGAACT
Enteroendocrine cells (EECs)	*CHGA*	CAGCTCGTCCACTCTTTCCG	CCTCTCGTCTCCTTGGAGGG
EECs (Enterochromaffin cells)	*THP1*	ACGTCGAAAGTATTTTGCGGA	ACGGTTCCCCAGGTCTTAATC
EECs (MX cells)	*GHRL*	CAGGGGTTCAGTACCAGCAG	CCTCTTTGGCCTCTTCCCAG
EECs (MX cells)	*MLN*	ATGGTATCCCGTAAGGCTGTG	CTGGAGTTCGCCATAGGTGAA
EECs (L cells)	*GCG*	ACATTGCCAAACGTCACGATG	TCTGCGGCCAAGTTCTTCAA
EECs (L cells)	*NTS*	TGCTTTAGATGGCTTTAGCTTGG	TTCCTGGATTAACTCCCAGTGT
EECs (K cells)	*GIP*	GGATCTCATGCTAAGGTGAGC	GTCTTGTTGGTGAATCTTGTCCA
EECs (D cells)	*SST*	ACCCAACCAGACGGAGAATGA	GCCGGGTTTGAGTTAGCAGA

Table 8
Primary antibodies for immunofluorescence staining

Cell type	Marker	Species	Dilution	Supplier	Catalog
Stem cells	OLFM4, on paraffin sections (*see* **Note 6**)	Rabbit	1:5000	Thermofisher Scientific	PA5-85041
Enterocytes	APOA1	Rabbit	1:100	Thermofisher Scientific	PA5-88109
M cells	GP2	Mouse	1:100	mblbio	D277-A48
BEST4/CA7+ cells	BEST4	Rabbit	1:500	ATLAS antibodies	HPA058564
Paneth cells	LYZ	Rabbit	1:100	GeneTex	GTX72913
Goblet cells	MUC2	Rabbit	1:100	Santa Cruz	sc-15334
Tuft cells	AVIL	Rabbit	1:100	Sigma	HPA058864
Tuft cells	POU2F3	Rabbit	1:100	Sigma	HPA019652
Tuft cells	GNAT3	Goat	1:100	LSBio	LS-B4942
Enteroendocrine cells (EECs)	CHGA	Rabbit	1:100	Invitrogen	ab16693
EECs (enterochromaffin cells)	5-HT (serotonin)	Rat	1:100	Abcam	ab66047
EECs (MX cells)	GHRL	Goat	1:100	Santa Cruz	sc-517596
EECs (L cells)	GLP-1	Goat	1:100	Santa Cruz	Santa Cruz, cat# sc-73508
EECs (D cells)	SST	Rabbit	1:100	Thermofisher Scientific	17512-1-AP

either IF medium (Table 2) or TpC medium (Table 3). Replace medium every 2–3 days.

2. Differentiation is completed after 10 days. Organoid cultures can, however, be expanded long-term in these conditions, maintaining both proliferating and differentiated cell types.

See **Note 9**. For BMP signaling activation, use human recombinant BMP2 (50 ng/mL, Peprotech) and BMP4 (50 ng/mL, Peprotech) and withdraw Noggin from the culture medium [22, 23].

3.3 Differentiation of Human Intestinal Organoids Towards Other Cell Types

1. Grow hSIOs in conventional expansion medium (Table 1) for 4–6 days until these have reached a size between 100–200 μm, it is crucial that the organoids remain cystic and do not start to spontaneously differentiate (*see* **Note 10** and Fig. 2B). If the organoids are too dense, replating might be desirable to prevent cell death in the center of the hydrogel droplet.
2. To replate organoids, aspirate culture medium. Use a 1000 μL pipette to add 1 mL cold AdDF +++ to the well. Disrupt the hydrogel by repeatedly aspirating and dispensing the gel droplets until the gel is dissolved. Collect in a 15 mL conical tube.
3. Pipette the mixture up and down ~10 times inside the tube to further disrupt the gel and resuspend up to 10 mL with cold AdDF +++.
4. Centrifuge organoids at 300×*g* for 5 min at 4°C. Remove supernatant.
5. If a clear separation between hSIOs and hydrogel cannot be observed, remove the supernatant, resuspend the pellet in 10 mL ice-cold ADF+++, and repeat centrifugation and aspiration steps.
6. Resuspend organoid pellet in hydrogel and plate in 10–20 μL drops. For IL medium plate about 20–30 organoids per hydrogel droplet of 10–20 μL, for the other protocols a higher density (50–200 organoids) can be used (see Fig. 2). Incubate the plate at 37°C for 15–30 min.
7. Once the hydrogel has solidified, add the pre-warmed differentiation medium (see below).

a. **Paneth cell differentiation: IL medium protocol**

1. Prepare IL patterning medium and maturation medium with IL-22 (see Table 4).
2. Add patterning medium to organoids, refresh medium every 2–3 days for 14 days.
3. After 14 days, replace the medium with pre-warmed maturation medium with IL-22. Refresh medium every 2–3 days, after 7 days, organoids should start to bud and produce Paneth cells.

b. **Tuft cell differentiation** [10]

1. Prepare hENR (Table 5) without EGF supplemented with 10 μM Notch inhibitor DAPT, 5 ng/mL human IL-4, 5 ng/mL human IL-13, and 0.5 nM Wnt surrogate.
2. Refresh medium every 2–3 days, differentiation should be completed after 4–7 days.

c. **M cell differentiation** [12, 24]

1. Prepare hENR (Table 5) with 100 ng/mL RANKL, 20 ng/mL TNFα, and 1 μM retinoic acid (RA).
2. Refresh medium every 2–3 days, differentiation should be completed after 5–8 days.

d. **Best4^{+}cell differentiation** [19]

1. Prepare hENR (Table 5) without EGF or Noggin supplemented with 10 ng/mL IFN-γ.
2. Refresh medium every 2–3 days, differentiation should be completed after 5–8 days.

3.4 qPCR-Mediated Validation of Differentiation

1. Remove medium from the desired wells and resuspend gel-embedded organoids in the provided lysis buffer. 100 μL of gel-embedded organoids (1 well of a 12-well plate) generates approximately 250–500 ng RNA.
2. Reverse transcribe RNA into cDNA using the GoScript Reverse Transcription System (or similar kits) following the manufacturer's protocol.
3. Run cDNA samples in duplicates or triplicates in a real-time PCR instrument (see Table 7 for a list of qPCR primers).

3.5 Immunofluorescence Mediated Validation of Differentiation

1. Remove culture medium from the desired wells. To release organoids from the hydrogel, collect organoids using 1 mL cold cell-recovery solution and transfer to a 15-mL tube precoated with 1× PBS + 2% v/v donkey serum (*see* **Notes 11 and 12**).
2. Incubate organoids 30 min at 4°C on a horizontal rotating platform until the hydrogel is dissociated.
3. Spin down organoids for 5 min at 300 g at 4°C. Wash once with 1× PBS.
4. Add 1 mL 4% formaldehyde to fix organoids for a minimum of 30 min at room temperature (RT).
5. Spin down organoids for 5 min at 300 g at RT. Wash organoids using 1 mL of 1× PBS, spin down again, and remove the supernatant.

6. Add 1 mL blocking solution: 2% v/v donkey serum in 1× PBS and incubate for 15 min at RT. Afterward, spin organoids down as before and remove supernatant.
7. Permeabilize organoids for 15 min at RT using 0.5% Triton X-100 in 1× PBS. Remove supernatant after spinning down.
8. Add primary antibody at indicated concentrations (see Table 8) in 100 μL 2% donkey serum in 1× PBS 0.1 % Triton X-100 in 1× PBS. Incubate organoids overnight on a horizontal rotating platform at 4°C.
9. Wash organoids three times using 1 mL of 2% donkey serum in 1× PBS 0.1 % Triton X-100 in 1× PBS.
10. Add secondary antibody, usually at a concentration of 1:1000 and DAPI (2 μg/mL) in 100 μL 2% donkey serum in 1× PBS 0.1 % Triton X-100 in 1× PBS for 1–2 h in the dark at RT in a rotating platform.
11. Spin down and wash three times. In the last step, carefully remove as much supernatant as possible.
12. To prevent the coverslip from disrupting organoids, apply four small amounts of Vaseline in the corners where the coverslip will be placed. To apply Vaseline, fill a syringe with the Vaseline and attach a 200 mL pipette tip using parafilm.
13. Cut the front end of a 200 μL pipette tip. Using this tip, resuspend organoids in 30 μL mounting solution and apply to the glass microscope slide.
14. Carefully place the coverslip on top, do not press.
15. Seal all coverslip edges using clear nail polish.
16. Image.

4 Notes

1. These differentiation protocols can be adapted to growth in 2D or in an apical-out system, to study microbial infection or drug metabolism [25–27].
2. Genetically engineered reporters are useful to visualize, quantify, and isolate specific cell populations [28].
3. Before any work with organoids derived from primary material is started, it is necessary that ethical approval is ensured and that experiments follow all relevant regulatory standards.
4. See Pleguezuelos-Manzano et al. [29] for the protocol to produce in-house Rspondin-1 and WNT3A-conditioned media.

5. The original IF protocol uses conditioned medium of WNT bound to a serum carrier protein called Afamin, which can be used instead of other WNT supplementation. Moreover, EGF addition to IF medium increases growth efficiency, especially during the first passage and after single-cell dissociation and electroporation, but should be removed to achieve efficient enteroendocrine differentiation.
6. OLFM4 presents a broader expression than LGR5, including early progenitor cells. LGR5 remains challenging to stain due to low expression levels.
7. Cells can be counted at this step using an automatic counter or counting chamber. An optimal plating density is 100 cells per μL, but this may vary per organoid line.
8. Rho kinase inhibitor (Rhoki) should be added to the medium during the first 2 days after splitting with organoids using TrypLE to prevent anoikis.
9. Activation of BMP signaling in organoids can be used to induce the expression of upper villus enterocyte, goblet, and enteroendocrine cells. This can be achieved by adding human recombinant BMP2 (50 ng/mL, Peprotech) and BMP4 (50 ng/mL, Peprotech) and withdrawing Noggin from the culture medium.
10. It is essential that organoids are cystic and growing well before starting the differentiation, as this is indicative of a pool of immature progenitor/stem cells that retain multipotency. If organoids do appear dense and display reduced growth, premature differentiation might have occurred. Lack of remaining undifferentiated cells will reduce the ability to further direct differentiation to specific lineages. In that case, organoids should be passaged again.
11. For immunofluorescence staining, it is recommended to remove all BME, as residual matrix may interfere with antibody staining.
12. Once organoids are removed from the gel, they can adhere to the sides of the tube and the pipette tips. To avoid loss of material, it is recommended to precoat the tubes and pipette tips with cold PBS + 2% donkey serum or PBS + 0.1% bovine serum antigen. Low-binding tubes can be used as well.

Acknowledgments

This work was supported by the Netherlands Organ-on-Chip Initiative, an NWO Gravitation project (024.003.001) funded by the Ministry of Education, Culture and Science of the government of the Netherlands and the European Research Council under ERC Advanced grant (Guthormones; nr 101020405). We thank Xuan Zheng, Robin Schreurs, Daisong Wang, and Joost Wijnakker for their help and advice.

References

1. Barker N et al (2007) Identification of stem cells in small intestine and colon by marker gene Lgr5. Nature 449:1003–1007
2. Sato T et al (2011) Paneth cells constitute the niche for Lgr5 stem cells in intestinal crypts. Nature 469(7330):415.
3. Bonis V, Rossell C, Gehart H (2021) The Intestinal Epithelium – Fluid Fate and Rigid Structure From Crypt Bottom to Villus Tip. Front Cell Dev Biol 9:661931
4. van Es JH et al (2005) Notch/γ-secretase inhibition turns proliferative cells in intestinal crypts and adenomas into goblet cells. Nature 435:959–963
5. VanDussen KL et al (2012) Notch signaling modulates proliferation and differentiation of intestinal crypt base columnar stem cells. Development 139:488–497
6. Bastide P et al (2007) Sox9 regulates cell proliferation and is required for Paneth cell differentiation in the intestinal epithelium. J Cell Biol 178:635–648
7. Yin X et al (2014) Niche-independent high-purity cultures of Lgr5+ intestinal stem cells and their progeny. Nat Methods 11:106–112
8. Basak O et al (2017) Induced quiescence of Lgr5+ stem cells in intestinal organoids enables differentiation of hormone-producing enteroendocrine cells. Cell Stem Cell 20:177–190.e4
9. von Moltke J, Ji M, Liang H-E, Locksley RM (2016) Tuft-cell-derived IL-25 regulates an intestinal ILC2–epithelial response circuit. Nature 529:221–225
10. Huang L et al (2024) Tuft cells act as regenerative stem cells in the human intestine. Nature 634:929–935
11. de Lau W et al (2012) Peyer's patch M cells derived from Lgr5(+) stem cells require SpiB and are induced by RankL in cultured 'miniguts'. Mol Cell Biol 32:3639–3647
12. Wood MB, Rios D, Williams IR (2016) TNF-α augments RANKL-dependent intestinal M cell differentiation in enteroid cultures. Am J Physiol Cell Physiol 311:C498–C507
13. Sato T et al (2009) Single Lgr5 stem cells build crypt-villus structures in vitro without a mesenchymal niche. Nature 459:262–265
14. Sato T et al (2011) Long-term expansion of epithelial organoids from human colon, adenoma, adenocarcinoma, and Barrett's epithelium. Gastroenterology 141:1762–1772
15. Martinez-Silgado A, Beumer J, Clevers H (2023) Directed differentiation of murine and human small intestinal organoids toward all mature lineages. Methods Mol Biol 2650:107–122
16. Fujii M et al (2018) Human intestinal organoids maintain self-renewal capacity and cellular diversity in niche-inspired culture condition. Cell Stem Cell 23:787–793.e6
17. He G-W et al (2022) Optimized human intestinal organoid model reveals interleukin-22-dependency of paneth cell formation. Cell Stem Cell 29:1333–1345.e6
18. Yang L et al (2025) A tunable human intestinal organoid system achieves controlled balance between self-renewal and differentiation. Nat Commun 16: 315
19. Wang D et al (2025) Interferon-responsive intestinal BEST4/CA7+ cells are targets of bacterial diarrheal toxins. Cell Stem Cell 32:598–612.e5
20. Bar-Ephraim YE, Kretzschmar K, Clevers H (2020) Organoids in immunological research. Nat Rev Immunol 20:279–293
21. Puschhof J et al (2021) Intestinal organoid cocultures with microbes. Nat Protoc 16:4633–4649
22. Beumer J et al (2018) Enteroendocrine cells switch hormone expression along the crypt-to-villus BMP signalling gradient. Nat Cell Biol 20(8):909-916.
23. Beumer J et al (2022) BMP gradient along the intestinal villus axis controls zonated enterocyte and goblet cell states. Cell Rep 38(9):110438
24. Ding S et al (2020) Retinoic acid and lymphotoxin signaling promote differentiation of human intestinal M cells. Gastroenterology 159:214–226.e1
25. Ettayebi K et al (2016) Replication of human noroviruses in stem cell-derived human enteroids. Science 353:1387–1393
26. Kourula S et al. (2023) Intestinal organoids as an in vitro platform to characterize disposition, metabolism, and safety profile of small molecules. Eur J Pharm Sci 188:106481

27. Nickerson KP et al (2021) A versatile human intestinal organoid-derived epithelial monolayer model for the study of enteric pathogens. Microbiol Spectr 9:e0000321
28. Artegiani B et al (2020) Fast and efficient generation of knock-in human organoids using homology-independent CRISPR–Cas9 precision genome editing. Nat Cell Biol 22:321–331
29. Pleguezuelos-Manzano C et al (2020) Establishment and culture of human intestinal organoids derived from adult stem cells. Curr Protoc Immunol 130:e106

Check for updates

Chapter 12

Mass Spectrometry-Based Proteomics of Intestinal Organoids

Roei Cohen-Almagor, Samna Sagadevan and Paloma Ordóñez-Morán

Abstract

Proteomics, the large-scale study of proteins, enables the identification, quantification, and functional characterization of proteins, revealing post-translational modifications and protein interactions that are not apparent from transcriptomic data. Human organoids, which recapitulate the structural and functional complexity of native epithelial tissues, provide powerful tools to study disease mechanisms and personalize therapies. However, their culture poses challenges for efficient protein extraction and reproducible analysis. Here, we present a proteomics workflow optimized to maximize protein recovery from Matrigel-encased organoids. Samples were processed using S-Trap microcolumns to minimize losses, followed by liquid chromatography–mass spectrometry (LC–MS) in data-independent acquisition (DIA/SWATH-MS) mode for comprehensive, untargeted quantification. Library-free computational analysis using DIA-NN, combined with differential expression analysis, enabled sensitive detection of key proteins in intestinal organoids.

Keywords Liquid chromatography–mass spectrometry, DIA-NN, Amica, Intestinal organoids, Proteomics, Protein quantification

1 Introduction

Proteins are the most functionally relevant class of molecules in almost all biochemical processes, serving as dynamic mediators of cell signaling, structural integrity, and metabolic regulation in the human body. Studying the complete protein composition of a sample, encompassing the identification, quantification, localization, turnover, and regulation in a biological system, otherwise known as proteomics, has emerged as a cornerstone of modern molecular biology [1]. Over the past two decades, advances in mass spectrometry (MS) and bioinformatics have transformed proteomics, enabling high-throughput, precise analyses of complex biological samples. Among these approaches, bottom-up proteomics is a key methodology that involves the enzymatic digestion of proteins into peptides. Generally, individual peptides are separated by liquid chromatography

Paloma Ordóñez-Morán (ed.), *Intestinal Stem Cells*, Methods in Molecular Biology, vol. 3059,
https://doi.org/10.1007/978-1-0716-5412-5_12,

(LC) and analyzed by LC/MS. This technique measures the chromatographic elution time and the unique mass-to-charge (m/z) ratios of peptides to infer protein identification and abundance [2].

Human organoids are three-dimensional, epithelial, self-organizing models that can recapitulate the morphology and molecular features of their tissue of origin [3]. Since organoids are derived from adult stem cells, they retain their self-renewal and multilineage differentiation capacity [4]. Consequently, they represent a superior translational model that combines the scalability of two-dimensional cell lines with the heterogeneity of xenograft models. However, their growth within a matrix (e.g., Matrigel), coupled with cellular heterogeneity and limited cellular yield, poses significant challenges for proteomic workflows.

Here, we describe our protocol for the proteomic analysis of intestinal organoids. Furthermore, we discuss solutions to overcome challenges associated with protein extraction from organoids.

2 Materials

Prepare all solutions using Milli-Q water to reduce the risk of contaminants. Prepare and store reagents at room temperature unless indicated otherwise. Follow waste disposal regulations diligently.

2.1 Organoids RIPA Buffer Lysis

Reagents:

1. Ice-cold phosphate-buffered saline.
2. RIPA buffer (25 mM Tris-HCl, 150 mM NaCl, 1% NP-40, 1% sodium deoxycholate, 0.1% SDS).
3. Protease and phosphatase inhibitor cocktail (*see* **Note 1**).

Consumables:

1. 1.5 mL microcentrifuge tubes
2. Bench top centrifuge (pre-chilled) (*see* **Note 2**).
3. LoBind microcentrifuge tubes (*see* **Note 3**).

2.2 Protein Quantification

Reagents:

1. Bovine serum albumin (BSA; ≥98% purity).
2. Bradford reagent.

Consumables:

1. 0.6 mL and 1.5 mL microcentrifuge tubes.
2. 96-well clear plates (non-sterile).

Equipment:

1. Magnetic stirrer and stir bars.
2. Microplate reader (e.g., TECAN Spark).
3. Weighing balance (0.1 mg precision).
4. Multichannel pipette reservoirs.
5. Vortex mixer.

2.3 Proteomic Analysis

Software platforms:

1. DIA-NN (1.9.1).
2. AMICA (v3.0.1).

3 Methods

Carry out all procedures at room temperature unless otherwise specified.

3.1 Organoid RIPA Buffer Lysis

1. After reaching the appropriate confluency (Fig. 1), harvest the organoids using ice-cold PBS, aspirating gently in a micropipette to disrupt the Matrigel (*see* **Note 4**).
2. Transfer dissociated cell lysate to a 1.5 mL microcentrifuge tube.
3. Centrifuge at 400×*g* for 5 min. Discard supernatant (*see* **Note 5**).

   Optional: Freeze pellet in −80°C. Thaw on ice before resuming.

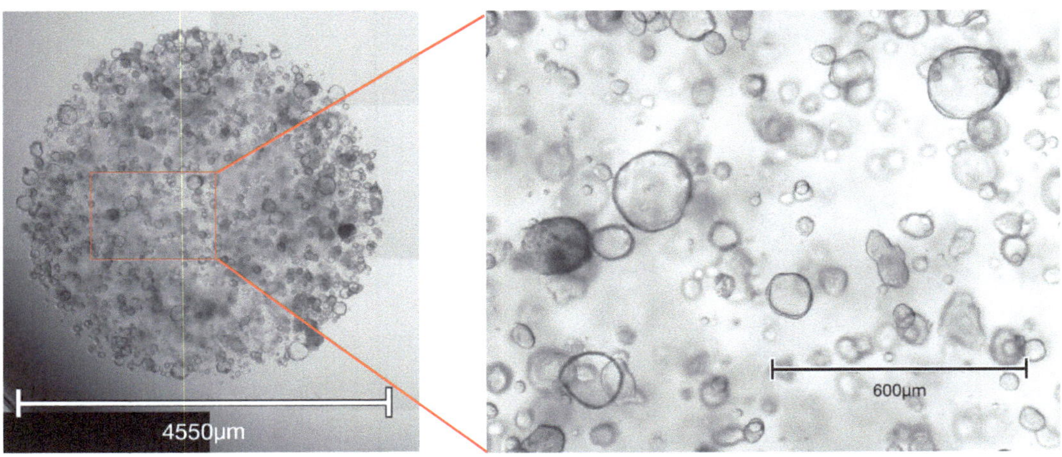

Fig. 1 Brightfield images of intestinal organoids before RIPA buffer lysis (images taken at 4× magnification on EVOS M700)

4. Resuspend pellet in 100 μL ice-cold RIPA buffer supplemented with 1× protease and phosphatase inhibitors (typically 1 μL per 100 μL of RIPA buffer)
5. Vortex briefly and incubate on ice for 30 min. Vortex every 5 min.
6. Centrifuge at 13,000 × *g* for 10 min at 4°C. Transfer supernatant to a LoBind tube (*see* **Note 6**).

3.2 Protein Quantification

1. Dissolve 0.5 g BSA in 8 mL pure H_2O. Mix with a magnetic stirrer.
2. Add 2 mL of the pure H_2O to the side of the flask.
3. Aliquot (1 mL) and store at −20°C.
4. Prepare ten eppendorfs with 200 μL H_2O each. Add 200 μL of BSA stock to Tube 1 (2000 μg/mL).
5. Serially transfer 200 μL to subsequent tubes to generate 1000–15.6 μg/mL standards. Tube 10 is blank (H_2O).
6. Dilute protein lysate in H_2O (5 μL lysate + 45 μL H_2O). Vortex gently and keep on ice.
7. Dilute Bradford reagent 1:5 in H_2O (e.g., 3 mL reagent + 12 mL H_2O) (*see* **Note 7**).
8. Add 10 μL standards/samples in triplicate to a 96-well plate (Fig. 2).
9. Add 200 μL diluted Bradford reagent per well using a multichannel pipette.

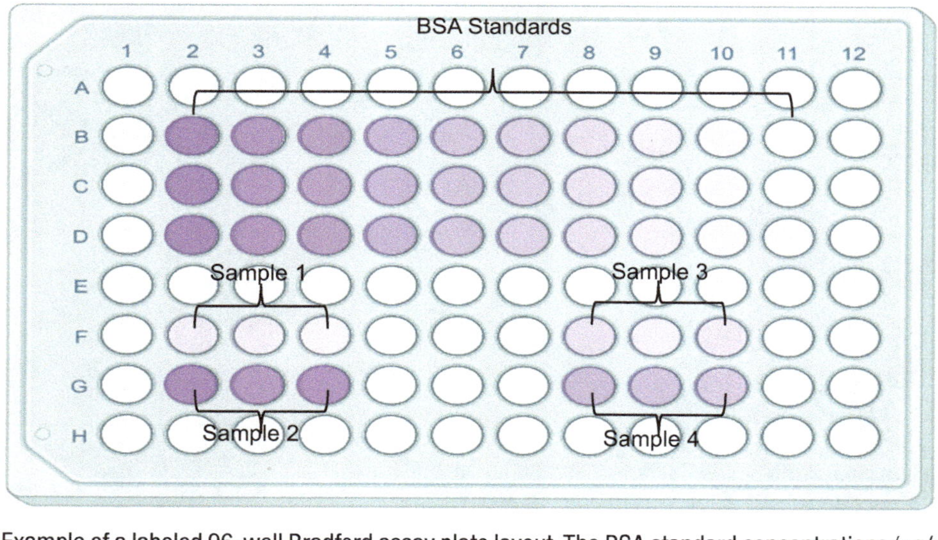

Fig. 2 Example of a labeled 96-well Bradford assay plate layout. The BSA standard concentrations (μg/mL) are as follows from left to right: 4000, 2000, 1000, 500, 250, 125, 61, 31, 15, and 0

10. Measure absorbance at 620 nm. (TECAN used, shake 3–5 s, scatter at 620 nm
11. Calculate the mean absorbance of the standard at each concentration
12. Subtract blank absorbance from standards/samples.
13. Plot log (standard concentration) vs. absorbance of standards to generate a standard curve.
14. Calculate the sample concentration by using the equation:
    a) Equation: ((1/x)*((avg-blank of sample/y))).
    b) Antilog the equation.
    c) Determine the dilution factor by multiplying the antilog by 10.
    d) Determine the concentration by dividing the dilution factor by 1000.
    e) To determine the μL of the sample to achieve 30 g of protein: 1/concentration*30.

3.3 Proteomic Analysis

1. After performing peptide preparation and LC/MS analysis according to your manufacturer's instructions, import SWATH data (.wiff files) into DIA-NN 1.9.1
2. Use the FASTA digest for library-free and deep learning-based spectra options as reference libraries (Swissprot Human Feb 2024 canonical and isoform and the cRAP proteome for contaminants).
3. Set parameters: FAST Digest for library-free search and deep learning, one missed cleavage, heuristic protein interference on, max variable mods 1 [4].
4. In the R-based software platform, AMICA (v3.0.1) [6], select the DIA file input mode.
5. Upload the DIA-NN matrix (pgmatrix.tsv).
6. Prepare a tab-separated experimental design with two columns: groups and samples.
    - groups—the experimental group name (e.g., Control, Treated).
    - samples—the sample identifiers, which must match the column names in pgmatrix.tsv, remaining in the same order.
7. Create a tab-separated contrast matrix that specifies which group comparisons to perform. The column names can be freely chosen. The sample groups in each row will be compared (*see* **Note 8**).

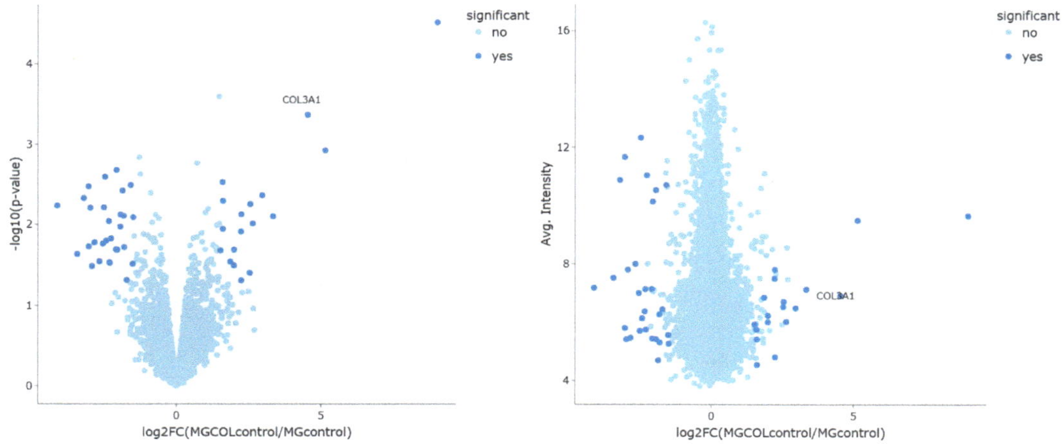

Fig. 3 Example of differential protein abundance analysis in AMICA by comparing two different condiitions: MG (organoids cultured in Matrigel) versus MGCOL (organoids cultured in Matrigel mixed with collagen I in 1:1 ratio). The volcano plot (left) shows significantly altered proteins (dark blue) based on $\log_2$ fold change and $-\log_{-10}$ (p-value). Proteins, such as COL3A1, can be highlighted and explored. The MA plot (right) displays average protein intensity against $\log_2$ fold change, illustrating expression distribution across the proteome

8. Upload the experimental design and contrast matrix in AMICA's upload panel (*see* **Note 9**).
9. Under Analysis parameters, select appropriate options in the advanced settings depending on your analysis. Click Analyze to run LIMMA-based differential abundance analysis [5]. Download and save the amica output containing change intensity and *p*-values.
10. Explore the Quality Control and Differential abundance ribbons to generate PCA, intensity distributions, MA, volcano plots, etc (Fig. 3). Alternatively, import the AMICA output into R for manual downstream analysis.

4 Notes

1. Keep on lysate and reagents on ice throughout the duration of the experiment to slow/prevent protein degradation and minimize enzyme activity.
2. Pre-chill centrifuges to 4°C >20 min before spinning to maintain lysate stability.
3. Use LoBind tubes to minimize protein adsorption during lysate storage.
4. Verify detachment of organoids from Matrigel by microscopy to ensure a higher protein yield.

5. Wash with ice-cold PBS and centrifuge again if residual Matrigel is still visible within the pellet. Excess Matrigel could skew proteins present in the final lysate.
6. If you accidentally collect the cell pellet, vortex, leave on ice for 5 min, vortex, repeat the centrifugation, and collect the supernatant.
7. Include a plate layout diagram on the plate cover to track replicates.
8. If you wish to change the sign of the fold changes, the position of the groups needs to be switched in the file (e.g., the columns will be swapped to Group 2–Group 1 instead of Group 1–Group 2).
9. Ensure there are no whitespaces and sample groups are exact in the experimental design and contrast matrix before uploading to prevent upload errors.

References

1. Sajic T, Liu Y, Aebersold R (2015) Using data-independent, high-resolution mass spectrometry in protein biomarker research: perspectives and clinical applications. Proteomics Clin Appl. 9(3–4):307–321. https://doi.org/10.1002/prca.201400117
2. Lindhorst PH, Hummon AB (2020) Proteomics of colorectal cancer: tumors, organoids, and cell cultures—a minireview. Front Mol Biosci 7:604492. https://doi.org/10.3389/fmolb.2020.604492
3. Wensink GE, Elias SG, Mullenders J, Koopman M, Boj SF, Kranenburg OW, Roodhart JML (2021) Patient-derived organoids as a predictive biomarker for treatment response in cancer patients. NPJ Precis Oncol 5(1):30. https://doi.org/10.1038/s41698-021-00168-1
4. Nakano K, Oki E, Yamazaki M et al (2025) Colorectal cancer cell line-derived organoid model with stem cell properties captures the regrowing state of residual cancer cells after neoadjuvant chemotherapy. Cell Death Discov 11:282. https://doi.org/10.1038/s41420-025-02567-w
5. Demichev V, Messner CB, Vernardis SI, Lilley KS, Ralser M (2020) DIA-NN: neural networks and interference correction enable deep proteome coverage in high throughput. Nat Methods 17(1):41–44. https://doi.org/10.1038/s41592-019-0638-x
6. Didusch S, Madern M, Hartl M, Baccarini M (2022) Amica: an interactive and user-friendly web-platform for the analysis of proteomics data. BMC Genomics 23(1):817. https://doi.org/10.1186/s12864-022-09058-7

Check for updates

Chapter 13

Organoid Culture Model for the Study of Adult Slow-Cycling Putative Intestinal Stem Cells

Maria Eugenia Gulino and Yashwant R. Mahida

Abstract

The study of intestinal stem cells is crucial for developing therapies for gut regeneration. Using Tet-Op histone 2 B (H2B)–green fluorescent protein (GFP) fusion protein transgenic mouse, we have developed a three-dimensional small intestinal (SI) organoid culture model in which cells with different cell cycle durations can be distinguished, allowing real-time visualization of the kinetics of slow-cycling H2B–GFP-retaining putative SI epithelial stem cells. The model allows real-time monitoring of the behavior of H2B–GFP-retaining putative SI stem cells under basal conditions, of their response to manipulations aimed at studying their regulation, and further functional characterization after isolation from SI organoids.

Keywords Stem cells, Organoids, Intestine, Regeneration, Paneth cells, Three dimensions

1 Introduction

Intestinal stem cells (ISCs) are crucial for the regeneration of the intestinal epithelium, which is constantly exposed to environmental stress. Two cell populations with stem cell properties exist in the murine small intestinal crypt. The first population, composed of cells at four cell position from the base of the crypt [1, 2], is capable of multilineage differentiation [3–5] and self-maintenance [3, 4]. These cells are slow cycling in a homeostatic state [6] but actively proliferate to support epithelial regeneration after injury [7]. The second population, located at crypt positions 1–4, oversees the rapid homeostatic turnover of the small intestinal epithelium [8].

A proportion of these cells is resistant to damage [9] and able to contribute to intestinal epithelium regeneration after injury [10].

Researchers hypothesize that these two stem cell populations play complementary functional roles, ensuring rapid physiological renewal and adaptable repair after damage [11].

Previous investigations into the hierarchical organization of the intestinal epithelium have relied on markers exhibiting suboptimal

Paloma Ordóñez-Morán (ed.), *Intestinal Stem Cells*, Methods in Molecular Biology, vol. 3059,
https://doi.org/10.1007/978-1-0716-5412-5_13,

specificity for the slow-cycling ISC population [12], thus underscoring the necessity for further characterization.

Since intestinal epithelial organoids preserve important physiological characteristics of the gut and have great potential for modeling tissue biology [13], we leveraged this model to further investigate slow-cycling small intestinal (SI) stem cells [14].

In this chapter, we provide details for the establishment of a 3D organoid culture model for the investigation of adult slow-cycling putative intestinal stem cells. Whole small intestinal crypts isolated from Tet-Op-H2B-GFP mice [15] effectively expand as three-dimensional structures, mirroring intestinal crypt-villus arrangement in the presence of intestinal stem cell niche signals (epidermal growth factor, the bone morphogenetic protein signal inhibitor Noggin, and the Wnt agonist/ligand of Lgr5, R-spondin1) as well as Matrigel (an extracellular matrix rich in collagen and laminin that mimics the basement membrane).

In these small intestinal organoids, the expression of the H2B–GFP transgene is transiently and ubiquitously induced by doxycycline administration (pulse). Discontinuation of doxycycline (chase) is followed by dilution of H2B–GFP from vigorously cycling crypt cells and longer retention in slow-cycling cells, which are designated "label"—retaining cells.

The protocol is suitable for the study of the activity of H2B-GFP-retaining putative adult intestinal stem cells in steady state conditions, for the in vitro monitoring of their early response to possible manipulations (e.g., genotoxic injury and modulation of ISCs regulatory pathways), and to conduct functional studies after isolating them from intestinal organoids.

2 Materials

2.1 Animals

Tet-OP-H2B–GFP mice should be generated as previously described (Foudi et al. [15]).

All the experiments must be performed as per regulatory requirements.

2.2 Isolation of Small Intestinal Crypts

1. Murine small intestinal tissue.
2. 500 mL Dulbecco's Modified Eagle Medium (DMEM) containing 2% fetal bovine serum, 100 µg/mL streptomycin, 100 U/mL penicillin, 50 µg/mL gentamicin, 2 mM L-glutamine, 20 mM HEPES, and 0.25 IU/mL insulin.
3. Dulbecco's phosphate-buffered saline without Ca^{2+} and Mg^{2+} (DPBS).
4. Weiser chelation stock solution, 3.95 g Na_2HPO_4, 5.45 g KH_2PO_4, 28.1 g NaCl, 0.6 g KCl, 75 g sucrose, 50 g

D-sorbitol, 0.4 g DL-1, 4-dithiothreitol, and q.s. to 1 L with sterile water.

5. Small intestinal crypt isolation solution [16], 200 mL Weiser stock solution, 700 mL sterile water, 2.26 g ethylenediaminetetraacetic acid (EDTA) [5 mM], 1.9 g ethyleneglycoltetraacetic acid (EGTA) [5 mM], q.s. to 1 L with sterile water, and PH 7.3.
6. 50 mL tubes.
7. Petri dish.
8. Scalpel blade.
9. 10 mL syringe with needle.
10. T25 flask.
11. 10 mL serological pipettes.
12. Phase contrast microscope.
13. 200 μm mesh.
14. Pot for crypt suspension collection.
15. 40 μm strainer.

2.3 Small Intestinal Organoid Culture

1. Small intestinal organoid culture medium, 500 mL Advanced DMEM F-12, 20 ng/mL recombinant mouse EGF, 100 ng/mL recombinant mouse Noggin, and 80–125 ng/mL recombinant mouse R-spondin 1 (*see* **Notes 1 and 2**), 100 U/mL penicillin, 100 μg/mL streptomycin, 50 μg/mL gentamicin, and 2 mM L-glutamine.
2. Tissue culture plate: 24-well, flat-bottom.
3. 96-well plate.
4. Ice.
5. Matrigel, basement membrane matrix.
6. Phase contrast microscope.

2.4 Passaging Small Intestinal Organoids

1. 15 mL tubes.
2. Ice.
3. Matrigel, basement membrane matrix.

2.5 Doxycycline Preparation

1. Doxycycline stock solution: 50 μg/mL doxycycline in Advanced DMEM F-12 (without EGF, Noggin, and R-spondin 1, but with penicillin, streptomycin, and gentamicin). Filter the solution to ensure sterility.
2. Doxycycline working solution: 800 ng/mL doxycycline in complete Advanced DMEM F-12, to prepare immediately before use.

2.6 Small Intestinal Organoid Paraffin Blocks

1. Matrigel.
2. 1 mL syringe.
3. 10% buffered formalin.
4. 70% ethanol.
5. Paraffin wax.

2.7 Immunohisto-chemistry

1. Xylene.
2. Ethanol.
3. Distilled water.
4. Antigen unmasking solution.
5. Microwave.
6. PBS.
7. 0.3% H_2O_2 solution (in methanol).
8. Normal serum of the same species as the secondary antibody's host.
9. 0.05% Tween in PBS.
10. Anti-GFP antibody.
11. Peroxidase-labeled secondary antibody.
12. Peroxidase substrate.
13. Haematoxylin.
14. Eosin.
15. Coverslips.
16. Mounting medium.

2.8 Small Intestinal Organoid Crypts Scoring

1. Light microscope equipped with an oil immersion objective 100×.
2. Immersion oil.

3 Methods

3.1 Animals

Obtain the whole small intestine from 6- to 12-week-old female Tet-OP-H2B–GFP mice.

3.2 Isolation of Small Intestinal Crypts

1. Sacrifice a mouse, harvest the entire small intestine, and place it into a 50 mL tube containing 40 mL cold DMEM.
2. Place the intestine into a 100 mm sterile dry Petri dish and cut it into two parts of equal length with a scalpel blade. Transfer each half to a Petri dish containing cold DPBS and flush it with 10 mL cold DPBS using a 10 mL syringe with a needle to remove luminal contents.

3. Divide each intestinal segment into two parts by cutting it longitudinally with a scalpel, then cut each longitudinal section transversely to obtain 3–4 mm wide pieces of small intestine.
4. Collect the pieces of small intestine with a scalpel, transfer them to a 50 mL tube, and rinse with 20 mL of cold DPBS three times.
5. Transfer the pieces of small intestine to a T25 flask and incubate with 12 mL of cold crypt isolation solution [containing Weiser solution, 5 mM ethylenediaminetetraacetic acid (EDTA), and 5 mM ethylene glycol tetraacetic acid (EGTA)] at 4°C for 30 min to achieve crypt release.
6. After incubation, shake the T25 flask at intermediate strength to initially release villous epithelium in as large pieces as possible. Let the pieces of intestine settle by gravity and remove the cloudy supernatant (containing detached epithelium) with a 10 mL serological pipette (first wash).
7. Repeat Step 6 another about six times. At the last wash, shake the T25 flask vigorously to allow maximum crypts extraction. Using a phase contrast microscope, check the outcome of each wash (*see* **Note 3**).
8. Filter out the intestinal pieces through a 200 μm mesh and collect the crypt suspension into a pot.
9. Filter the crypt suspension through a 40 μm strainer to collect crypts and rinse the strainer with 5 mL DMEM.
10. Transfer the 40 μm strainer to a pot containing 20 mL of 37°C DMEM and tumble and shake to release crypts into the medium, which turns cloudy. The isolated crypts are ready to be seeded.

3.3 Small Intestinal Organoid Culture

1. Count crypt numbers under a phase contrast microscope (*see* **Note 4**).
2. To plate 350 intestinal crypts in a well of a 24-well plate, transfer the required volume of crypt suspension into a 15 mL tube and centrifuge at 1100 rpm for 3 min.
3. Remove the supernatant and place the tube with the pelleted crypts on ice.
4. Using a P200 pipette and a cold tip, add 35 μl of cold Matrigel to the tube and gently resuspend the crypts. Place the tube back on ice.
5. Deposit the drop of 35 μl Matrigel with resuspended crypts into the center of the well. Repeat the procedure for all the required wells.

6. Place the culture plate into a CO_2 incubator (5% CO_2, 37°C) for 5 min, to allow Matrigel to solidify.
7. Add 500 μl Advanced DMEM/F-12 culture medium supplemented with EGF, Noggin, and R-spondin 1 to each well. Replace the medium three times per week and culture for 8–10 days (Fig. 1) (*see* **Notes 5 and 6**).
8. In order to prevent cell debris from building up in the lumens, passage the organoids around the tenth day of culture (*see* **Note 7**).

3.4 Passaging Small Intestinal Organoids

1. Before passaging, thaw Matrigel overnight at 4°C.
2. Using a 1000 μl pipette, scrape the surface of one well in opposite directions in order to fragment the drop of Matrigel into small pieces.

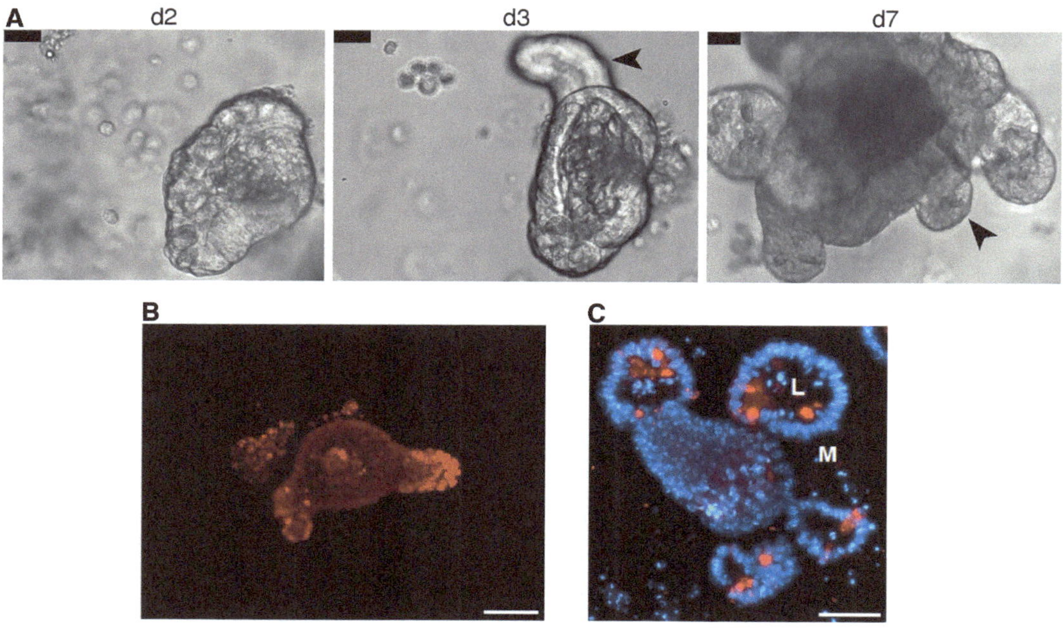

Fig. 1 SI organoid growth and compartmentalization. (**A**) Phase contrast images depict SI organoids at days 2 (d2), 3 (d3), and 7 (d7) of culture. During the first week of culture, both the number of crypts (indicated by notched arrows, d3 and d7) and the size of the SI organoids increase gradually. (**B**) Day 2 culture. The DNA dye DRAQ5 shows that, in branched SI organoids, cell proliferation is confined to crypts, revealing that the compartmentalization observed in the small intestine in vivo is maintained in vitro. (**C**) Only organoid crypts include Paneth cells (red, lysozyme), which mirror their topology in vivo. While lysozyme resides in the apical part of the cytoplasm and points toward the crypt lumen (L), the nuclei of Paneth cells (blue, DAPI) point toward Matrigel (M). Paneth cells are therefore polarized in vitro. This polarization is similar to that of Paneth cells in vivo, where the lysozyme granules face the intestinal lumen and the nucleus faces the basement membrane. (**A**) Scale bar: 25 μm; (**B**, **C**) scale bars: 50 μm

3. Using a 1000 µl pipette, pipette the content of the well (medium and fragmented Matrigel containing organoids) around 25 times to release the organoids from the Matrigel.
4. Using a 200 µl pipette, aspirate and dispense slowly around twenty times to mechanically dissociate the organoids into individual crypts.
5. Continuing to use a 200 µl pipette, slowly aspirate and dispense the content of the well for ten or fifteen more times, allowing an extensive fragmentation of the organoids into individual crypt-domains. Organoid dissociation can be evaluated under a microscope (*see* **Note 8**).
6. Once individual crypt-like structures are obtained, place the content of the well into a 15 mL tube and centrifuge at 1100 rpm for 3 min.
7. Remove the supernatant (medium and Matrigel), place the tube on ice, and working on ice, resuspend the crypt pellet in 120 µl of fresh Matrigel (40 µl per well to be plated, three wells in total), according to a 1:3 split ratio.
8. Deposit 40 µl Matrigel with resuspended crypts as a drop in the middle of one well of a 24-well plate.
9. Follow the same procedure (Steps 1–9) for all the required wells.
10. Place the plate inside the culture incubator at 37°C and in the presence of 5% CO_2 for 5 min to allow matrigel to solidify.
11. Add 500 µl of small intestinal organoid culture medium to each well.
12. Change the medium every other day and passage the intestinal organoids within 10 days of culture.

3.5 Induction of H2B-GFP Transgene Expression in Small Intestinal Organoids and Chase for Detection of H2B-GFP-Retaining (Slow-Cycling) Cells

Work on at least twice-passaged intestinal organoids, so that organoid growth and differentiation are more predictable and inter-organoid variability is reduced.

1. After 24 h from the last passage (day 1 culture), add 500 µl Advanced DMEM/F-12 + EGF, Noggin, and R-spondin 1, supplemented with 800 ng/mL doxycycline to each well for 24 h (pulse) (Fig. 2A, B).
2. After 24 h of pulse, change the medium with fresh 500 µl Advanced DMEM/F-12 + EGF, Noggin, and R-spondin 1, omitting doxycycline (0-day chase) (Fig. 2A).
3. The chase lasts 6 days. At day 6 of culture, isolated H2B-GFP-retaining cells will be detectable within the crypt domains (Figs. 2C and 3) (*see* **Notes 9–12**). During the chase, replace the medium with fresh Advanced DMEM/F-12 + EGF, Noggin, and R-spondin 1, every other day.

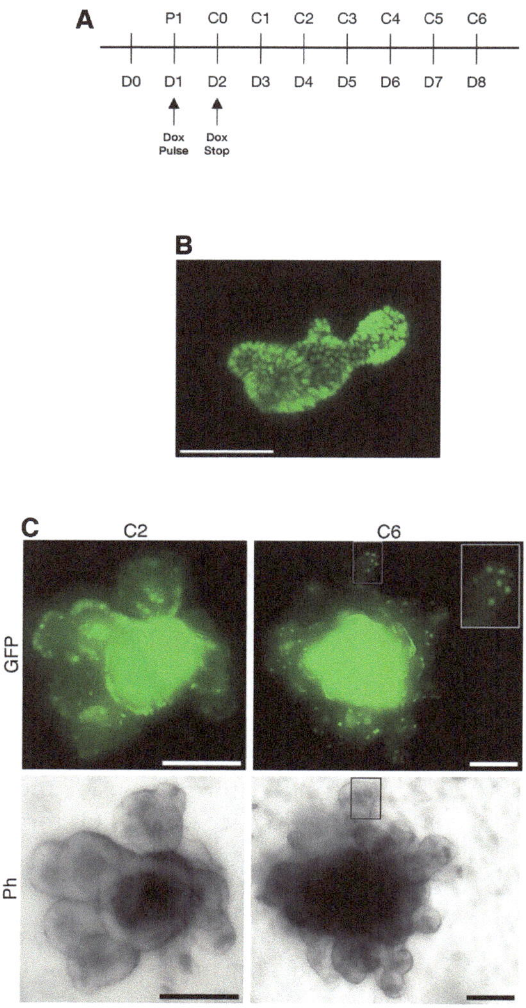

Fig. 2 Generation of H2B–GFP-retaining cells in SI organoids. (**A**) Line diagram displaying the experimental protocol to generate H2B–GFP-reatining cells in SI organoids. P1, 24h doxycycline pulse. D, day of culture. C, day of chase. (**B**) The ubiquitous expression of the H2B–GFP transgene following a 24h doxycycline pulse is shown by confocal imaging and 3D reconstruction of a live SI organoid. (**C**) Fluorescence microscopy images of live SI organoids. In the 2-day chase organoid (C2), the dilution process is at the beginning, as evidenced by the presence of continuous rows of GFP-positive nuclei (green) observed in certain crypts and within the main body of the organoid. Conversely, in the 6-day chase organoid (C6), only a limited number of H2B–GFP-retaining cells persist within the organoid crypts (insert). Extended focus for all the images. The accompanying phase contrast images (Ph) show how the distribution of H2B-GFP-retaining cells varies along the intestinal epithelium as the chase progresses, as well as the localization of H2B-GFP-retaining cells in the organoid crypts at 6-day chase. The black insert in the C6 Ph image corresponds to the white insert in the C6 GFP image. Scale bars: 100 μm

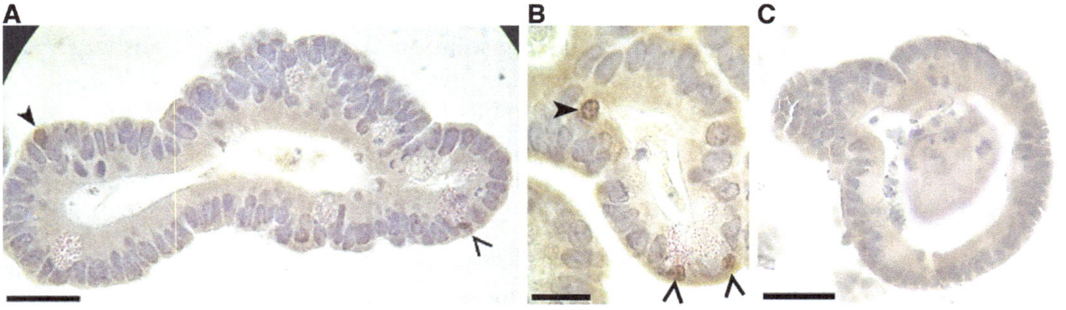

Fig. 3 H2B–GFP-retaining cells persisting in SI organoids at 6-day chase. (**A**) At least two crypts are distinctly identifiable, with Paneth cells serving as a crypt landmark. Immunohistochemistry for GFP, in conjunction with eosin staining for lysozyme granules, shows one H2B–GFP$^+$ non-Paneth cell (slow-cycling putative small intestinal stem cell) (notched arrow) and one H2B–GFP$^+$Paneth cell (open arrow) within the crypts. H2B–GFP-positive cells are not present in the intercrypt epithelium. H2B–GFP-negative Paneth cells are newly generated cells resulting from the divisions of H2B–GFP-positive precursor cells. To capture the entire organoid, the camera field of view is reduced in size. (**B**) SI organoid crypt, 6-day chase. Two H2B–GFP$^+$Paneth cells (open arrows) and one H2B–GFP$^+$ non-Paneth cell (slow-cycling putative small intestinal stem cells) (notched arrow) are detectable. (**C**) Control organoid. Immunohistochemistry shows that an organoid that was not treated with doxycycline does not include any cells that express or retain H2B-GFP. (**A** and **C**) Scale bars: 50 μm; (**B**) scale bar: 25 μm

3.6 Small Intestinal Organoid Paraffin Blocks

To prepare one paraffin block, recover intestinal organoids from three wells.

1. With a P1000 pipette, fragment the Matrigel present in each well and gently aspirate and dispense to release intact organoids from the Matrigel.
2. Transfer the contents of three wells into a 15 mL tube and centrifuge at 235 × *g* for 3 min.
3. Discard the supernatant and resuspend the organoid pellet in 50 μl of fresh Matrigel (working on ice).
4. Load the intestinal organoids embedded in Matrigel into a 1 mL syringe in which the tip and proximal part are cut to leave a cylinder.
5. Place the syringe in an incubator at 37°C until the Matrigel solidifies.
6. Extrude the block into 10% buffered formalin and fix for 3 h.
7. After fixation, wash the block in PBS.
8. Transfer the block to 70% ethanol.
9. Process in paraffin wax as per tissue.

3.7 Immunohistochemistry

Use 4 μm sections of SI organoids for immunohistochemistry.

1. Deparaffinize and rehydrate the slides.

2. Microwave the sections for total 15 minutes in citrate-based antigen unmasking solution. After cooling and washing incubate them in a 0.3% hydrogen peroxide solution (in methanol) for 30 min to block the endogenous peroxidase.
3. To prevent nonspecific binding, incubate the slides for 20 min in normal serum of the same species as the secondary antibody's host.
4. Wash the sections twice in 0.05% Tween 20 in PBS.
5. Incubate overnight at 4°C with a primary antibody directed against green fluorescent protein (GFP).
6. Wash in 0.05% Tween 20 in PBS and incubate for 1 h at room temperature with the secondary antibody conjugated with peroxidase.
7. Detect peroxidase activity using the appropriate substrate.
8. Wash the sections in tap water for 5 min.
9. Counterstain the nuclei with hematoxylin.
10. Incubate organoid sections with eosin for 2 min to stain the cytoplasm and Paneth cell granules.
11. Dehydrate the stained sections in an ascending ethanol series, clear in xylene, and mount with coverslips using mounting medium.

3.8 Organoid Crypt Scoring

Following immunoistochemistry for GFP detection, the GFP-expressing cells can be analyzed and quantified using a light microscope. Slow-cycling putative stem cells can be distinguished from GFP-retaining Paneth cells via identification of cytoplasmic granules in the latter. To identify organoid crypts to score, use Paneth cells as a crypt landmark.

1. Score at least 10 well-oriented organoid crypts per experiment (three independent experiments) by employing a light microscope and using 100× magnification combined with oil immersion. Count both the H2B-GFP-retaining non-Paneth cells (H2B-GFP$^+$ NP) and H2B-GFP-retaining Paneth cells (H2B–GFP$^+$ P; identified by the presence of eosinophilic cytoplasmic granules) persisting at 6-day chase.
2. Determine percentages out of the total H2B-GFP$^+$ population (H2B-GFP$^+$ NP + H2B-GFP$^+$ P cells).

4 Notes

1. Since the biological activity of commercially available R-spondin 1 may vary, the optimal concentration for culturing organoids should be determined for each batch. An

optimal R-spondin 1 concentration synergizes the Wnt pathway so that the small intestinal organoids form approximately 10 finger-like crypt domains at one week (Fig. 1). If the concentration is inappropriate, the organoids may develop inadequately indicating insufficient growth (small organoids with few and faint branches) or exhibit disproportionate growth (large organoids with swollen branches rather than finger-like structures), indicating overstimulation of the Wnt pathway.

2. The growth factors reported in this protocol are commercially available (rmR-spondin 1 R&D Systems, rmEGF, and rmNoggin from Peprotech).
3. After each wash, before removing the cloudy supernatant, check the result under the microscope. You should observe a progressive reduction of the amount of villous epithelium and, at the same time, an increasing presence of intestinal crypts in the washing solution. Crypt epithelium will begin to be released during the mid-washes. The product of the final wash should be a suspension of relatively pure crypts. If, after the final wash, you do not obtain a crypt-enriched suspension, add two more washes.
4. Using a 96-well plate, under the microscope, count in triplicate the number of crypts contained in 25 μl of crypt suspension. Average the counts and determine the volume of crypt suspension needed for plating 350 crypts/well.
5. The standard plating efficiency expected from the protocol is 20%. This means that 20% of the plated SI crypts turn into established SI organoids.
6. Plating efficiency is strictly dependent on the quality of SI crypt preparation and on the plating technique. SI crypts must not be damaged. A source of damage for the crypts can be the extended timing between their extraction from the intestinal wall and their embedding in Matrigel. Furthermore, mechanical damage can be induced by pipetting repeatedly in order to resuspend crypts in Matrigel before plating.
7. SI organoids consist of a single layer of epithelial cells surrounding a central lumen. Apoptotic epithelial cells are released into the lumen and start to accumulate there, leading to organoid deterioration after approximately 10 days of culture. Additionally, the lack of vascularization limits the growth capacity of the organoids in vitro. In line with these observations, it is necessary to passage the organoids within or shortly after 10 days of growth, by breaking them into smaller pieces that can be replated in fresh culture medium.
8. The critical point of this step resides in the number of times the organoids are aspirated and dispensed. Excessive pipetting should be avoided in the final stage for preventing the disruption of the crypt-like structures.

9. In order to evaluate the extent of H2B-GFP expression in live, three-dimensional, SI organoids exposed to doxycycline (after pulse and during chase), both fluorescence and confocal microscopy can be employed (Fig. 2B, C). For confocal live imaging, grow the SI organoids on 35 mm glass-bottom dishes. Use laser line 488 nm to excite GFP and relative filter set for signal detection.
10. At the 6-day chase, both slow-cycling H2B-GFP-retaining non-Paneth cells (slow-cycling putative intestinal stem cells) and post-mitotic long-lived H2B-GFP-retaining Paneth ones [17, 18] persist within the organoid crypt (Fig. 3). To distinguish H2B-GP-retaining non-Paneth cells from H2B-GFP-retaining Paneth cells, combine the immunohistochemistry for GFP detection with eosin staining: eosin stains in pink lysozyme granules that are present exclusively in the Paneth cell cytoplasm, thus rendering them clearly recognizable.
11. Using the H2B-GFP TetOP mouse technology, antibodies for the isolation of "label"-retaining cells are unnecessary: the"label"-retaining cells can be isolated from the small intestinal organoids simply on the basis of their intrinsic GFP fluorescence through fluorescence-activated sorting (FACS), thus allowing recovery of viable slowly-cycling cells suitable for functional studies and gene expression analysis. Note that after the period of chase, two types of H2B-GFP-retaining cells persist within the organoid crypt: slow-cycling H2B-GFP-retaining non-Paneth cells (putative intestinal stem cells) and post-mitotic long-lived H2B-GFP-retaining Paneth ones [17, 18] (Fig. 3). For FACS, the discrimination between the two populations can be undertaken with the use of Ulex Europaeus I lectin (UEA-I) conjugated with a fluorochrome (for example, UEA-I lectin-Atto550 conjugated) which does not require fixation. UEA-I has been widely used as a specific marker for small intestinal Paneth cells [19, 20].
12. Although the (H2B–)GFP signal decays only gradually after the sixth day of chase, the need to passage the SI organoids prevents long-term studies of slow-cycling cells in the described organoid culture model.

References

1. Cairnie AB, Lamerton LF, Steel GG (1965) Cell proliferation studies in the intestinal epithelium of the rat. I. Determination of the kinetic parameters. Exp Cell Res 39:528–538. https://doi.org/10.1016/0014-4827(65)90055-8
2. Potten CS, Kovacs L, Hamilton E (1974) Continuous labelling studies on mouse skin and intestine. Cell Tissue Kinet 7:271–283. https://doi.org/10.1111/j.1365-2184.1974.tb00907.x
3. Sangiorgi E, Capecchi MR (2008) Bmi1 is expressed in vivo in intestinal stem cells. Nat Genet 40:915–920. https://doi.org/10.1038/ng.165

4. Takeda N, Jain R, LeBoeuf MR et al (2011) Interconversion between intestinal stem cell populations in distinct niches. Science 334:1420–1424. https://doi.org/10.1126/science.1213214
5. Powell AE, Wang Y, Li Y et al (2012) The pan ErbB negative regulator Lrig1 is an intestinal stem cell marker that functions as a tumor suppressor. Cell 149:146–158. https://doi.org/10.1016/j.cell.2012.02.042
6. Montgomery RK, Carlone DL, Richmond CA et al (2010) Mouse telomerase reverse transcriptase (mTert) expression marks slowly cycling intestinal stem cells. Proc Natl Acad Sci U S A 108:179–184
7. Yan KS, Chia LA, Li X et al (2011) The intestinal stem cell markers Bmi1 and Lgr5 identify two functionally distinct populations. Proc Natl Acad Sci USA 109:466–471. https://doi.org/10.1073/pnas.1118857109
8. Barker N, van Es JH, Kuipers J et al (2007) Identification of stem cells in small intestine and colon by marker gene Lgr5. Nature 449:1003–1007. https://doi.org/10.1038/nature06196
9. Metcalfe C, Kljavin NM, Ybarra R, de Sauvage FJ (2013) Lgr5+ stem cells are indispensable for radiation-induced intestinal regeneration. Cell Stem Cell 14:149–159. https://doi.org/10.1016/j.stem.2013.11.008
10. Barriga FM, Montagni E, Mana M et al (2017) Mex3a marks a slowly dividing subpopulation of Lgr5+ intestinal stem cells. Cell Stem Cell 20:801–816.e7. https://doi.org/10.1016/j.stem.2017.02.007
11. Li L, Clevers H (2010) Coexistence of quiescent and active adult stem cells in mammals. Science 327:542–545. https://doi.org/10.1126/science.1180794
12. Itzkovitz S, Lyubimova A, Blat IC et al (2011) Single molecule transcript counting of stem cell markers in the mouse intestine. Nat Cell Biol 14:106–114. https://doi.org/10.1038/ncb2384
13. Clevers H (2016) Modeling development and disease with organoids. Cell 165:1586–1597. https://doi.org/10.1016/j.cell.2016.05.082
14. Gulino ME, Ordóñez-Morán P, Mahida YR (2024) Establishment of a 3D organoid culture model for the investigation of adult slow-cycling putative intestinal stem cells. Histochem Cell Biol 162(5):351–362. https://doi.org/10.1007/s00418-024-02312-x Epub 2024 Jul 29. PMID: 39073425
15. Foudi A, Hochedlinger K, Van Buren D et al (2009) Analysis of histone 2B-GFP retention reveals slowly cycling hematopoietic stem cells. Nat Biotechnol 27:84–90. https://doi.org/10.1038/nbt.1517
16. Weiser MM (1973) Intestinal epithelial cell surface membrane glyco- protein synthesis. I. An indicator of cellular differentiation. J Biol Chem 248:2536–2541
17. Ireland H, Houghton C, Howard L, Winton DJ (2005) Cellular inheritance of a Cre-activated reporter gene to determine Paneth cell longevity in the murine small intestine. Dev Dyn 233:1332–1336. https://doi.org/10.1002/dvdy.20446
18. Hughes KR, Gândara RMC, Javkar T et al (2012) Heterogeneity in histone 2B-green fluorescent protein-retaining putative small intestinal stem cells at cell position 4 and their absence in the colon. Am J Physiol Gastrointest Liver Physiol 303:G1188–G1201. https://doi.org/10.1152/ajpgi.00080.2012
19. Falk P, Roth KA, Gordon JI (1994) Lectins are sensitive tools for defining the differentiation programs of mouse gut epithelial cell lineages. Am J Physiol 266(6 Pt 1):G987–G1003. https://doi.org/10.1152/ajpgi.1994.266.6.G987 PMID: 8023947
20. Wong VW, Stange DE, Page ME, Buczacki S, Wabik A, Itami S, van de Wetering M, Poulsom R, Wright NA, Trotter MW, Watt FM, Winton DJ, Clevers H, Jensen KB (2012) Lrig1 controls intestinal stem-cell homeostasis by negative regulation of ErbB signalling. Nat Cell Biol 14(4):401–408. https://doi.org/10.1038/ncb2464 PMID: 22388892; PMCID: PMC3378643

Check for updates

Chapter 14

Intestinal Organoids Culture in Synthetic Hydrogels

Magda Martinez-Espuga, Alvaro Mata and Paloma Ordóñez-Morán

Abstract

Derived from adult intestinal stem cells, organoids form intricate three-dimensional structures that closely mimic the cellular diversity and spatial organization of the human gut epithelium. Patient-derived organoids (PDOs) can recapitulate patient-specific tumor heterogeneity, making them highly valuable for colorectal cancer research and personalized medicine. However, their long-term in vitro culture still depends mainly on animal-derived materials (e.g., Matrigel) with undefined composition, batch variability, and limited tunability. To overcome these limitations, we introduced the use of synthetic peptide-based hydrogels, particularly peptide amphiphiles alone and in combination with extracellular matrix macromolecules, as a reproducible and tunable alternative. Here, we present a protocol for embedding PDOs in synthetic peptide-based hydrogels, and describe how to extract their RNA for transcriptomic analysis. This workflow enables the culture of intestinal PDOs in synthetic peptide-based hydrogels in a reproducible way, preserving key proliferation and differentiation lineage markers characteristic of PDOs.

Keywords Organoids, Synthetic peptide-based hydrogels, Matrigel, Colorectal cancer, Differentiation, 3D, Intestine

1 Introduction

Patient-derived organoids (PDOs) are complex three-dimensional (3D) models that more accurately replicate tumor heterogeneity and spatial organization of the tumor from which they are derived [1, 2]. Their ability to preserve patient-specific characteristics makes them highly valuable for modeling colorectal cancer (CRC) and advancing personalized medicine [3–5]. Recreating the tumor microenvironment (TME) is fundamental in PDO culture, as it supports the preservation of cellular heterogeneity, enables cell-cell/matrix interactions, and ensures more faithful modeling of tumor dynamics and therapeutic responses.

Although PDOs represent one of the most advanced preclinical tumor models, their use is still hindered by practical challenges. These include deriving organoids from limited patient-tumor samples,

Paloma Ordóñez-Morán (ed.), *Intestinal Stem Cells*, Methods in Molecular Biology, vol. 3059,
https://doi.org/10.1007/978-1-0716-5412-5_14,

optimizing culture conditions, embedding them in suitable matrices, recovering them for passaging or analysis, and systematically evaluating their behavior [6]. In addition to these limitations, organoid culture still relies on Matrigel, a basement membrane-like matrix derived from a mouse sarcoma tumor [7]. While Matrigel has been widely used as the gold-standard scaffold for organoid culture, it presents many limitations, such as batch-to-batch variability, undefined composition, and limited tunability [8]. These limitations hinder experimental reproducibility and prevent systematic dissection of how individual extracellular matrix (ECM) components regulate organoid growth, differentiation, and response to therapy.

To address these limitations, we explore the use of synthetic peptide-derived hydrogels, particularly peptide amphiphiles (PAs), as a defined, reproducible, and tunable alternative to Matrigel [9, 10]. These hydrogels are engineered to more reliably mimic the mechanical, biochemical, and architectural complexity of the CRC-TME, providing a physiologically relevant culture system for PDOs. By incorporating well-defined molecular moieties and tunable mechanical properties, synthetic hydrogels offer an opportunity not only to replace animal-derived scaffolds but also to systematically investigate the role of specific cues that drive cancer progression, stem cell maintenance, and lineage differentiation [11–13].

In this chapter, we compare two different synthetic peptide-derived hydrogels versus Matrigel: the first one composed of PA and the second PA mixed with ECM macromolecules, including collagen type I, laminin alpha 4, fibronectin, and hyaluronan. This comparison allows us to evaluate how defined synthetic hydrogels, either with minimal or ECM-enriched components, influence organoid viability, morphology, and differentiation capacity relative to Matrigel. The work is divided into a series of experimental steps designed to systematically evaluate the performance of synthetic peptide-derived hydrogels in comparison with Matrigel. First, PDOs were embedded in synthetic peptide-based hydrogels and Matrigel. We assessed their viability, proliferation, and morphology over time. Second, we optimized a protocol for hydrogel disassembly and RNA extraction for transcriptomic analysis. Finally, we compare gene expression profiles of PDOs cultured in PA-derived hydrogels with those in Matrigel (Fig. 1).

This work aims to establish a defined, reproducible, and physiologically relevant platform that more accurately reflects the mechanical, biochemical, and architectural complexity of the CRC microenvironment. Synthetic hydrogels represent a promising next-generation platform that can be engineered to recapitulate the dynamic TME, offering greater reproducibility for cancer modeling and enhancing the potential for personalized medicine.

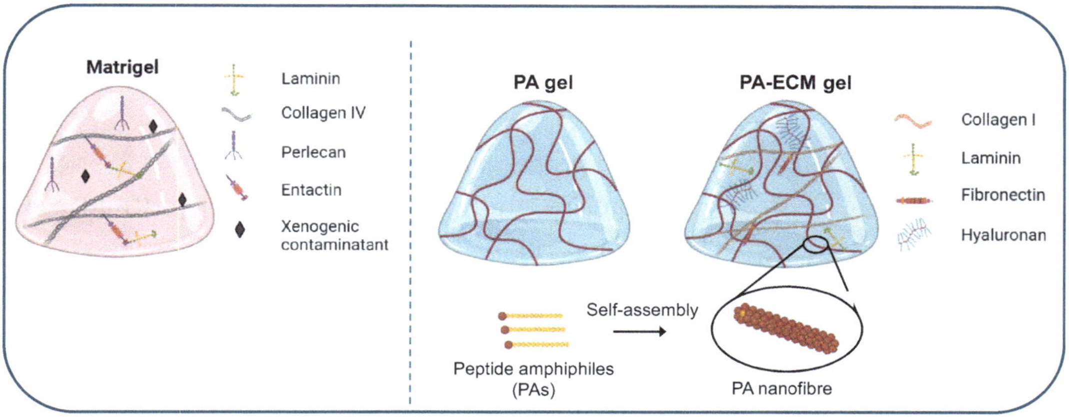

Fig. 1 Human organoids embedded in Matrigel, PA, and PA-ECM hydrogels scheme.

2 Materials

2.1 Organoid Culture in Matrigel

1. Intestinal organoid media:
   - Advanced Dulbecco's Modified Eagle Medium (DMEM) supplemented with F12 (Advanced DMEM/F12).
   - GlutaMAX.
   - Penicillin (100 U/mL) and streptomycin (100 µg/mL).
   - Primocin.
   - *N*-Acetyl-L-cysteine (NAC) [500× stock; 81.5 mg/mL in distilled water (500 mM)].
   - N21 supplement.
   - B27 supplements.
   - Nicotinamide.
   - CHIR99021.
   - PGE_2.
   - SB431542.
   - A83-01.
   - Y-27632 [1000× stock; 10 mM in PBS].
   - Growth factors: EGF, Noggin, and R-spondin-1.
2. Matrigel™, basement membrane matrix, growth factor reduced (GFR), phenol red-free (Corning).
3. Cold Dulbecco's phosphate-buffered saline without Ca^{2+} and Mg^{2+} (DPBS; Gibco).
4. Flat-bottom 48-well cell culture plate.
5. 15 mL centrifuge tube.

6. Centrifuge.
7. Incubator 37°C.
8. Ice.

2.2 Synthetic Peptide-Based Hydrogels

1. Lyophilized peptide amphiphile (PA).
2. HEPES buffer (10 mM HEPES, 3 mM KCl, 150 mM NaCl, pH 7.2–7.4).
3. Human collagen type I (2.5 mg/mL).
4. Low molecular weight hyaluronan (2.5 mg/mL).
5. Recombinant human laminin α4 (250 μg/mL).
6. Human fibronectin (250 μg/mL).
7. $CaCl_2$ solution (50 mM, in HEPES buffer).
8. Flat-bottom 48-well cell culture plate.

2.3 Disassembly of Synthetic Peptide-Based Hydrogels

1. Centrifuge.
2. Cold advanced DMEM/F12.
3. TrypLE select enzyme (10×), no phenol red.
4. Eppendorf (1.5 mL).
5. Incubator at 37°C.

2.4 RNA Extraction

1. RNeasy Mini Spin Columns.
2. Collection tubes (1.5 mL).
3. Collection tubes (2 mL).
4. RLT buffer.
5. RW1 buffer.
6. RPE buffer.
7. RNase-free water.
8. β-Mercaptoetanol.
9. Ethanol 70%.

2.5 cDNA Synthesis

1. RNA template (total RNA).
2. Oligo dT master mix (18 and 24) (100 μM).
3. dNTPs (10 mM each).
4. 0.1 M DTT.
5. 250 mM Tris-HCl pH 8.3; 375 mM KCl, 15 mM $MgCl_2$ 5× RT buffer.
6. Reverse transcriptase (Superscript II, 200 U/μl).
7. Ribonuclease inhibitor (RNasin, 40 U/μl).
8. Nuclease-free water.
9. Thermal cycler.

Table 1
Primers used for gene expression analysis

Gene	Forward (5′->3′)	Reverse (3′->5′)
GAPDH	GTCTCCTCTGACTTCAACA	AGCCAAATTCGTTGTCATAC
VILLIN	AAAGTGGACGTGTTCAATGCTAA	AATCTTCAATGGACAGGTGTTCC
MUC2	ACCCGCACTATGTCACCTTC	GGACAGGACACCTTGTCGTT
LYZ1	ACTACAATGCTGGAGACAGAAGC	GCACAAGCTACAGCATCAGCGA
PCNA	AGGCACTCAAGGACCTCATCA	GAGTCCATGCTCTGCAGGTTT

2.6 cDNA Amplification by qPCR

1. cDNA pool: our samples and negative controls (samples without Superscript enzyme).
2. PCR primers for the genes of interest and a housekeeping gene (Table 1).
3. SYBR Green: includes dNTPs, Taq Polymerase, and $MgCl_2$.
4. MicroAmp Fast Optical 96-well plate.
5. MicroAmp Optimal Adhesive Film.
6. Quantitative real-time thermal cycler.

3 Methods

3.1 Establishment of Organoid Lines in Matrigel

1. Expand epithelial cells isolated from tumor biopsies in Matrigel over 15–21 days to establish organoid lines.
2. Dissociate organoids into single cells and re-embed in 25 μL of Matrigel (pre-thawed on ice) in a 48-well plate.
3. Incubate for 15 min at 37°C to allow Matrigel to solidify,
4. Add 200 μL intestinal organoid media supplemented with Y-27632.
5. Replace the intestinal organoid media every 48–72 h.

3.2 Embedding of Organoids in Matrigel and Synthetic Peptide-Derived Hydrogels

1. Thaw Matrigel on ice. Pre-warm 48-well plates at 37°C.
2. Remove the intestinal organoid media carefully, without disrupting Matrigel domes containing organoids.
3. Add 300 μL cold DPBS and pipette up and down, disrupting the domes.
4. Transfer the resulting suspension to a 15 ml centrifuge tube and add 10 mL of DPBS.
5. Centrifuge at 300 × *g* for 5 min. Remove the supernatant carefully.

6. Add 10 mL of cold Advanced DMEM/F12 and mechanically disrupt (pipetting) the organoids.
7. Centrifuge at 300 × *g* for 5 min. Remove the supernatant carefully.
8. Resuspend the organoids in HEPES buffer (5 μL/ gel).
   a. For Matrigel, add cold Matrigel (20 μL/gel) to the HEPES buffer with organoids.
   b. For synthetic peptide-derived hydrogels:
      i. Add 7.5 μL of gelling solution (Control gelling solution or ECM gelling solution) to a 48-well plate.
      ii. Mix the PA solution with the HEPES buffer solution containing organoids (12.5 μL PA solution/ 5 μL organoids in HEPES).
      iii. Add the PA solution with HEPES containing organoids into the gelling solution (*see* **Note 1**).
9. Incubate for 15 min to allow hydrogels to solidify.
10. Add 200 μL of intestinal organoid media supplemented.
11. Change intestinal organoid media every 48 h.
12. Check organoid viability and growth under a brightfield microscope by assessing morphology and monitoring changes over time (Fig. 2) (*see* **Note 2**).

3.3 Disassembly of Matrigel and Synthetic Peptide-Based Hydrogels

1. Carefully remove the intestinal organoid media without disrupting the domes of Matrigel and synthetic peptide-based hydrogels.
2. Disruption of Matrigel and synthetic peptide-based hydrogels:
   a. For Matrigel:
      i. Add cold Advanced DMEM/F12 and mechanically disrupt (pipetting) the matrix with organoids.

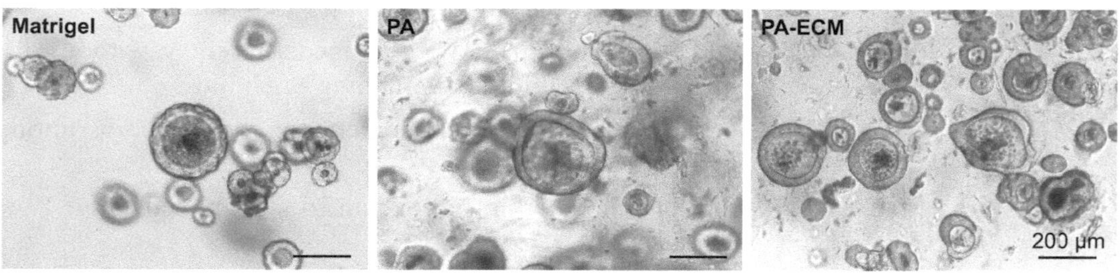

Fig. 2 Human organoids embedded in Matrigel, PA, and PA-ECM hydrogels after 7 days in culture. Scale bar, 200 μm

ii. Centrifuge at 300 × *g* for 5 min. Remove the supernatant carefully without disrupting the pellet of organoids (*see* **Note 3**).

iii. Add 1 mL of Qiazol and transfer to a 1.5 mL Eppendorf.

b. For synthetic peptide-based hydrogels:

i. Add 1 mL of Qiazol and transfer to a 1.5 mL Eppendorf.

STOPPING POINT: Samples can be stored at 4°C overnight or at −20°C for up to 6 months.

3.4 RNA Extraction and Purification

1. Thaw the Eppendorf containing organoids in Qiazol for Matrigel and synthetic peptide-based hydrogel samples.
2. Leave for 5–10 min at RT.
3. Add 200 μL of chloroform per 1 mL of Qiazol and shake vigorously for 15 s.
4. Incubate at RT for 2–3 min.
5. Centrifuge 12,000 × *g* for 15 min at 4°C.
6. Transfer the upper aqueous phase to a new tube. Avoid the interface.
7. Add 500 μL of isopropanol and vortex.
8. Incubate for 10 min at RT.
9. Centrifuge 12,000 × *g* for 10 min at 4°C.
10. Remove the supernatant and dry the ethanol (15–30 min).
11. Follow the RNeasy® MinElute® Cleanup protocol for purification:

    a. Add 100 μL of RNAse-free water.

    b. Add 350 μL Buffer RLT and mix well.

    c. Add 250 μL of 96% ethanol to the diluted RNA and mix well by pipetting. Do not centrifuge.

    d. Transfer the sample (700 μL) to an RNeasy MinElute spin column placed in a 2 ml collection tube and close the lid.

    e. Centrifuge for 15 s at 10,000 rpm. Discard the flow-through.

    f. Place the RNeasy MinElute spin column in a new 2 ml collection tube.

    g. Add 500 μL of Buffer RPE to the RNeasy MinElute spin column. Close the lid gently and centrifuge for 2 min at 10,000 rpm to wash the spin column membrane. Discard the flow-through.

    h. Add 500 μL of 80% ethanol to the RNeasy MinElute spin column. Close the lid gently and centrifuge for 2 min at 10,000 rpm to wash the spin column membrane. Discard the flow-through and collection tube

i. Place the RNeasy MinElute spin column in a new 2 mL collection tube. Open the lid of the spin column, and centrifuge at full speed for 5 min. Discard the flow-through and collection tube.

j. Place the RNeasy MinElute spin column in a new 1.5 ml collection tube. Add 20 μL RNase-free water directly to the center of the spin column membrane. Close the lid gently and centrifuge for 1 min at full speed to elute the RNA.

3.5 cDNA Synthesis

1. Mix 1 μg of total RNA with 1 μL of oligodT master mix and 1 μL dNTPs master mix.
2. Heated the tubes to 65°C for 5 min to denature secondary structures.
3. Place the tubes on ice immediately to prevent reformation of secondary structures.
4. Centrifuge briefly to collect the solution at the bottom of the tube.
5. Add 2 μL of 0.1 M DTT, 4 μL of 5× Reaction Buffer, 1 μL Ribonuclease inhibitor (RNasin), and 1 μL of Superscript II Reverse Transcriptase to 12 μL of the annealed RNA/primer mix.
6. Incubate at 40°C for 50 min.
7. Inactivate the reverse transcriptase by heating at 70°C for 15 min.
8. Store the synthesized cDNA at −20°C.

3.6 Real-Time Quantitative PCR (RT-qPCR)

1. Prepare a 1/20 dilution of cDNA mixture in RNase-free water.
2. Choose the set of primers that will amplify your gene of interest. We selected for our purposes:

Cell function	Markers
Absorptive lineage	*VILLIN*
Secretory lineage	*MUC2, LYZ1*
Proliferation	*PCNA*

3. For gene expression analysis, in a 96-well plate, prepare a PCR master-mix containing gene primers mix, SYBR green, and RNase-free water. Find an example below for the housekeeping gene:

Master mix 1 (*GAPDH*)	Sample (1×)	Total
GAPDH primers mix	1.5 μL	1.5 μL × number of wells
SYBR green	7.5 μL	7.5 μL × number of wells
H_2O	3 μL	3 μL × number of wells
Total volume	12 μL	

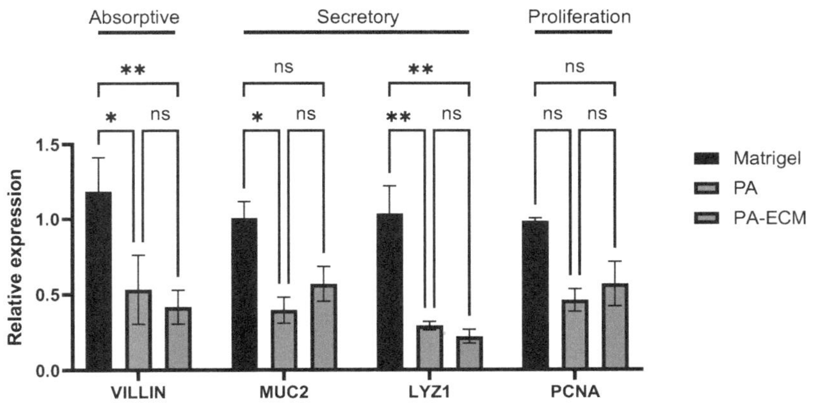

Fig. 3 RT-qPCR measurement of mRNA expression in human organoids embedded in Matrigel, PA, and PA-ECM hydrogels after 7 days in culture. Absorptive lineage marker: *VILLIN*. Secretory lineage markers: *MUC2*, *LYZ1*. Proliferation marker: *PCNA*

4. Dispense 12 μL of the master mix into each well of the 96-well plates.
5. Dispense 3 μL of cDNA into each well of the 96-well plates.
6. Place the optical strip lids on the wells.
7. The thermal cycler should be programmed as described below:

   - 98°C for 30 s (initial denaturation)

     Repeat for 40 cycles:
     - 98°C—10 s (denaturation).
     - 57 °C—30 s (primer annealing).
     - 72°C—60 s (extension).

   - 72°C for 10 min (to fill in the protruding ends of newly synthesized PCR products).

8. Ct values (threshold cycle) were obtained after RT-qPCR.
9. RT-qPCR data based on the e $2^{-\Delta\Delta Ct}$ method were analyzed (Fig. 3) (*see* **Note 4**).

4 Notes

1. Insert the bottom of the 200 μL tip into the gelling solution and rapidly dispense the PA solution with HEPES containing organoids; the gelation occurs very quickly.

2. Remove quickly and carefully the supernatant containing the dissolved Matrigel. Matrigel is thermo-sensible; if left too long at room temperature (RT), it will solidify. If this happens, repeat the process, adding cold Advanced DMEM/F12 and centrifuge at $300 \times g$ for 5 min.
3. When evaluating the suitability of different hydrogels for organoid culture, it is important to visualize their 3D structure, cell viability, and differentiation capacity, using techniques such as brightfield microscopy. Successful organoid formation was observed across all hydrogel conditions, including Matrigel, PA and PA-ECM. Across these matrices, organoids presented a cystic morphology characterized by a central lumen. Over time, stem cells self-organize and differentiate into various specialized cell types. As proliferation and differentiation progress, the cellular wall thickens, debris accumulates in the lumen and budding structures start to form on the surface of these cysts, features that are characteristic of epithelial organoid maturation (Fig. 2). After 7 days in culture, no major morphological differences were observed across hydrogel conditions. These results demonstrate that synthetic peptide-based hydrogels support organoid growth, proliferation, and differentiation comparable to Matrigel.
4. To further demonstrate that cells within organoids proliferate and differentiate across different hydrogel conditions, we analyzed the expression of gene-specific markers of proliferation, secretory, and absorptive lineages (Fig. 3). VILLIN was selected as a marker of absorptive lineage encoding for an actin-binding protein localized in the apical brush border of enterocytes [14]. It is considered an early marker of intestinal absorptive cells. Relative expression of *VILLIN* was significantly higher in Matrigel compared to both PA and PA-ECM hydrogels, with no significant differences between these two synthetic hydrogels. *MUC2* and *LYZ1* were selected as markers of secretory lineage, representing goblet and Paneth cells, respectively. *MUC2* is the major intestinal mucin expressed in goblet cells [15], while *LYZ1* is a lysozyme specifically produced by Paneth cells [16]. Both markers showed significantly higher expression in Matrigel compared to PA alone, while for *LYZ1*, significant differences were also observed between Matrigel and PA-ECM. No significant differences were found between PA and PA-ECM hydrogels for these secretory markers. *PCNA* was selected as it encodes a proliferating cell nuclear antigen directly involved in DNA synthesis [17]. No significant differences were observed in proliferation across the different hydrogel conditions.

References

1. LeSavage BL, Suhar RA, Broguiere N, Lutolf MP, Heilshorn SC (2022) Next-generation cancer organoids. Nat Mater 21:143–159. https://doi.org/10.1038/s41563-021-01057-5
2. Fatehullah A, Tan SH, Barker N (2016) Organoids as an in vitro model of human development and disease. Nat Cell Biol 18:246–254. https://doi.org/10.1038/ncb3312
3. Thorel L et al (2024) Patient-derived tumor organoids: a new avenue for preclinical research and precision medicine in oncology. Exp Mol Med 56:1531–1551. https://doi.org/10.1038/s12276-024-01272-5
4. Sato T et al (2009) Single Lgr5 stem cells build crypt-villus structures in vitro without a mesenchymal niche. Nature 459:262–265
5. Sato T et al (2011) Long-term expansion of epithelial organoids from human colon, adenoma, adenocarcinoma, and Barrett's epithelium. Gastroenterology 141:1762–1772
6. Andrews MG, Kriegstein AR (2022) Challenges of organoid research. Annu Rev Neurosci 45:23–39
7. Aisenbrey EA, Murphy WL (2020) Synthetic alternatives to Matrigel. Nat Rev Mater 5:539–551. https://doi.org/10.1038/s41578-020-0199-8
8. Kozlowski MT, Crook CJ, Ku HT (2021) Towards organoid culture without Matrigel. Commun Biol 4(1):1387. https://doi.org/10.1038/s42003-021-02910-8
9. Gan Z, Qin X, Liu H, Liu J, Qin J (2023) Recent advances in defined hydrogels in organoid research. Bioactive Mater 28:386–401. https://doi.org/10.1016/j.bioactmat.2023.06.004
10. Webber MJ, Appel EA, Meijer EW, Langer R (2015) Supramolecular biomaterials. Nat Mater 15:13–26. https://doi.org/10.1038/nmat4474
11. Hartley A, Williams PM, Mata A (2025) A comparison of the mechanical properties of ECM components and synthetic self-assembling peptides. Adv Healthcare Mater 14(11):e2402385. https://doi.org/10.1002/adhm.202402385
12. Hartgerink JD, Beniash E, Stupp SI (2022) Peptide-amphiphile nanofibers: a versatile scaffold for the preparation of self-assembling materials. Proc Natl Acad Sci U S A 99(8):5133. https://doi.org/10.1073/pnas.072699999
13. Sato K, Hendricks MP, Palmer LC, Stupp SI (2018) Peptide supramolecular materials for therapeutics. Chem Soc Rev 47:7539–7551
14. West AB et al (1988) Localization of villin, a cytoskeletal protein specific to microvilli, in human ileum and colon and in colonic neoplasms. Gastroenterology 94:343–352
15. Velcich A et al (2002) Colorectal cancer in mice genetically deficient in the mucin Muc2. Science (1979) 295:1726–1729
16. Bel S et al (2017) Paneth cells secrete lysozyme via secretory autophagy during bacterial infection of the intestine. Science (1979) 357:1047–1052
17. Moldovan G-L, Pfander B, Jentsch S (2007) PCNA, the maestro of the replication fork. Cell 129:665–679

Chapter 15

Live Fluorescence Lifetime Imaging Microscopy (FLIM) of the Epithelial Polarity and Cell-Specific Segmentation in the Small Intestinal Organoids

Hang Zhou, Irina A. Okkelman, Amber De Kinder, Bert Devriendt, Max Nobis and Ruslan I. Dmitriev

Abstract

Apical-basal polarity is essential for gastrointestinal epithelium function, where the apical membrane mediates nutrient absorption, host-microbiota interactions, and pathogen defense. However, conventional intestinal organoid culture displays an "apical-in" topology, limiting functional research with respect to the apical membrane. F-actin labeling with fluorescent phalloidin, a well-established apical marker, is also incompatible with live imaging. To address these limitations, in this chapter we describe an extracellular matrix (ECM) removal protocol to achieve "apical-out" topology in porcine intestinal organoid model and further validated WGA and Nile Red live imaging tracers enabling discrimination between apical-out (AO) and basal-out (BO) organoids. Subsequently, using integrated computational pipelines (automatic nuclei segmentation and single-nucleus phasor-based lifetime analysis), we can quantify organoid proliferative states at single-nucleus resolution following BrdU and Hoechst 33342 staining. Using the analysis of Hoechst 33342 fluorescence lifetime, we demonstrate that both organoid topologies exhibit intrinsic proliferative heterogeneity, but overall proliferative capacity is topology-dependent: apical-out organoids displayed decreased proliferation, whereas basal-out organoids remain highly proliferative.

Keywords Polarity reversion, Intestinal organoid, Live microscopy, FLIM, Cell proliferation

1 Introduction

The intestinal epithelium displays strong apical-basal polarity. Its apical membrane, interfacing with the luminal environment, is involved in nutrient uptake, host-microbiota interactions, and pathogen defense. Conversely, the basal membrane adheres to the underlying basement membrane, facilitating nutrient transport from the lumen to the bloodstream and mediating intercellular signaling of neighboring cells [1, 2]. However, investigating the polarity-specific features remains challenging: in vivo approaches constrain dynamic observation and manipulation of these polarized membranes, while

Paloma Ordóñez-Morán (ed.), *Intestinal Stem Cells*, Methods in Molecular Biology, vol. 3059,
https://doi.org/10.1007/978-1-0716-5412-5_15,

2D cultures are incapable to recapitulate epithelial cytoarchitecture, precluding the functional study of gastrointestinal polarity. To overcome these limitations, intestinal organoids have emerged as a promising ex vivo model that mimics epithelial morphology and function [3, 4]. These self-organized 3D structures, derived from either adult stem cells (ASCs) or pluripotent stem cells (PSCs), reconstitute the polarized epithelium that exhibits apical membrane facing lumen and basolateral membrane anchored to the basement membrane extract (e.g., Matrigel, Geltrex, Cultrex, and related EHS extracts) [4, 5]. However, "apical-in"/"basal-out" topology of conventional intestinal organoid culture hinders the apical membrane direct access to the luminal content (i.e., nutrients, microbiota, and pathogens), limiting the research of the apical membrane function. While this limitation has been addressed by some labor-intensive techniques, such as (i) microinjection and (ii) partial disaggregation to crypts and re-assembly [5, 6], depletion of the extracellular matrix (ECM) and transitioning to suspension culture, as pioneered by Amieva's group, represents a more elegant alternative offering polarity reversion. This method produces "apical-out" organoids that retain differentiation capacity [6]. It has been validated across multiple species, including human PSC [7], porcine [8, 9], bovine [10, 11], chicken [12], and murine [8] organoids.

The apical membranes of gastrointestinal epithelium feature abundant F-actin-based microvilli with lateral attachment to the plasma membrane, which is interconnected by a subapical terminal web of actin and myosin. This enables validation of the apical membrane using phalloidin labeling, an established high-affinity F-actin probe [6]. However, the inherent toxicity of phalloidin and impermeability to live cells impede its application in real-time apical membrane labeling. To address this limitation for tracking polarity in live organoids, new tracers capable of discriminating apical from basal membrane are critically needed to enable real-time visualization of topological structure without disturbing native epithelial physiology.

Following the identification of live topology, polarity-dependent physiological functions, including proliferation, require deeper investigation. In the intestinal organoids, crypt-based cell proliferation occurs prominently in stem cells and is transiently sustained in transit-amplifying cells. The conventional proliferation assays, including MTT, BrdU, EdU, and genetically encoded biosensors labeling, are time-consuming, often incompatible with live imaging, or require genetic manipulation, hindering downstream dynamic observation of proliferative cell states throughout the crypt-villus axis. To overcome these limitations, a live-compatible Fluorescence Lifetime Imaging Microscopy (FLIM) proliferation method was developed to identify the S-phase in intestinal organoids [13]. As an advanced imaging methodology, FLIM enables multiparametric and multi-dimensional 5D imaging (X,Y,Z, t, and luminescence lifetime (tau, τ) dimensions), allowing quantitative analysis of cell physiology in live organoids. Our prior work revealed that incorporation of BrdU into dsDNA during S phase quenches the Hoechst 33342 signal and reduces its fluorescence lifetime in nuclei. Such a lifetime shift of Hoechst 33342 enables quantitative distinction of S-phase from non-proliferating cells in apical-out and basal-out organoids [13, 14]. Moreover, to achieve single-cell resolution analysis of proliferation, deep-learning facilitated nucleus segmentation and fluorescence lifetime analysis pipelines are essential to segment nuclei robustly and quantify Hoechst 33342 lifetime shifts within the individual nuclei. This should allow the precise mapping of DNA synthesis phases to apical-out and basal-out organoids.

In this protocol, we describe new tracers indicating apical membrane orientation in live organoids following polarity reversal. Fluorophore-conjugated wheat germ agglutinin (WGA) intensely labels outward-facing apical membrane by binding to cell surface proteoglycans in the apical-out organoids, while Nile Red visualizes intracellular lipid droplets distribution in both apical-out and basal-out topology organoids. We subsequently employ the latest published segmentation model, Cellpose-SAM [15], to generate a nuclei mask, followed by the application of masks to the corresponding lifetime data using open-source phasor-based lifetime analysis software GSLab [16] to achieve single-nucleus proliferation analysis. This protocol enables quantitative, topology-resolved profiling of proliferation at single-cell resolution in live organoids.

2 Materials

2.1 Small Intestinal Organoids

Porcine small intestinal organoids were established from *jejunum* (Porgj_2 organoid line, *Sus scrofa domesticus*, female, 6–7 weeks) [8].

2.2 Reagents, Consumables, and Equipment

1. Matrigel (growth factors reduced, phenol red-free). Aliquot 500 μL in 1.5 mL Eppendorf tubes and store at −20°C. Defrost the aliquot at 4°C and store at 4°C.
2. Washing medium: DMEM-F12 HAM supplemented with 12 mM GlutaMAX. Store at 4°C.
3. Culture medium: IntestiCult Organoid Growth Medium (Human): aliquot IntestiCult™ Organoid Growth Medium (OGM) Human Basal Medium and OGM Supplement by 5 mL into separate sterile 15 mL tubes. Store aliquots at −20°C. Prior to use, thaw aliquots of OGM and supplements overnight at 4°C and mix in a proportion 1:1 to make a final OGM media. Store at 4°C, avoid repeated freeze–thaw cycles.
4. Imaging medium: DMEM without phenol red and sodium bicarbonate, supplemented with 10 mM D-glucose, 2 mM GlutaMAX, 1 mM sodium pyruvate, 10 mM HEPES, pH 7.2. Store at 4°C.

5. DMEM (no glucose, no pyruvate, no glutamine, and no phenol red).
6. 1 M HEPES solution pH 7.2, 100× concentrate, sterile.
7. 200 mM GlutaMAX supplement, 100× concentrate, sterile.
8. 100 mM sodium pyruvate solution, 100× concentrate, sterile
9. 1 M D-Glucose solution, 100× concentrate, sterile
10. 0.5% Lipidure-coating solution, lipidure powder was dissolved in 95% molecular grade ethanol. (Amsbio, AMS.52000034GB1G)
11. 0.5 M EDTA solution, pH 8.0, store at room temperature.
12. Phosphate-buffered saline (PBS) solution, store at room temperature.
13. Dimethyl sulfoxide (DMSO), "Hybri-Max" grade.
14. 3.7% paraformaldehyde (PFA) in PBS solution
15. 0.1% bovine serum albumin (BSA) in PBS solution
16. 0.1% Triton X-100 in PBS solution
17. Tris-buffered saline with Tween 20 (TBST)
18. 5-Bromo-2′-deoxyuridine (BrdU), 100 mM stock solution in DMSO, Store in aliquots at −20°C.
19. Electronic pipette controller
20. Serological pipettes
21. 1.5 mL 15 mL centrifuge tube
22. Flat-bottom 24-well cell culture plate
23. 8-well glass bottom imaging chamber (#1.5 thickness glass coverslip bottom)
24. Pipettes for 10 μL, 200 μL, and 1000 μL
25. Low retention tips for 10 μL, 200 μL, and 1000 μL
26. Humidified CO_2/37°C incubator.
27. Swinging bucket rotor-equipped centrifuge with cooling option 4°C.
28. Nutating Shaker

2.3 Microscopy Setup

TCSPC-based confocal and multiphoton FLIM systems from Leica Microsystems, Becker & Hickl, PicoQuant, or from other vendors can provide appropriate excitation sources (i.e., pulsed diode, white light, and two-photon lasers), acousto-optical modulators, detectors, and respective software. Below are examples of instruments used in this work:

1. Confocal FLIM based on an inverted STELLARIS 8 Falcon microscope (Leica Microsystems), equipped with a white-light

laser (440–790 nm), HC PL Apo 10×/0.4 air, HC Fluotar 25×/0.95 W, HC PL Apo 40×/1.25 GLYC corr, HC PL Apo 63×/1.4 oil objectives, HyD X, HyD R, and HyD S detectors, temperature / CO_2-controlled incubator, and dedicated LAS X acquisition and analysis software (ver. 4.6.0 and ver. 4.8.1)

2. Two-photon FLIM based on an inverted STELLARIS 8 DIVE Multiphoton Microscope (Leica Microsystems), equipped with a tunable infrared laser (IR Laser Mai Tai HP, 690-1040 nm), HC PL Apo 10×/0.4 air, HC Fluotar 25×/0.95 W, HC PL Apo 40×/1.25 GLYC corr, HC PL Apo 63×/1.4 oil objectives, 4Tune HyD NDD detectors, temperature / CO_2-controlled incubator (black, for light protection), and dedicated LAS X acquisition and analysis software (ver. 4.6.0 and ver. 4.8.1)

2.4 Fluorescent Probes

All probes are prepared in small aliquots in dark plastic vials and stored at −20°C. After defrosting, WGA conjugates are stored at 4°C to avoid protein aggregation caused by repeated freeze-thaw cycles. The details about probes used in this chapter are listed in Table 1.

2.5 Analysis Software

1. Cellpose-SAM: A deep learning platform for automated cell and nucleus segmentation, used for nucleus segmentation in Hoechst 33342-labeled intestinal organoid [15].
2. GSLab: An open-source software to achieve advanced phasor-based fluorescence lifetime analysis, implemented for phasor plot presentation and single-nucleus fluorescence lifetime analysis [16].

3 Methods

3.1 Preparation of Porcine Organoids for Polarity Reversion

1. Prior to polarity reversion, organoids are cultured for 3–4 days at a density of 150–300 organoids per 50 μL Matrigel dome in 24-well plate, supplemented with 500 μL human IntestiCult organoid growth medium per well.
2. Dislodge Matrigel domes using a pipette with 1000 μL tip attached by a 10 μL tip to its end. Pipette the mixture of Matrigel and culture medium up and down 20–50 times with this tip to ensure robust mechanical disruption.
3. Transfer all content into a 15 mL tube with 10 mL washing medium, centrifuge at 300 ×*g* for 5 min at 4°C.
4. Aspirate supernatant gently without disturbing the pellet at the bottom of the tube.
5. Resuspend pellet with liquid ice-cold Matrigel using 100–150 μL per original dome to split organoids at 1:2 to 1:3.

Table 1
Fluorescent probes used in the protocol

Probe	Physiological parameter	Stock concentration (solvent)	Staining concentration	Excitation/emission (nm)	Fluorescence lifetime (τ, ns), [citation]
Alexa Fluor 488-conjugated Wheat Germ Agglutinin (AF488-WGA)	Binds *N*-acetyl-D-glucosamine and sialic acid; mucous membranes	1 mg/mL (PBS)	4 μg/mL	490/495–520	1.2–2.6 ns [8]
Nile Red	Solvatochromic dye. Lipids, LDs, cytoplasm.	1 mM (DMSO)	2 μM	490/588–620	2–4 ns
Phalloidin-Texas Red	F-actin labeling in fixed cells	66 μM	0.132 μM	591/608–625	3–3.3 ns [8]
Hoechst 33342	Nuclear staining	1 mM	200 nM	(Two photon)740/411–491	0.8–2.6 ns (depends on BrdU concentration and pulse duration)

6. Seed organoid/Matrigel mixture in a pre-warmed 24-well plate at 50 μL/well, then incubate for 5 min at 37°C to solidify Matrigel domes.
7. Add 500 μL/well growth medium to cover Matrigel domes. Culture organoids for 1–3 days (*see* **Note 1**).

3.2 Polarity Reversion of Small Intestinal Organoids

1. Prepare low-adhesion 24-well plates by adding 500 μL 0.5% (w/v) Lipidure coating solution (dissolved in ethanol) in each well and incubate for 2 min at RT.
2. Aspirate coating solution from each well using 1000 μL tip and transfer back to the stock glass bottle for reuse. Dry the coated low-adhesion plate with lids open under laminar flow hood for 10–30 min until no ethanol is left.
3. Remove the growth media from organoid cultures. Rinse each well twice with 500 μL of ice-cold 5 mM EDTA in PBS (prepared from 0.5 M EDTA stock, 1:100 dilution).
4. Disrupt each Matrigel dome gently in 500 μL residual EDTA solution using a 1000 μL tip, then collect organoids from all wells in a 15 mL tube.
5. Add an additional 5 mM EDTA solution to the tube to achieve a ratio of 12 mL EDTA solution per 4 domes (each dome contains ~50 μL Matrigel).
6. Place this tube on an orbital shaker for 1 h at 4°C to fully dissolve the Matrigel.
7. Centrifuge the tube at 300 × *g* for 5 min at 4°C. Discard most of the supernatant using a 1000 μL pipette tip. Resuspend the organoid pellet gently in the remaining EDTA solution (*see* **Note 2**).
8. Add 10 mL of ice-cold washing medium. Centrifuge again and discard the washing medium completely using a 1000 μL pipette tip.
9. Resuspend organoids in growth medium at a ratio of 500 μL for each dome collected previously.
10. Seed 500 μL suspension per well in the prepared low-adhesion 24-well plates and culture overnight to achieve epithelium polarity reversion.

3.3 Validation of Apical-Out Topology by Co-Staining of WGA-Labeled Organoids with Fluorescent Phalloidin

1. Incubate 500 μL organoid suspension with WGA (final concentration: 4 μg/mL) in the dark at 37°C for 1 h following polarity reversion.
2. Transfer the organoid suspension to a 1.5 mL low-adhesive Eppendorf tube completely (0.5% (w/v) lipidure coating) and centrifuge at 300 × *g* for 2 min.

3. Aspirate the supernatant gently without disturbing the pellet. Resuspend organoids in 1 mL room-temperature PBS, then continue centrifugation (300×*g*, 2 min). Repeat two additional wash cycles to eliminate unbound WGA.
4. Fix the organoid with pre-warmed 4% PFA for 10 min at room temperature. Centrifuge (300×*g*, 2 min) and wash with PBS three to five times to remove residual PFA.
5. Permeabilize organoids with 0.1% Triton X-100 in PBS for 10 min at room temperature. Wash three times with 1× TBST.
6. Block organoids with 5% FBS in TBST blocking solution for 1 h at room temperature.
7. Stain organoids with Texas Red-conjugated phalloidin (final concentration: 0.132 μM) in blocking solution for 1 h at room temperature, protected from light.
8. Wash organoids three to five times with 1× TBST to remove unbound phalloidin conjugate. Transfer organoids to #1.5 H glass-bottom μ-chamber and perform imaging following protocols in Step 3.3 at room temperature (*see* **Note 3**).

3.4 Live Imaging of Polarity Reversed Organoids Using WGA and Nile Red

1. After overnight culture, add Alexa Fluor 488-conjugated Wheat Germ Agglutinin (WGA, final concentration: 4 μg/mL) and Nile Red (final concentration: 2 μM) into each well containing 500 μL of organoid suspension. Incubate the microplate at 37°C for 1 h.
2. Gently tilt the 24-well microplate at angle of 45° for 2–5 min, allowing organoids to sediment to the lower edge of each well via gravity. Gently tap the microplate's lower edge against the laminar flow surface to facilitate sedimentation.
3. Aspirate 90% of culture medium from the upper edge of each well without disturbing the organoid sediment.
4. Put the microplate back to a flat position gently and resuspend the organoids with 500 μL of imaging medium in each well.
5. Wash three times with imaging medium to effectively remove unbound WGA and Nile Red probes, minimizing residual background fluorescence.
6. Transfer the organoid suspension to an 8-well imaging μ-chamber (No. 1.5 H glass coverslip bottom). Limit the loading volume to 200–300 μL for each well to avoid tracers' contamination between neighbor wells.
7. Prior to imaging, equilibrate the microscope incubator chamber to 37°C.
8. Keep the loaded μ-chamber in the incubator for 10–20 min to minimize cell stress before live imaging.

9. Initialize the confocal microscopy system, choose 40×/1.25 NA objective, and add immersion fluid (e.g., glycerol, refractive index: 1.47).
10. Start image acquisition software and set excitation and emission for WGA (exc. 490 nm/em. 498–520 nm) and Nile Red (exc. 490 nm/em. 588–620 nm).
11. Secure the μ-chamber on the motorized stage, locate organoids by adjusting the Z- focus and XY-axis coordinates using the transmission light mode.
12. Activate preview mode and use the zoom function to achieve a field of view where organoids occupy ≥80% of the imaging area.
13. Adjust laser excitation power (WGA and Nile Red: 490 nm), image resolution (256-, 512-, or 1024-pixel solution), scan speed (100–400 Hz), pinhole (1.0–1.2 airy units), and frame repetitions (1–3 frame repetition rate) to achieve robust fluorescent signals (*see* **Note 4**).
14. Capture images using optimized parameters. Save data files.

3.5 Live Imaging of Proliferation State in Polarity Reversed Organoid

A thymidine analog 5-bromo-2′-deoxyuridine (BrdU) is used here as a marker for cell proliferation. It incorporates into dsDNA during S-phase in cell cycle (DNA synthesis) and quenches Hoechst 33342 fluorescence via a heavy atom effect [17]. The lifetime of Hoechst 33342 is reduced upon BrdU incorporation into genomic DNA. Here, we report labeling of the apical-out organoids with BrdU and Hoechst 33342 to achieve live tracing of proliferating cells.

1. During polarity reversion, incubate organoids with 5-bromo-2′-deoxyuridine (BrdU, final concentration: 2.5 μM) in culture medium at 37°C overnight.
2. Incubate organoids with Hoechst 33342 (final concentration: 200 nM) at 37°C for 1 h.
3. Wash organoids three times with pre-warmed imaging medium (as in Step 3.4) in order to remove unbound BrdU and Hoechst 33342. Subsequently, transfer the organoids to an 8-well μ-chamber (#1.5 H glass bottom).
4. Place μ-chamber in microscope incubator chamber, pre-equilibrated to 37°C for 10–20 min in order to minimize organoid stress.
5. Initialize two-photon microscopy and select a 25×/0.95 NA objective with distilled water as immersion fluid.
6. Fix the μ-chamber on the microscopy stage. Using transmission light, adjust XY coordinates and Z-focus to find the organoids in the μ-chamber.

7. Start image acquisition software, turn on two-photon laser, and wait for 10 min for warm-up. Set excitation wavelength for Hoechst 33342 (exc. 700 nm), wait for laser tuning ("locked" mode). Upon laser stabilization at 740 nm, set the emission spectrum: 410–490 nm (*see* **Note 5**).
8. Initialize in the preview mode with FLIM enabled, adjust laser excitation power (Hoechst 33342: 740 nm), optimize image resolution (256-, 512-, or 1024-pixel solution), and scan speed (100 – 400 Hz). Digitally zoom in to ensure that organoids occupy ≥80% of the imaging area. Switch to FLIM imaging software interface, ensuring peak intensity counts over 10^5 photons on the global decay curve for a reliable Hoechst 33342 fluorescence lifetime estimation (*see* **Note 6**).
9. Capture images with optimal imaging settings.

3.6 Nuclear Segmentation for Single Cell Proliferation Analysis

1. Start microscopy acquisition software (e.g., LAS X) and load the Hoechst 33342-labeled intestinal organoid images. Right-click the selected images to export all images as TIFF format.
2. Import the exported TIFF images into Cellpose software, set the pixel size to 20, and run the pretrained nucleus segmentation model—Cellpose-SAM. Check the mask precision by visually inspecting the mask boundaries against the nuclear outlines. Manually improve the ROI selection by correcting segmentation errors and eliminating non-organoid nuclei signals such as apoptotic epithelial cells in BO lumen and dissociated cells near AO apical membrane. Upon confirming ideal ROIs, export generated binary masks as PNG files to a dedicated folder (*see* **Note 7**).
3. Reopen and select the original organoid images within the FLIM analysis software (e.g., LAS X FLIM/FCS software). On the export tab, choose all images in TIFF format with the "Save Phasor GS" option. Select "Fixed range," set "Factor or Range" to "per Gray Level" and enter 1. Export all images to a dedicated folder, including TIFF format images and a metadata folder containing lifetime information (G, S coordinates). (*see* **Notes 8 and 9**)
4. Install and initialize the GSLab platform within a MATLAB environment. Launch the graphic interface.
5. Within GSLab, right-click on the file tab and select the "Leica FALCON SnG." Navigate to the folder containing both TIFF files and lifetime metadata from Step 3.6.3. Select this folder to import all images and corresponding lifetime information into GSLab. The software will display images and parse G/S coordinates as pixels on phasor plots for each image (*see* **Note 10**).

6. Click "Phasor plot" tab and set median filter as 3 × 3. Transition to image tab, utilize the intensity histogram (x-axis: photon counts, y-axis: pixel frequency) to mitigate background fluorescence. It is optional to activate "Relative" function to normalize photon counts to 0–1 range. Adjust the lower bound threshold slider to exclude low-photon pixels corresponding to background signals.
7. Click "load mask" and import the Cellpose-generated PNG masks from 3.6.2. Execute "apply mk" to restrict phasor analysis exclusively to segmented nuclei regions.
8. Confirm that G, S coordinates on the phasor plot represent intranuclear pixels after applying the mask.
9. Navigate to Export Tab, click on "Output Folder" to select a directory for data storage. Tick "Masked obj data as csv," Click "Export HR" to save metrics in a CSV file.
10. Open the exported lifetime.csv file, collect mean Int (photons/pixel), τ_{ϕ} (phase lifetime), and $\tau_{(m)}$ (modulation lifetime) values per nucleus. Determine the proliferative state of each cell by τ value as previously described [13] and complement with column plots to quantify proliferating cells in organoids.

4 Anticipated Results

4.1 Live apical-out and basal-out organoids display distinct staining patterns with WGA and Nile Red

Following the method described by Amieva et al. [6], we successfully adapted the polarity reversion protocol to porcine small intestinal organoids. To identify live imaging-compatible topological tracers, we first validated wheat germ agglutinin (WGA) via co-staining with Phalloidin (Fig. 1). In basal-out organoids (BO), phalloidin labels the apical membrane interfacing the lumen. WGA labels and internalizes through the basal epithelial surface, outlining peripheral cellular contours without co-localizing with apical F-actin signal. (Fig. 1, **top**). Conversely, in apical-out organoid (AO), WGA entry was blocked by the exposed apical membrane, generating intense boundary fluorescent labeling that is precisely co-localized with phalloidin-stained apical F-actin (white merged signal from green WGA and magenta phalloidin) (Fig. 1, **bottom**). It should be noted that prolonged WGA incubation (>18 h) causes internalization and more profound intracellular accumulation, which can hinder characteristic polarity distinction from BO (data not shown). However, short-term incubation demonstrates WGA as a robust live-imaging tracer for real-time discrimination of intestinal organoid polarity.

We further evaluated Nile Red capacity to identify AO topology in live small intestinal organoids through characteristic lipid droplets patterning and further validated this via WGA co-staining. Multiple

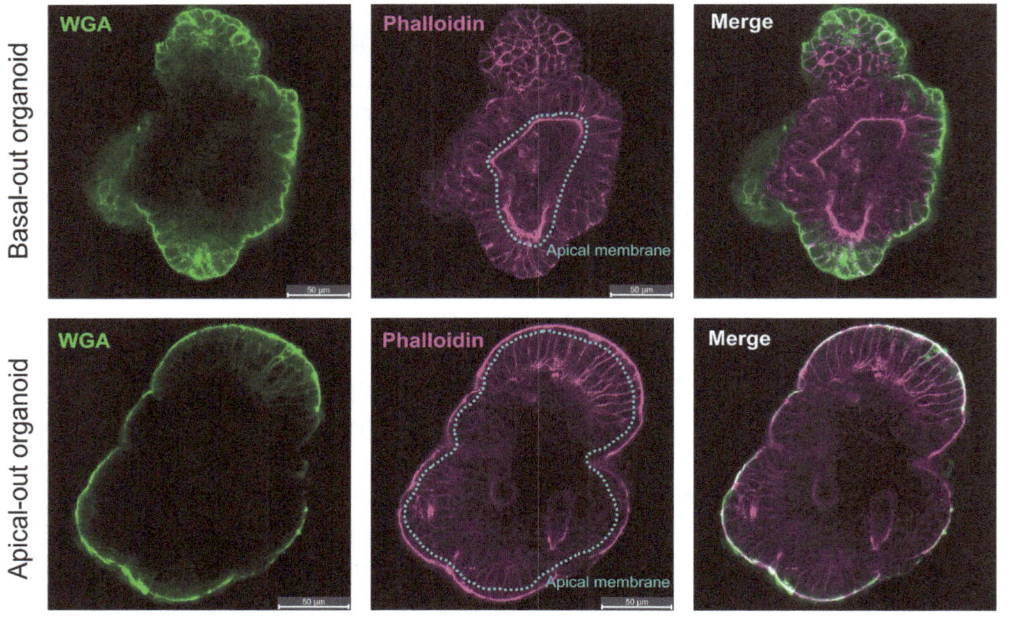

Fig. 1 WGA labeling can reliably identify apical-out and basal-out topology, validated with Phalloidin co-staining. (**Top**) In basal-out organoids, WGA permeates through the basal membrane and Phalloidin outlines bright F-actin signal at the apical membrane. (**Bottom**) Apical-out organoids exhibit intense WGA fluorescence merged with Phalloidin co-localizing at the exterior apical epithelium membrane. Scale bars: 50 μm

WGA and Nile Red co-stained organoids were imaged (Fig. 2A, **Top left**), which in high-resolution view clearly differentiated AO and BO topologies [8] (Fig. 2A **Top right**). Fluorescent WGA consistently exhibited topology-specific labeling patterns as previously described [8] (Fig. 2A **Bottom left**). Nile Red staining revealed topology-dependent lipid droplet distribution: AO organoids exhibited abundant droplets concentrated at the interior basal membrane, whereas BO organoids showed sparse droplets localization on the exterior membrane surface (Fig. 2A **Bottom right**). We next performed fluorescence lifetime imaging to characterize WGA and Nile Red lifetime distribution in AO and BO organoids: WGA showed heterogeneous lifetimes (color coding line range from 1.15 to 1.88 ns on the phasor plot) in organoids, predominantly in AO organoids, where a discrete apical membrane domain displayed distinctive lifetime (short: blue-coded, long: red-coded) (Fig. 2B, **top**). Such spatial heterogeneity suggests a distinct regional composition of N-acetylglucosamine and sialic acid groups at the apical membrane surface. FLIM analysis of the Nile Red signal (Fig. 2B, **bottom**) revealed dual phasor ROIs distinguishing lipid droplets (blue ROI: 3.3 ns) from bulk lipids (yellow ROI: 2.9 ns). High-intensity lipid droplet-rich regions exhibited longer lifetime than the low-intensity lipid cytoplasmic localization, enabling clear separation of the lipid state: lipid droplets (blue) versus dispersed lipids (yellow).

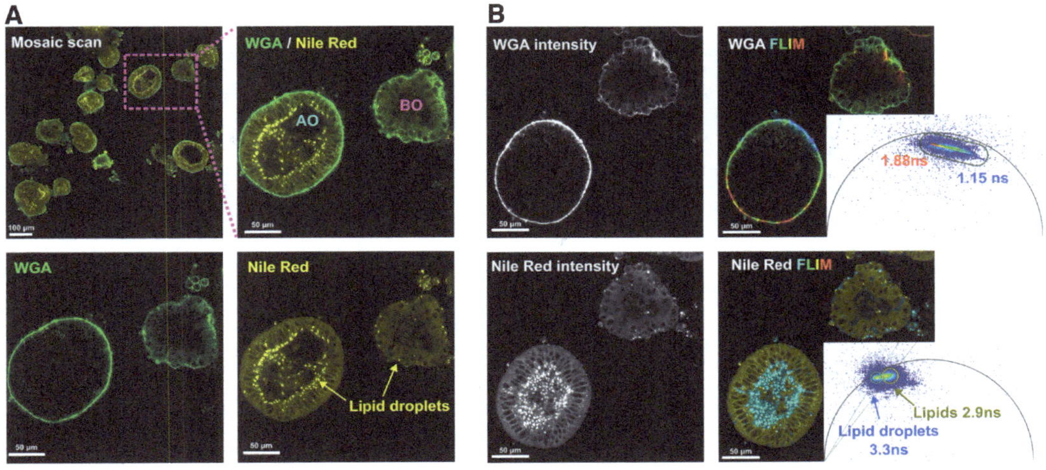

Fig. 2 Topology-dependent spatial distribution and fluorescence lifetime of WGA and Nile Red in live porcine organoids. (**A**) Mosaic scan of organoids after an overnight polarity reversion procedure, co-stained with AF488-WGA and Nile Red (**Top left**). Insets: Representative images of AO and BO organoids showing WGA (green), Nile Red (yellow), and merged channels. Yellow arrows indicated lipid droplets. Scale bar: 100 μm (mosaic), 50 μm (**insets**). (**B**) FLIM of WGA and Nile Red in AO and BO organoids. Top: representative WGA intensity and FLIM images. WGA phasor plot incorporated in FLIM image, with a color-coding bar to depict WGA lifetime from 1.15 ns (blue) to 1.88 ns (red). Bottom: representative Nile Red intensity and FLIM images. Nile Red phasor plot embedded in FLIM image, with two regions of interest (ROI) to display lifetime differences of lipids and lipid droplets. Blue: lipid droplets (mean ROI lifetime: 3.3 ns). Yellow: dispersed lipids (mean ROI lifetime: 2.9 ns). Scale bar: 50 μm

4.2 Single Nucleus FLIM Analysis Reveals Distinct Proliferative States in Apical-Out and Basal-Out Organoids

Hoechst 33342 staining revealed topology-specific nucleus positioning: nuclei consistently aligned with the basal membrane, peripherally in the basal-out organoids, while their alignment was more "internal" in apical-out organoids (Fig. 3A). This basal membrane-associated spatial distribution suggests the potential of the nuclear intracellular location to indicate epithelium polarity. To quantify single-nucleus proliferation via FLIM, organoids were labeled by Hoechst 33342, after overnight incubation with 5-bromo-2′-deoxyuridine (BrdU). During S-phase, BrdU incorporation into genomic DNA quenched Hoechst 33342, shortening its fluorescence lifetime, as detected by FLIM, thereby establishing Hoechst 33342 lifetime shift as an indicator for proliferative status [13]. Here, we applied this methodology to assess single-nucleus proliferation in intestinal organoids, revealing both intrinsic heterogeneity and topologically driven proliferative divergence. We first generated preliminary nucleus masks using a Cellpose-SAM, followed by manual refinement to correct errors and eliminate non-organoid nucleus signals (e.g., luminal debris in BO or dissociated cells adjacent to AO apical membrane) (Fig. 3A). Subsequently, nuclei-specific fluorescence lifetime data were converted into G-S coordinates and displayed on a phasor plot via the GSLab software, revealing a distinct distribution of coordinates

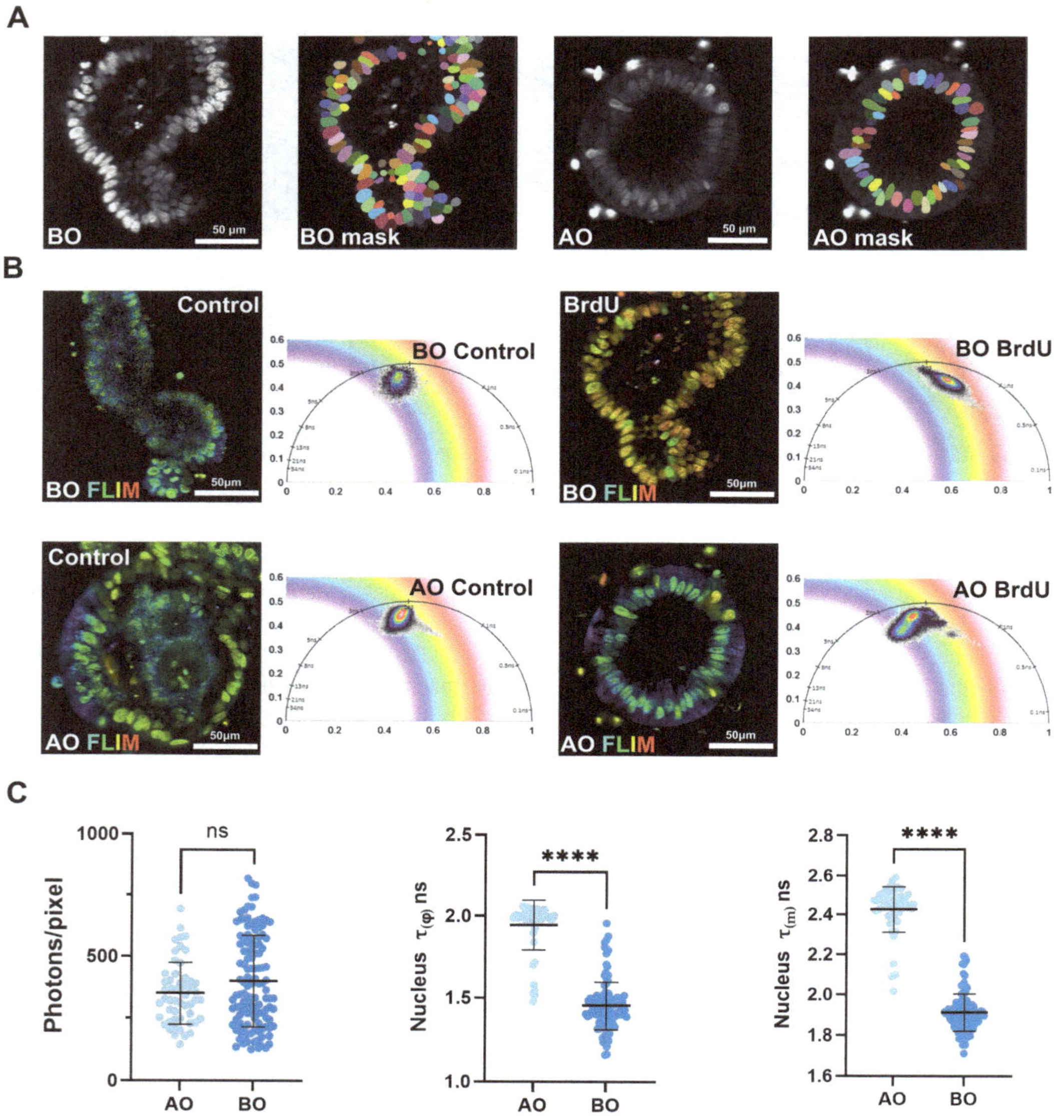

Fig. 3 Single-cell proliferation analysis of AO and BO organoids via nuclei segmentation and FLIM. (**A**) Representative images of Hoechst 33342-labeled nuclei and Cellpose-generated segmentation masks for apical-out (AO: 53 ROIs) and basal-out (BO: 121 ROIs) organoids. Scale bars: 50 μm. (**B**) Nuclear fluorescence lifetimes visualized with phasor-based pseudo-coloring. FLIM images are colored by mapping their pixel distribution to a rainbow gradient where hue represents $\tau_{(m)}$ magnitude on phasor plot. Following BrdU labeling, BO nuclei exhibit shorter $\tau_{(m)}$ (red-shifted), while AO nuclei retain longer $\tau_{(m)}$ (blue-shifted) compared to control groups. (**C**) Single-nucleus parameters in BrdU-treated AO and BO, quantified from masked FLIM data. Left: Photons per pixel for each nucleus. Middle: mean nucleus $\tau_{(\varphi)}$. Right: mean nucleus $\tau_{(m)}$. Significance: Mann–Whitney (photons/pixel); unpaired t-tests (mean nucleus $\tau_{(\varphi)}$, $\tau_{(m)}$); ****$p < 0.0001$

corresponding to the AO and BO nuclei (Fig. 3B). Pseudo coloring of FLIM images was achieved with GSLab's TauMod function (phasor tab), which directly mapped G-S coordinates to a spectral gradient where the hue represents $\tau_{(m)}$ magnitude. Each image pixel retained its corresponding G-S coordinates hue within the rainbow gradient. Through such mapping, both AO and BO control organoids (untreated with BrdU) showed extended Hoechst 33342 lifetimes (AO mean $\tau_{(\varphi)}$: 1.88 ns, mean $\tau_{(m)}$: 2.32 ns; BO mean $\tau_{(\varphi)}$: 2.00 ns, mean $\tau_{(m)}$: 2.40 ns), establishing a Hoechst 33342 lifetimes baseline for cell proliferation validation via BrdU treatment. Conversely, the majority of BrdU-treated BO nuclei exhibited shorter fluorescence lifetimes (red shift), whereas BrdU-exposed AO nuclei retained longer fluorescence lifetime (blue shift), demonstrating that polarity reversion depleted proliferative capability of intestinal organoids (Fig. 3B). For single-nucleus FLIM analysis, we implemented nucleus segmentation masks (AO: 53 ROIs, BO: 121 ROIs) to the corresponding FLIM images in GSLab to obtain nuclei-segmented areas lifetime. Photon/pixel plots revealed comparable photon counts in BrdU-treated AO and BO (200 – 1000 photons/pixel; non-significant: $p > 0.05$, Fig. 3C, **left**). Both nuclear $\tau_{(\varphi)}$ and $\tau_{(m)}$ (Fig. 3C, **middle, right**) demonstrated significant topological difference: post-polarity reversal, nearly 90% of AO nuclei showed longer lifetime, indicating quiescent state (mean $\tau_{(\varphi)}$: 1.95 ns, mean $\tau_{(m)}$: 2.43 ns), with rare nuclei (4-5 nuclei in organoid) exhibiting shorter lifetimes, suggesting their location within a proliferative niche ($\tau_{(\varphi)}$: 1.5 – 1.7 ns, $\tau_{(m)}$: 2.0 – 2.2 ns). Conversely, BO nuclei predominantly displayed short lifetime (mean $\tau_{(\varphi)}$: 1.45 ns, mean $\tau_{(m)}$: 1.91 ns), with minor subpopulations of cells showing quiescence. This observed divergence between AO and BO reflects their distinct growth contexts: basal-out organoids, analyzed after 1–2 days in culture, were still in an active expansion period and sustained proliferating cell activity. At the same time, the predominance of nuclei with long lifetimes in AO suggested that the majority of the cells were in a quiescent state, as was also evidenced by the gradual cell death of AO organoids after 7 days of culture (data not shown). Further, the small proliferative cell subset in AO organoids deserves future investigation to understand its cellular identity and functional properties (potentially stem-like populations). Collectively, the presented single-nucleus FLIM pipeline establishes a live-cell methodology for proliferation analysis, which is adaptable to other organoids and 3D ex vivo systems.

5 Notes

1. Selecting the culture duration (1–3 days) determines the percentage of apical-out organoids that can be obtained. In porcine intestinal organoids, we observed more apical-out organoids after 3 days in culture. After only a 1-day culture, some organoids kept their basal-out orientation or partially returned their membrane polarity to BO.

2. Resuspending the organoid pellet in a residual EDTA solution is essential to avoid formation of large organoid aggregates. These aggregates are hard to separate by pipetting after washing with medium. Do not resuspend organoids using a serological pipette. Organoids adhere to glass and cell culture plastic surfaces easily, which will cause their loss during pipetting. Instead, resuspend organoids gently using low-retention pipette tips.
3. When using phalloidin staining to validate organoid topology displayed by live WGA staining, we recommend implementing a sequential scanning approach for WGA (ex: 490 nm/em: 498–520 nm) and Texas Red-Phalloidin (ex: 586 nm/em: 600–630 nm) spectral channels to eliminate fluorescence crosstalk, as combined excitation of Alexa fluor 488-conjugated WGA (AF488-WGA) may bleed into emission spectrum of Texas Red-Phalloidin.
4. While adjusting imaging parameters in the Leica Stellaris system, activate two useful image display functions: (i) switch "Channel LUT" to "Grayscale" mode in order to monitor under/overexposure zones through intensity gradients. (ii) the "histogram" function displays the histogram of grayscale values for the current acquisition, enabling real-time intensity monitoring to avoid insufficient intensity signal or signal saturation. These functions guarantee robustness of the fluorescence signals and prevent saturation artifacts created by pixels over 255 intensity values at an 8-bit pixel depth.
5. Hoechst 33342 can be excited in both one- and two-photon microscopies. On a conventional confocal FLIM microscope, we recommend using 405 nm pulsed laser. However, the two-photon excitation range is broader, allowing choosing between 690 and 740 nm, or even longer (>800 nm) wavelengths. This enables multiplexing with other fluorophores, e.g., 740 nm can excite both Hoechst 33342 and endogenous NAD(P)H, providing the possibility to visualize cell metabolism with respect to localization of the proliferating, stem, and differentiated cells.
6. When adjusting the two-photon laser power in LAS X, note that the laser panel percentage values (%) reflect the actual power delivery (mW) non-linearly. To mitigate phototoxicity and photobleaching, we recommend to iteratively increase power in 1% increments while monitoring photon counts. Additionally, two to three frame repetitions or line repetitions can be applied to accumulate a sufficient number of photons without the need to increase the high laser power applied to the sample, achieving robust decay curves for the lifetime estimation while preserving organoid viability.
7. Cellpose is an open-source, deep learning-based tool for cell and nucleus segmentation. To segment nuclei for intestinal

organoid proliferation analysis, follow the latest installation and use guide from Cellpose GitHub (https://github.com/MouseLand/cellpose).

8. For FLIM data acquired from non-Leica systems (e.g., Zeiss, PicoQuant, Becker & Hickl, LaVision etc.), export lifetime images with a GSLab-compatible format from the vendor-provided software, following the file loading guidance in GSLab GitHub repository's documentation (https://github.com/AlexVallmitjana/GSLab).
9. Typically, phasor plot data were applied with threshold and filters (median or wavelet) in LAS X FALCON FLIM software. To ensure analytical consistency in the subsequent GSLab workflow, we suggest exporting raw lifetime data by disabling all filters and setting threshold to 1 before export.
10. Lifetime calibration is essential for FLIM data analysis on phasor plots. GSLab incorporates lifetime calibration tools for accurate phasor analysis, calibrating mono-exponential lifetime data from reference probes (e.g., fluorescein, Atto 488, etc.). However, when analyzing data from Leica FLIM system, manual calibration may not be necessary as these instruments have an integrated auto-calibration using internal IRF during the acquisition.

Acknowledgments

This research was supported by the Special Research Fund (BOF) grants (BOF/STA/202009/003, BOF/BAF/1y/25/1/004), Research Foundation Flanders (FWO, I001922N, I004124N), and the European Union, fliMAGIN3D-DN Horizon Europe-MSCA-DN No. 101073507 grants. We would like to thank Colinda Scheele for supporting access to two-photon measurements on SP8 Falcon Dive system.

References

1. Klunder LJ, Faber KN, Dijkstra G, Ijzendoorn SCDv (2017) Mechanisms of cell polarity–controlled epithelial homeostasis and immunity in the intestine. Cold Spring Harb Perspect Biol 9(7):a027888
2. Schneeberger K, Roth S, Nieuwenhuis EES, Middendorp S (2018) Intestinal epithelial cell polarity defects in disease: lessons from microvillus inclusion disease. Dis Models Mech 11(2):dmm031088
3. Fatehullah A, Tan SH, Barker N (2016) Organoids as an in vitro model of human development and disease. Nat Cell Biol 18(3):246–254
4. Roberto de Barros N, Wang C, Maity S, Peirsman A, Nasiri R, Herland A et al. (2023) Engineered organoids for biomedical applications. Adv Drug Deliv Rev 203:115142
5. Puschhof J, Pleguezuelos-Manzano C, Clevers H (2021) Organoids and organs-on-chips: insights into human gut-microbe interactions. Cell Host Microbe 29(6):867–878
6. Co JY, Margalef-Català M, Li X, Mah AT, Kuo CJ, Monack DM et al. (2019) Controlling

epithelial polarity: a human enteroid model for host-pathogen interactions. Cell Reports 26(9):2509–2520.e4

7. Kakni P, López-Iglesias C, Truckenmüller R, Habibović P, Giselbrecht S (2022) Reversing epithelial polarity in pluripotent stem cell-derived intestinal organoids. Front Bioeng Biotechnol 10:879024
8. Okkelman IA, Zhou H, Borisov SM, Debruyne AC, Lefebvre AEYT, Zoccoler ML, Chen L, Devriendt B, Dmitriev RI (2025) Visualizing the internalization and biological impact of nanoplastics in live intestinal organoids by Fluorescence Lifetime Imaging Microscopy (FLIM). Light: Sci Appl 14(1):272. https://doi.org/10.1038/s41377-025-01949-0
9. Li Y, Yang N, Chen J, Huang X, Zhang N, Yang S et al (2020) Next-generation porcine intestinal organoids: an apical-out organoid model for swine enteric virus infection and immune response investigations. J Virol 94(21):e01006-20. https://doi.org/10.1128/jvi.01006-20
10. Blake R, Jensen K, Mabbott N, Hope J, Stevens J (2022) The development of 3D bovine intestinal organoid derived models to investigate *Mycobacterium avium* ssp paratuberculosis pathogenesis. Front Vet Sci 9:921160
11. Varegg MS, Robertson LJ, Hovd LBN, Stokstad M, Jiménez-Meléndez A (2025) Apical-out bovine intestinal organoids as an infection model for Cryptosporidium parvum. Curr Res Parasitol Vector-Borne Dis 8:100284
12. Nash TJ, Morris KM, Mabbott NA, Vervelde L (2021) Inside-out chicken enteroids with leukocyte component as a model to study host–pathogen interactions. Commun Biol 4(1):377
13. Okkelman IA, Dmitriev RI, Foley T, Papkovsky DB (2016) Use of fluorescence lifetime imaging microscopy (FLIM) as a timer of cell cycle S phase. PLoS ONE 11(12):e0167385
14. Okkelman IA, Foley T, Papkovsky DB, Dmitriev RI (2017) Multi-parametric imaging of hypoxia and cell cycle in intestinal organoid culture. Adv Exp Med Biol 1035:85–103
15. Pachitariu M, Rariden M, Stringer C (2025) Cellpose-SAM: superhuman generalization for cellular segmentation. bioRxiv. 2025.04.28.651001
16. Vallmitjana A, Torrado B, Durkin AF, Dvornikov A, Rajil N, Ranjit S et al (2025) GSLab: open-source platform for advanced phasor analysis in fluorescence microscopy. Bioinformatics 41(4):btaf162
17. Kubbies M, Rabinovitch P (1983) Flow cytometric analysis of factors which influence the BrdUrd-Hoechst quenching effect in cultivated human fibroblasts and lymphocytes. Cytometry 3(4):276–281

Check for updates

Chapter 16

Whole-Mount Immunostaining of Human Intestinal Organoids

Shivam Patel and Paloma Ordóñez-Morán

Abstract

Human intestinal organoids (HIOs) recapitulate the architecture and cell diversity of intestinal epithelium, hence providing a system for modeling processes like regeneration and tumorigenesis. Here, we describe a detailed whole-mount immunostaining protocol for HIOs to visualize proliferative cell population. Briefly, this protocol includes preparation of HIOs, fixation and permeabilization of tissue, blocking of non-specific binding, antibody crosslinking, and visualization using a confocal microscope. Additionally, the protocol is broadly adaptable for investigating different antibodies, enabling the exploration of other cell signaling processes in HIO models. By eliminating embedding and sectioning steps, this method is both time-efficient and preserves spatial tissue architecture.

Keywords Intestine, Organoids, Immunostaining, Whole-mount, Permeabilization, Colon

1 Introduction

HIOs have emerged as an excellent tool to study intestinal stem cells [1, 2] and diseases like colorectal cancer [3] or Inflammatory bowel disease [4] in vitro. HIOs can be generated from either adult intestinal stem cell expressing LGR5+ cells [5] or pluripotent stem cells [6] of a human individual. Self-organizing 3D-culture of intestinal organoids recapitulates features of intestinal tissue physiology, such as multilineage differentiation [7]. Due to these properties, intestinal organoids can be used to model regenerative development and explore novel therapeutic strategies.

Despite these advantages, analysis of HIOs often relies on conventional histological techniques that involve embedding, sectioning, and 2D-staining. This process can disrupt the structural integrity of organoids and can often result in loss of spatial information. Also, due to the size of HIOs and chemical properties of scaffolding gels, embedding can be challenging, time-consuming, and it might require additional optimization steps. In contrast, whole-mount

Paloma Ordóñez-Morán (ed.), *Intestinal Stem Cells*, Methods in Molecular Biology, vol. 3059,
https://doi.org/10.1007/978-1-0716-5412-5_16,

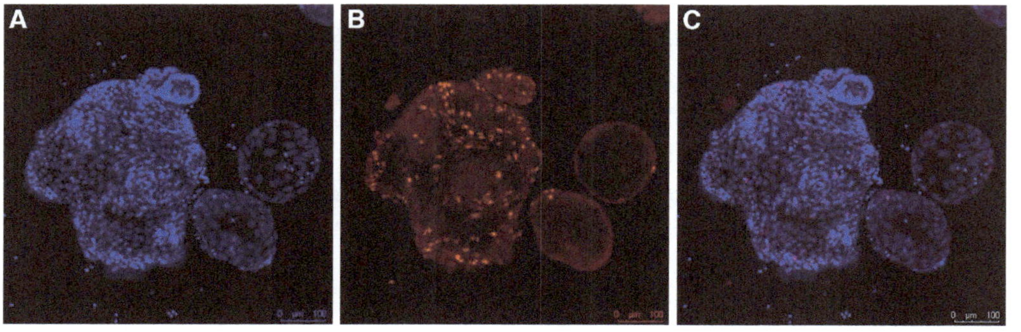

Fig. 1 Lecia confocal SPE immunofluorescence of HIOs showing (**A**) DAPI (Blue-Nucleus), (**B**) Ki67 (Red-Proliferating cells), and (**C**) Merged DAPI and Ki67 images

immunostaining allows visualization of intact of HIOs by preserving their architecture. In addition, spatial patterns of proliferation [5], differentiation [5, 8], and cell signaling [9] can be studied using whole-mount immunostaining approach.

Whole-mount immunostaining method involves fixation, permeabilization, and antibody labeling without embedding HIOs into paraffin blocks. In this chapter, we described a detailed protocol for whole-mount immunostainings of HIOs using Ki67 antibody (Fig. 1). This resource aims to support researchers in studying processes like proliferation, intestinal stem cell dynamics and/or epithelial integrity.

2 Materials

2.1 Equipment and Consumables

1. HIOs culture in 48-well plate.
2. 15 mL Falcon tubes.
3. 1.5 mL Protein LoBind® tubes.
4. Centrifugation machine for 15 mL and 1.5 mL tubes.
5. Tube rolling platform.

2.2 Chemicals

1. Expansion media of organoids (composition shown in Table 1).
2. 4% paraformaldehyde (PFA) solution.
3. PBS1X solution.
4. 0.5% Triton X-100.
5. Blocking solution (10 mg/mL BSA in PBS solvent).
6. Primary antibody mix, 1:250 dilution of anti-Ki67 rabbit monoclonal antibody in blocking solution.
7. Secondary antibody mix, 1:500 dilution of anti-rabbit IgG and 1:500 DAPI (1 mg/mL) mix in blocking solution.

Table 1
Expansion media components with their volume and concentration for HIOs culture

Reagent name	Master stock concentration	Final concentration	Amount for making 500 mL expansion media
Advance DMEM/F-12	–	–	457.05 mL
GlutaMAX™	200 mM	2 mM	5 mL
Penicillin–streptomycin	10,000 U/mL	100 U/mL	5 mL
HEPES	1 M	10 mM	5 mL
N21	50×	1×	10 mL
B27	50×	1×	10 mL
Nicotinamide	1 M	10 mM	5 mL
N-acetyl-L-cystine	500 mM	1.25 mM	1.25 mL
PGE2 (prostaglandin E2)	5 mM	1 μM	100 μL
SB202190 (p38 inhibitor)	30 mM	3 μM	50 μL
A83-01(ALK5 inhibitor)	10 mM	1 μM	50 μL
Wnt3a	100 μg/mL	50 ng/mL	250 μL
EGF	100 μg/mL	50 ng/mL	250 μL
Noggin	100 μg/mL	100 ng/mL	500 μL
R-spondin1	100 μg/mL	100 ng/mL	500 μL

3 Methods

3.1 Preparation of Intestinal Organoids

1. Start with established HIO cultures grown in 48-well plate using appropriate expansion or differentiation media. For this protocol, we used intestinal organoids expansion media with Wnt3a, R-spondin1, Noggin, and EGF (*see* **Note 1**).
2. Before staining, coat each 15-mL Falcon tube with blocking solution (*see* **Note 2**).
3. From 48-well plate, remove culture media carefully without disturbing the Matrigel domes. Add 1 mL ice-cold PBS on top of the Matrigel dome using pre-coated pipette tips, mix to separate HIOs from Matrigel. Transfer HIOs into pre-coated 15 mL Falcon tube.
4. Centrifuge 15-mL Falcon tube at a speed of 500 × g for 5 min and remove the supernatant.
5. Repeat the centrifugation process one more time if the Matrigel is not dissolved (*see* **Note 3**).

3.2 Fixation of Organoids

1. Add 1 mL of 4% PFA solution to fix organoids. Incubate the tube at 4°C overnight on a constantly rolling platform (*see* **Note 4**).
2. Centrifuge 15-mL Falcon tube at 500 ×*g* for 5 min and remove supernatant.
3. Add 500 μL of PBS to the pellet of fixed HIOs and transfer them to Protein LoBind® tube coated with blocking solution.
4. Centrifuge Protein LoBind® tube at 500 ×*g* for 5 min and remove PBS without disturbing settled HIOs.

3.3 Permeabilization and Blocking

1. Add 750 μL of permeabilizing solution and mix gently using a pipette. Keep the tubes on the rolling platform for 1 h at room temperature.
2. Remove permeabilization solution by spinning protein-LoBind tube at 500 ×*g* for 5 min.
3. Add 500 μL of blocking solution to HIOs pellet and mix it gently using a micropipette. Incubate the tube at room temperature while constantly rolling for 1 h.
4. Centrifuge the tube for 5 min at 500 ×*g* to remove supernatant.

3.4 Antibody Crosslinking

1. Add 250 μL of primary antibody mix to HIOs pellet and incubate the tube on a rolling platform overnight at 4°C. (*see* **Note 5**).
2. The next day, centrifuge the tube at 500 ×*g* for 5 min and remove supernatant.
3. Add 500 μL of PBS to HIOs pellet for washing out unlinked primary antibody from the tube.
4. Incubate the tube at room temperature on a rolling platform for 10 min. Centrifuge the tube at 500 ×*g* for 5 min to remove supernatant.
5. Repeat the PBS washing step and centrifugation steps two more times
6. Add 250 μL of secondary antibody mixture to the pellet and mix it gently using a pipette. Incubate in the dark for 1 h at room temperature while constantly rolling (*see* **Note 6**).
7. Collect the stained HIOs pellet by centrifuging the tube at 500 ×*g* for 5 min. Repeat the PBS washing and centrifugation step two times.
8. Add 500 μL of Milli-Q purified water and centrifuge the tube at 500 ×*g* for 5 min.
9. Gently mix HIOs pellet in 200 μL of PBS and transfer it to 96-well plate with a glass bottom and black wall. Immediately, use this plate for image acquisition using a confocal microscope (*see* **Note 7**).

4 Notes

1. Larger and mature organoids are prone to getting ruptured due to repeated pipetting in this method. Avoid pipetting vigorously throughout this whole procedure of immunostaining.
2. Blocking solution prevents organoids from sticking to the surface, thus preventing loss. Additionally, researchers can also coat pipette tips with a blocking solution when transferring organoids into microtubes or Falcon tubes.
3. Keep this centrifugation to 4°C if Matrigel is not clearly separated.
4. Alternatively, incubate the tube on a roller for 1 h at room temperature.
5. Primary and secondary antibody mix dilutions can differ and require optimization whenever using a new antibody or acquired from a new company/supplier.
6. From this step onward, cover the tube with silver foil to prevent fluorophores from getting bleached.
7. Plate can be stored up to 2 weeks at 4°C in the dark. Although prolonged exposure to the microscope laser can decrease the fluorescence signal.

Acknowledgments

This work was supported by Medical Research Council (MRC) New Investigator Research Award under grant number (MR/W01825X/1) and GUTS UK (ECR2021_1).

References

1. Clevers H (2016) Modeling development and disease with organoids. Cell 165(7):1586–1597. https://doi.org/10.1016/j.cell.2016.05.082
2. Basak O, Beumer J, Wiebrands K, Seno H, van Oudenaarden A, Clevers H (2017) Induced quiescence of Lgr5+ stem cells in intestinal organoids enables differentiation of hormone-producing enteroendocrine cells. Cell Stem Cell 20(2):177–190.e4. https://doi.org/10.1016/j.stem.2016.11.001
3. Novellasdemunt L, Kucharska A, Baulies A, Hutton C, Vlachogiannis G, Repana D, Rowan A, Suárez-Bonnet A, Ciccarelli F, Valeri N, Li VSW (2023) USP7 inactivation suppresses APC-mutant intestinal hyperproliferation and tumor development. Stem Cell Rep 18(2):570–584. https://doi.org/10.1016/j.stemcr.2022.12.013
4. Dennison TW, Edgar RD, Payne F, Nayak KM, Ross ADB, Cenier A, Glemas C, Giachero F, Foster AR, Harris R, Kraiczy J, Salvestrini C, Stavrou G, Torrente F, Brook K, Trayers C, Elmentaite R, Youssef G, Tél B, Winton DJ, … Zilbauer M (2024) Patient-derived organoid biobank identifies epigenetic dysregulation of intestinal epithelial MHC-I as a novel mechanism in severe Crohn's Disease. Gut 73(9):1464–1477. https://doi.org/10.1136/gutjnl-2024-332043
5. Sato T, Vries RG, Snippert HJ, van de Wetering M, Barker N, Stange DE, van Es JH, Abo A, Kujala P, Peters PJ, Clevers H (2009) Single

Lgr5 stem cells build crypt-villus structures in vitro without a mesenchymal niche. Nature 459(7244):262–265. https://doi.org/10.1038/nature07935

6. McCracken KW, Howell JC, Wells JM, Spence JR (2011) Generating human intestinal tissue from pluripotent stem cells in vitro. Nat Protoc 6(12):1920–1928. https://doi.org/10.1038/nprot.2011.410
7. He GW, Lin L, DeMartino J, Zheng X, Staliarova N, Dayton T, Begthel H, van de Wetering WJ, Bodewes E, van Zon J, Tans S, Lopez-Iglesias C, Peters PJ, Wu W, Kotlarz D, Klein C, Margaritis T, Holstege F, Clevers H (2022) Optimized human intestinal organoid model reveals interleukin-22-dependency of paneth cell formation. Cell Stem Cell 29(9):1333–1345.e6. https://doi.org/10.1016/j.stem.2022.08.002
8. Huang L, Bernink JH, Giladi A, Krueger D, van Son GJF, Geurts MH, Busslinger G, Lin L, Begthel H, Zandvliet M, Buskens CJ, Bemelman WA, López-Iglesias C, Peters PJ, Clevers H (2024) Tuft cells act as regenerative stem cells in the human intestine. Nature 634(8035):929–935. https://doi.org/10.1038/s41586-024-07952-6
9. de Jong PR, Taniguchi K, Harris AR, Bertin S, Takahashi N, Duong J, Campos AD, Powis G, Corr M, Karin M, Raz E (2016) ERK5 signalling rescues intestinal epithelial turnover and tumour cell proliferation upon ERK1/2 abrogation. Nat Commun 7:11551. https://doi.org/10.1038/ncomms11551

Chapter 17

Monitoring Calcium Dynamics in Pig Intestinal Organoids with FLIM and FRET

Michele Cervellera, Hang Zhou, Bram van den Broek, Kees Jalink and Ruslan I. Dmitriev

Abstract

Intracellular calcium signaling plays a crucial role in intestinal epithelial function. In this chapter, we describe a method to monitor calcium dynamics in live pig small intestinal organoids using stably expressed genetically encoded FRET-based biosensor in combination with intensity-based ratiometric and fluorescence lifetime imaging (FLIM) microscopy readouts. The calcium biosensor Twitch-2B was introduced into organoids by electroporation. Despite the modest lifetime contrast, the biosensor provides a highly reliable ratiometric signal and is compatible with both conventional fluorescence and FLIM-based imaging platform. Following electroporation, organoids are allowed to form and subsequently were selected to establish stable biosensor-expressing lines. This is followed by the live microscopy of heterogeneous and cell-specific calcium responses upon pharmacological stimulation. The presented protocol offers a versatile approach for studies of calcium signaling in relevant 3D cell models, such as organoids and tissues-on-a-chip.

Keywords Calcium signaling, Intestinal organoids, Electroporation, Fluorescence lifetime imaging microscopy (FLIM), Förster resonance energy transfer (FRET), Live cell imaging

1 Introduction

Calcium (Ca^{2+}) signaling plays a fundamental role in coordinating different stimuli to modulate cell biology, physiology and inter- and intracellular communication. In 3D stem-cell-derived intestinal organoid model, dynamic changes in Ca^{2+} define and orchestrate a large number of responses, ranging from fluid and mucus secretion [1–3] to nutrient transport [4] and cell proliferation [5]. In particular, pig small intestinal organoids are emerging as valuable translational models in biomedical research and being considered anatomically and physiologically very similar to humans, displaying crypt-like morphology and multilineage differentiation [6–8].

Ca^{2+} signals are highly dynamic and can vary in amplitude, frequency, duration, and spatial specificity, dictating specific downstream

Paloma Ordóñez-Morán (ed.), *Intestinal Stem Cells*, Methods in Molecular Biology, vol. 3059,
https://doi.org/10.1007/978-1-0716-5412-5_17,

signaling: for instance, rapid highly localized Ca^{2+} spikes regulate fast responses, whereas repetitive global transients or intracellular Ca^{2+} waves control slower responses [9]. Highly controlled homeostatic mechanism keeps cytosolic Ca^{2+} at resting conditions within nanomolar (nM) range of concentrations, while rising to the micromolar range upon stimulation [10]. At the same time, different cell types can exhibit distinct Ca^{2+} signaling patterns according to their physiological role: goblet cells, which under cholinergic stimulation show rapid Ca^{2+} triggered exocytosis [11]; enterocytes, which mediate Ca^{2+} dependent fluid and ion transport; enteroendocrine cells, which respond to luminal cues with hormone secretion [4]; and intestinal stem cells, which integrate Ca^{2+} signals to regulate proliferation andregeneration [5]. Monitoring Ca^{2+} dynamics in live organoids can offer deeper insight about the intestinal epithelium responses to signals of different nature in a real time.

Imaging intracellular Ca^{2+} in the organoids poses several technical challenges. These include penetration depth of excitation and emission light, potentially different concentrations of fluorophore within the spatial context, as well as anisotropy of physical, biological, and chemical characteristics at different tissue depths and subcellular levels. Lastly, employed hydrogel matrices, such as Matrigel or other (bio)polymers, can also impose additional diffusion barriers for embedded organoids, which can hinder dye diffusion and limit the effectiveness of conventionaltransfection methods for fluorophore delivery [12–14].

Traditional Ca^{2+} imaging often relies on the use of small-molecule indicators, such as Fura-2 AM, Fluo- 2, or -4-AM, which can permeate cell membranes and report calcium levels via changes in fluorescence intensity or emission wavelength [4, 15]. However, dyes can often display uneven loading efficiency, time-dependent cytotoxicity, or inability to target specific cell compartments [16]. In the last decades, genetically encoded calcium indicators (GECIs) have transformed Ca^{2+} imaging, helping to visualize intracellular Ca^{2+} signals in live cells and model organisms. GECIs are engineered protein biosensors that change their fluorescence properties upon binding of Ca^{2+} [17]. The first sensor, Cameleon, reported in 1997, has been created by fusing calmodulin-binding peptide between two fluorescent proteins, cyan donor and a yellow acceptor. Subsequently, this design has provided a framework for development of Cameleons with improved sensing properties. The sensing principle of Ca^{2+} sensing exploited in Cameleon and many other subsequently designed biosensors is based on the Förster Resonance Energy Transfer (FRET) phenomenon: when Ca^{2+} binds the sensing moiety of the indicator, the donor and acceptor fluorescent proteins (FP) come into proximity, increasing FRET efficiency and altering the fluorescence intensity of the two FPs, enabling ratiometric measurements. Numerous FRET-based GECIs have been developed using troponin C (e.g., TN-XXL [18] and Twitch series [19]), as well as single FP indicators like Camgaroo and Pericam series [20] or Tq-Ca-FLITS [35]. These biosensors offer sensing over a broad range of Ca^{2+} affinity, dynamic range, and kinetics, so that many researchers can tailor to different experimental needs and cellular models, for long-term imaging and with minimal phototoxicity.

The energy transfer from a donor to an acceptor fluorophores in FRET-based Ca^{2+} indicators can be estimated both in intensity (ratiometric) and lifetime (FLIM-FRET) readouts, providing means for quantitative measurements. In this protocol, we demonstrate applications of both ratiometric (donor/acceptor) and fluorescence lifetime imaging microscopy (FLIM) readouts, using the Twitch-2B biosensor. In FLIM-FRET, the excited-state lifetime of the donor fluorophore is measured. Fluorescence lifetime is an intrinsic property of every fluorophore, reflecting how long it remains in the excited state. Upon FRET, the donor lifetime decreases due to nonradiative energy transfer to the acceptor [36]. Lifetime measurements are generally not influenced by intensity-related factors such as fluctuations in excitation power, photobleaching, or fluorophore concentration, providing an advantage over intensity-based FRET imaging. FLIM-FRET can be considered insensitive to variations in the biosensor expression level, fluorophore concentrations, uneven illumination (as long as sufficient photon budget for FLIM calculations is provided), photobleaching, and light scattering [37]. FLIM can be extremely valuable to separate true FRET signals from tissue autofluorescence, as these factors display different lifetime than the donor [38]. The use of FRET-FLIM provides a robust quantitative readout of FRET efficiency (and thus Ca^{2+} binding) that remains accurate even in thick, heterogeneous samples like organoids or live animal tissues (*see* **Note 3**). There are few FRET-FLIM calcium indicators with strong lifetime contrast currently available (Table 1). For this reason, we selected the Twitch-2B biosensor for proof-of-principle measurements in small intestinal organoids. This biosensor, despite its relatively modest fluorescence lifetime contrast, remains a robust ratiometric reporter. Its performance makes it suitable for showing the possibility to perform FLIM-FRET measurements and, at the same time, conventional intensity-based imaging on setups that do not support lifetime measurements.

As an alternative to genetically encoded biosensors, dye-based indicators can sometimes offer an attractive and somewhat cheaper alternative: such dyes as OGB-1 and Fura-2 can be added directly to cells, organoids, or applied intravitally, requiring much shorter preparation time before the actual measurements (Table 2). Thus, Fura-2 can be added to small intestinal organoids and provide ratiometric Ca^{2+} readout [4]. On the other hand, OGB-1, X-Rhod-1, and other dyes enable FLIM-based quantification [26, 27], being more robust than simple intensity-based dyes. However, these dyes, efficiently performing with, e.g., neural cells, tissues, and organoids, did not provide successful staining with mouse and pig intestinal organoids (data not shown).

Table 1
Overview of FRET-based Ca^{2+} biosensors with respect of their properties for FLIM readout

Biosensor	Sensing principle (FPs and Ca^{2+} sensing domain)	Reported cell models in literature	Readout
Twitch-2B [19]	Cyan FP-Opsanus troponin C—Yellow FP	Neurons (mouse brain); T-lymphocytes (in vivo)	Ratiometric FRET (~5-fold FRET ratio change), sub-optimal FLIM-FRET (τ_m 2.9~3.1 ns; lifetime shift 0.2 ns) K_d 200 nM
Tq-Ca-FLITS [35]	cpTurquoise2 FP + calmodulin (CaM) + M13 peptide	Endothelial cells; human intestinal organoids	FLIM (τ_m 1.8~3.01 ns; lifetime shift ~1.3 ns) K_d 265 nM
YC-nano [21, 22]	ECFP-CaM-M13-cpVenus (improved Yellow Cameleon-Nano variant)	Neuronal cultures (high-density networks)	Ratiometric FRET Tunable K_d, e.g. 50 nM for YC-Nano50 or 15 nM for YC-Nano15. τ_m not reported
D1GO-Cam [23]	cp173EGFP-CaM-M13-mKOκ (Green donor, Orange acceptor)	HeLa cells; pancreatic β-cells	Ratiometric FRET (1.5-fold ratio change). K_d 1.53 μM
TN-XXL [18, 24]	TN-XXL: ECFP-Troponin C- cpCitrine	Neurons in vivo and ex vivo (e.g., Drosophila, mouse cortex, live brain slices)	Ratiometric FRET (~2.5-fold FRET ratio change) K_d 800 nM
TN-L15 [25]	Cerulean—Troponin C—Citrine	Neurons (brain stem, hippocampus) of transgenic mice (in vivo and ex vivo slices)	FRET-FLIM K_d 1.29 μM (FLIM) τ_m 0.8 ~ 2.4 ns, lifetime shift ~1.6 ns
Camgaroo and Pericam series [20]	Single-FP intensiometric sensor (CaM based)	HeLa cells (mitochondria); neuronal and fly models	Intensity K_d (Camg1) 7 μM K_d (Camg2) 5.3 μM

Efficient delivery of genetically encoded Ca^{2+} indicators into organoids remains technically challenging, especially due to the physical barrier imposed by the 3D hydrogel matrices required for their growth. Conventional gene delivery methods, such as lipofection and viral transduction, often yield low efficiency. On the other hand, electroporation using brief high-voltage electrical pulses to transiently permeabilize cell membranes allows efficient delivery of plasmid and large DNA constructs [28–30]. Dissociated organoid fragments are

Table 2
Overview of Ca^{2+} responsive dyes reported with or potentially useful for organoid cultures

Dye	Type	Stock and working concentration	Excitation/emission (nm)	Read-out and dynamic range	Organoid application
Fura-2 AM (Thermo Fisher F1221)	Ratiometric Ca^{2+} dye	1 mM (in DMSO); 2–5 μ M working	335/505 (high Ca^{2+}) and 363/512 (zero Ca^{2+})	Ratiometric intensity shift. K_d 0.14 μM	Used in intestinal organoids for Ca^{2+} imaging [4]
OGB-1-AM (Thermo Fisher O6807)	Green fluorescent cytosolic Ca^{2+} dye	1 mM (DMSO + 20% Pluronic F-127); 2–5 μM	494/523	Intensity (> 14-fold), FLIM (τ_m 0.73 ~ 4 ns, lifetime shift ~3 ns) [27] K_d 170 nM	Uneven, no cytosolic staining with small intestinal organoids
X-Rhod-1-AM (Thermo X14210)	Red fluorescent cytosolic Ca^{2+} dye	1 mM (DMSO + 20% Pluronic F-127); 1–5 μM	580/602	Intensity (> 100-fold) and FLIM (lifetime shift ~1 ns) K_d 700 nM	Uneven, non cytosolic staining

electroporated and re-embedded to regenerate organoids expressing the biosensor, which can undergo a selection process to establish stably expressing fluorescent cell lines that can help streamlining imaging from the cell culture to the microscopy set-up of choice preserving tissue architecture. In this chapter, we describe imaging pipeline performed on a Stellaris 8 Falcon confocal microscope (Leica Microsystems). Although tailored to this system, the described step-by-step protocol is broadly adaptable to any FLIM-compatible microscope and confocal/widefield systems with spectral detection and CFP-YFP detection filter sets.

2 Materials

Prepare all solutions using ultrapure water (Milli-Q grade, sterile filtered 0.22 μm, 18 MΩ cm). Store all reagents at 4°C for no longer than 4 weeks, unless specified otherwise. To prevent contamination, perform all cell culture under laminar flow (class II Microbiological Safety Cabinet with HEPA filter) and aseptically. Wear gloves at all times and spray all used materials and surfaces with 70% ethanol. Unless provided sterile, autoclave all glass and plasticware (121°C, 20 min) and filter-sterilize media and solutions prior to use.

2.1 Pig Intestinal Organoid Culture

Pig small intestinal organoids (Porgj2 line) [31] established from jejunum (*Sus scrofa domesticus*, female, 6–7 weeks) were provided by Bert Devriendt (Ghent University) [6, 31].

2.2 Electroporation Set-Up

Electroporation was performed using the Lonza 4D-Nucleofector™ system, configured with the pre-programmed setting DS-138, optimized for primary cells and other hard-to-transfect cell types. While the exact parameters of DS-138 are not disclosed by the manufacturer, comparable programs typically apply an electrical potential of approximately 1600 V, with two pulses of 20 ms each. Electroporation was carried out using the P3 Primary Cell 4D-Nucleofector™ X Kit S (Lonza, Cat. No. V4XP-3024), which contains the P3 Primary Cell Solution, Supplement 1, and Nucleocuvette certified strips. All procedures were conducted under aseptic conditions within a Class II biosafety cabinet. Nucleocuvette strips supplied in the kit were used for a single electroporation run to ensure consistency and avoid contamination.

2.3 Microscope Set-Up

Confocal FLIM and intensity imaging were performed on a Stellaris 8 Falcon microscope (Leica Microsystems, Germany). The system includes a white-light laser (440–790 nm) operating at 80 MHz pulse frequency, with the 448 nm line used for donor excitation in Förster resonance energy transfer (FRET) experiments. Emission detection was carried out using two Power HyD X detectors, configured for simultaneous collection of donor and acceptor signals. The microscope was fitted with HC FLUOTAR VISIR 25×/0.95 water objective and XYZ direction motorized stage. A temperature- and CO_2-controlled incubation chamber (Okolab) was integrated with the microscope for live-cell imaging. In addition, widefield fluorescence imaging was used for colony screening and detection of positively transfected organoids. Example: inverted microscope Olympus IX81, equipped with CFP, GFP, and YFP filter cubes, 40×/0.6 UPlanFL N and 20×/0.45 UPLanFL N) air objectives, with motorized *Z*-axis control, CoolLED, an Hamamatsu ORCA-Flash 4.0LT$^+$ cMOS camera, glass warming plate Okolab, CellSens Dimension v.3 software [32].

2.4 Chemicals and Plasticware

Reagents and plasticware used for culture, passaging, electroporation, and imaging of pig intestinal organoids.

1. Matrigel (growth factors reduced, phenol red-free). Aliquot 500 μL in 1.5 mL Eppendorf tubes and store at −20°C. Defrost the aliquot at 4°C and store at 4°C after use.
2. DMEM supplemented with 25 mM glucose, 1 mM pyruvate, 2 mM GlutaMAX, with phenol red. Store at 4°C.
3. Complete organoid growth medium: IntestiCult Organoid Growth Medium Human (C. 100-0190). Aliquot 10 mL of IntestiCult™ Organoid Growth Medium (OGM) Human Basal Medium and 10 mL of OGM Supplement into separate

sterile 15 mL tubes. Store aliquots at −20°C. Prior to use, thaw and mix basal medium and supplement thoroughly by pipetting. Store at 4°C to avoid repeated freeze–thaw cycles.

4. Imaging medium: DMEM without phenol red (C. D5030-10X1L), supplemented with 10 mM D-glucose, 2 mM GlutaMax, 1 mM sodium pyruvate, 10 mM HEPES. Store at 4°C.
5. DMEM (no glucose, no pyruvate, no glutamine, no phenol red).
6. Opti-MEM supplement with 25 mM glucose, 1 mM pyruvate, 2 mM GlutaMAX, with phenol red. Store at 4°C.
7. 1 M HEPES solution pH 7.2, 100× concentrate, sterile.
8. 100 mM sodium pyruvate solution, 100× concentrate, sterile
9. 200 mM GlutaMAX supplement, 100× concentrate, sterile.
10. 1 M D-Glucose solution, 100× concentrate, sterile
11. 0.5% Lipidure-coating solution, Lipidure powder was dissolved in 95% molecular grade ethanol. (Amsbio, AMS.52000034GB1G)
12. Phosphate buffer saline (PBS) solution, store at room temperature.
13. Dimethyl sulfoxide (DMSO), "Hybri-Max" grade.
14. Y-27632 (ROCK inhibitor) 10 mM stock in PBS. Store at −20°C.
15. CHIR99021 3 mM stock in DMSO. Store at −20°C.
16. Puromycin 1 mg/mL stock in sterile water. Store in aliquots at −20°C.
17. Bradykinin 1 mM stock in water, store at 4°C, prepare fresh before use.
18. NR12S dye 1 mM stock in DMSO. Store at −20°C, protected from light.
19. Electronic pipette controller.
20. Serological pipette.
21. 1.5 mL and 15 mL sterile centrifuge tube, including lipidure-coated tubes.
22. Flat bottom 24-well cell culture plate, sterile, tissue culture grade.
23. 8-well glass bottom imaging chamber (#1.5 H glass coverslip bottom)
24. Pipettes for 10 μL, 200 μL, and 1000 μL
25. Low retention tips for 10 μL, 200 μL, and 1000 μL
26. Humidified CO_2/37°C incubator.
27. Swinging bucket rotor-equipped centrifuge with cooling option 4°C.

Table 3
Required live imaging probes and plasmid DNAs

Probe	Target	Stock concentration (solvent)	Working concentration	Excitation/ emission (nm)
NR12S (Bio-Techne C. 7509)	Membrane order	1 mM (DMSO)	0.5–1 μM	554/627

All probes were obtained commercially (Table 3), prepared according to manufacturer's instruction and aliquoted in dark plastic vials and stored at −20°C.

The calcium biosensor Twitch-2B (Addgene plasmid #49531) was subcloned into Tol2-Q2 vector. The clone was linearized at *Bam*HI-*Eco*RI restriction sites and ligated into the Q2 vector for stable integration into the genome via co-electroporation with a transposase plasmid. The vector contains the following key features: EF1α promoter followed by intron A and Kozak sequence for efficient expression and puromycin selection cassette. Plasmid DNA was prepared using PureLink™ HiPure Plasmid Midiprep Kit (C. K210004) and Sanger sequencing of the cloning site.

All compounds were reconstituted and stored according to guidelines. Stock solutions were prepared in water or DMSO. Note that final concentration can be optimized according to organoid size and to reduce toxicity if the aim is to study the concentration-dependent drug effect (Table 4).

2.5 Data Acquisition and Analysis Software

1. Leica LAS X software (version 4.8.1.29271 or earlier, e.g., 4.6.0), which provides dedicated modules for FLIM and intensity-based image acquisition and lifetime decay fitting.
2. ImageJ/Fiji with FAST-HIPPOS plugin and Cellpose Advanced integration (working Cellpose environment in Python). For more information and detail of usage, refer to https://imagej.net/plugins/fast-hippos.
3. R (version 4.4.4 or higher) and custom R scripts.

3 Method

3.1 Defrosting the Pig Small Intestinal Organoids

1. Retrieve a cryovial from liquid nitrogen storage or −80°C freezer and thaw rapidly in a 37°C water bath until a small ice crystal remains.
2. Transfer carefully, with the use of a 1000 μL tip, the thawed content to a 15 mL tube prefilled with cold DMEM.
3. Centrifuge at 300 × *g* for 5 min at 4°C in a prechilled centrifuge.

Table 4

Recommended pharmacological compounds for studying calcium dynamics in organoids

Compound	Mechanism of action	Calcium-induced mechanism	Concentration used in this study	Observed response time	Observed organoid contraction
ATP	Activates purinergic receptors	Both IP_3-mediated Ca^{2+} release and direct calcium influx	100 µM	Rapid <10 s	Strong
Bradykinin	Binds B2 receptor (GPCR)	PLC → IP_3 → ER Ca^{2+} release	20 µM	~30–60 s, oscillatory	No, at this concentration
Carbachol	Activates M3 receptors	PLC → IP_3 → ER Ca^{2+} release	10 µM	<10 s	Strong
Forskolin	Activates adenylate cyclase, ↑cAMP	Modulation through PKA	25 µM	Rapid <10 s	Strong
Glucose	Metabolic trigger	Voltage-gated calcium channels	10–20 mM	At least 1 min	Moderate
Histamine	H1 receptor (GPCR)	PLC → IP_3 → ER Ca^{2+} release	1 µM	< 1 min	Variable
Ionomycin	Ca^{2+} ionophore	Direct elevation of cytosolic Ca^{2+} from intra- and extracellular pools	1–10 µM	Slow (concentration-dependent response)	Collapse of epithelial membrane structure

4. Carefully remove supernatant and resuspend the pellet in 50–200 µL of ice-cold Matrigel, depending on pellet size.
5. Plate 50 µL Matrigel domes into wells of a prewarmed 24- or 12-well plate.
6. Incubate the plate at 37°C for 5 min to allow Matrigel to solidify.
7. Cover each dome with 500 µL complete organoid growth medium, optionally supplemented with 10 µM Y-27632 for the first 48 h to enhance post-thaw survival.
8. Incubate at 37°C, 5% CO_2, and refresh medium every 2–3 days.

3.2 Electroporation

Organoids Dissociation

1. Use 1-day cultured intestinal organoids post splitting in complete growth medium, with a good confluency (200–400 organoids per dome) and small organoid size (50–100 µm per organoid) before electroporation.
2. Pre-warm all media and reagents to room temperature or 37°C or according to use.
3. Thaw plasmids from −20°C and allow to equilibrate to room temperature.
4. Prepare 0.5% lipidure-coated tubes (1.5 mL or 15 mL) for dissociation and recovery steps to minimize cell adhesion and loss.
5. Turn on the 4D-Nucleofector™ X Unit and warm the 24-well plate that will be used for re-plating domes.
6. Remove medium from 2–3 wells in a 24-well plate without disturbing the Matrigel domes.
7. Wash each well gently with 500 µL prewarmed Opti-MEM.
8. Add 500 µL TrypLE + 10 µM Y-27632 to each well.
9. Break up the Matrigel dome by pipetting five to ten times with a 1000 µL tip.
10. Transfer the suspension to a 15 mL Lipidure-coated tube.
11. Incubate for 5 min at 37°C (in the incubator), helping mechanical dissociation by pipetting 10–15 times with 1000 µL tip for mechanical dissociation. Repeat one to two times. Check under a bench microscope that all large organoid aggregates have been dissociated into single-cell suspension.
12. Centrifuge at 300 × *g* for 2 min.
13. Resuspend the pellet in 5 mL Opti-MEM supplemented with 10 µM Y-27632 and 3 µM CHIR99021.

14. Count cells accurately and prepare a 10^5 cell suspension for electroporation for each well of the Nucleovette strip.
15. Centrifuge again at 300 × *g* for 5 min.

Electroporation

16. Resuspend cell pellet in 17.5 μL P3 Primary Cell Nucleofector ™ Solution and Supplement 1 in a 4.6:1 ratio according to manufacturer's instructions.
17. In a separate tube, prepare 4 μg biosensor plasmid + 2 μg transposase plasmid (1 μg/μL each).
18. Add 7.5 μL plasmid mix to the 17.5 μL cell suspension; mix gently and avoid air bubbles.
19. Transfer 25 μL of the mixture into one well of a 16-well Nucleovette strip assuring there are no air bubbles.
20. Move the strip out of the biological safety cabinet after closing the lid, place it in the 4D-Nucleofector™ X Unit and start program DS-138 toward the desired wells.

Post-electroporation organoid culture

21. Immediately after electroporation, transfer cells into lipidure-coated 1.5 mL tubes containing 1 mL Opti-MEM + 10 μM Y-27632 + 3 μM CHIR99021.
22. Centrifuge at 300 × *g* for 5 min.
23. Leave 20–30 μL of supernatant to avoid pellet loss; resuspend gently in 50 μL of ice-cold Matrigel.
24. Plates as domes in prewarmed 24-well plates.
25. Incubate 5 min at 37°C to solidify the Matrigel and then cover with 500 μL complete organoid growth medium + 10 μM Y-27632 + 3 μM CHIR99021.
26. Inspect organoid morphology and viability daily under transmission light microscope and refresh complete organoid growth medium after 48 h post-electroporation.
27. Try to avoid overgrowth (organoids should not exceed 100–150 μm in diameter) as larger structures become more resistant to puromycin selection.
28. Look for positively fluorescent organoids under an inverted fluorescence microscope, e.g., in the CFP or GFP fluorescence channels, depending on the reporter.
29. If fluorescence is detectable, begin selection of positive clones by adding 2 μL of puromycin stock solution (1 μg/μL) to each 500 μL of culture medium (final concentration: 4 μg/μL) (*see* **Note 4**).

30. Continue selection for at least 2–3 days, check your wells daily and if required, refresh puromycin-containing medium.
31. Identify domes with healthy, expanding fluorescent organoids 4–7 days after puromycin selection. If needed, organoids showing fluorescent signals of the reporter can be manually isolated with a fine-tipped pipette or pulled glass capillary after identification under inverted microscope and then plated and expanded through standard passaging.
32. Passage transfected and selected organoids as they reach confluency, using a 1:2 to 1:5 split ratio depending on growth rate.

3.3 Passaging, Seeding, and Staining of Newly Generated Organoid Lines for Microscopy

3.3.1 Passaging of Selected Fluorescent Organoids

1. Organoids population 4–7 days after puromycin selection must show a homogeneous fluorescence.
2. Thaw Matrigel overnight on ice.
3. Prewarm 24-well plates, microscopy dishes, and complete organoid growth medium to 37°C for at least 30 min. Start and prechill centrifuge at 4°C.
4. Mechanically disrupt Matrigel and organoids by pipetting 15–30 times with a 10 μL tip attached to a 1 mL tip.
5. Transfer the suspension to a lipidure-coated 15 mL tube.
6. Rinse each well with prewarmed complete organoid growth medium and add to the same tube. Adjust to 10 mL total volume.
7. Centrifuge at 500 × *g* for 5 min at 4°C.
8. Remove carefully supernatant without disturbing the pellet.
9. Keep the tube on ice and resuspend the pellet in fresh cold Matrigel at 1:2 to 1:5 dilution of the original dome volume. For subculturing or expansion seed 50 μL per well in a prewarmed 24-well plate, incubate 5 min at 37°C, and cover with complete organoid growth medium.

Seeding organoid for imaging

10. Dispense 20 μL of the Matrigel-organoid suspension into the center of each well of a prewarmed glass-bottom microscopy dish.
11. Incubate the dish at 37°C for 5 min to solidify the Matrigel.
12. Overlay with 200 μL of complete organoid growth medium.

13. Incubate for 2–3 days to allow recovery and regrowth before staining or imaging.

Dye staining procedures

14. Prepare stock solutions at the previously described stock solutions in DMSO and store protected from light (store at −20°C for long-term storage).
15. Dilute the stocks in prewarmed complete organoid growth medium the required final concentration.
16. Remove medium from each well and replace it with 200 μL staining solution.
17. Incubate for indicated time by manufacturers i.e. ranging from 20 min to 1 h for the described dyes at 37°C, protected from light.
18. Wash at least three times with warm CO_2-independent imaging medium (e.g., with 10 mM HEPES buffer, pH 7.2) to remove excess dye.

3.4 Intensity and FLIM-Based Imaging of Calcium Dynamics

The microscope and associated components, including laser and incubation chamber, must be switched on for at least 30 min prior to imaging to ensure system stability and equilibration of temperature-controlled chamber to 37°C, as well as humidity control. Below is the description of the experiment performed in Leica LAS X software (version 4.8.1.29271) with spectral detection settings configured according to the CFP and Citrine spectral properties.

1. FRET donor excitation is performed within the 440–455 nm range, using a laser line modulated at 40–80 MHz; in our setup, we use the 448 nm line commonly available on commercial lasers and 80 MHz modulation for consistency and reproducibility in other systems too. Operate the system in *xyt* scan mode with a 600 Hz unidirectional scan speed. Set the pinhole size at 2.50 Airy units. Use the donor excitation at 448 nm (white light laser) at 30% power output (*see* **Note 1**). Acquire frames with a resolution of 512 × 512 pixels with no line or frame averaging.
2. Configure the emission detection for simultaneous collection of the FRET donor and acceptor using Power HyD detectors in counting mode. Define the detection channels at 452–490 nm (HyD X 1, for CFP) and 515-555 nm (HyD X 2, for YFP). The collection of emission from both fluorophores on the same line or optical path will help avoiding scanning artifacts and preserve temporal resolution of the FRET-FLIM measurements.
3. To image membrane boundaries and facilitate automatic segmentation, use NR12S (excited separately using 598 nm laser with power ranging from 10% to 30% and with zoom factor set

at 1.0 or higher). Collect its emission within a different spectral channel (e.g., using HyD R detector), covering emission range of 610–745 nm. This acquisition must be conducted prior to FRET channels detection on the same field of view to prevent spectral leak-through into the YFP detector, because of the overlapping emission spectra.

4. After initialization of timelapse sequence, record at least ten baseline frames to establish prestimulation resting calcium levels. 4 μL of 1 mM bradykinin in water must be added manually to the well with the Matrigel dome in 200 μL imaging volume, accounting for a final concentration of about 20 μM. The drug addition must be performed, assuring uniform resuspension by manual pipetting several times, making sure to avoid disturbing the sample and to avoid drift of the field of view or focus plane changes. Optionally, multiple stimuli can be added sequentially (*see* **Note 5**).
5. Finish the experiment and save the data.

3.5 Intensity Data Processing

1. Use Leica LAS X software (v4.8.1) to export intensity data as TIFF files. Each channel, CFP (channel 1) and YFP (channel 2), was exported separately using multipage TIFF format, with lossless compression and raw images settings to preserve data quality.
2. Open the two TIFF stacks on Fiji/ImageJ and assign distinct LUTs to avoid misidentification. To merge the channels into single ratiometric time-lapse, navigate to *Image* > *Color* > *Merge Channels*, assigning CFP as channel 1 and YFP as channel 2. The *Create Composite option should be enabled.*
3. Convert the merged image into a hyperstack via *Image* > *Hyperstack* > *Stack to Hyperstack*, and specify number of channels (2), slice (1), and total time frames according to the file, with the display mode set to *Composite.*
4. Alternatively, when FLIM is not used, the *.lif* file can be directly opened in Fiji using the Bio-Formats importer to generate a hyperstack, offering a faster and simpler solution.
5. Save the resulting 2 channels, timelapse hyperstack as TIFF and use it as an input on your own dedicated Fiji/ImageJ macro or for a publicly available plugin like FAST-HIPPOS.

3.6 FLIM Data Fitting and Export

1. Open the FLIM module of LAS X and fit donor decay curve values to fixed values of 0.4 ns ("high FRET" state) and 2.8 ns ("low FRET" state). These values are determined empirically after the thorough testing of the biosensor in different cellular models and gave the best fit to observed photon arrival time distribution assuring reproducibility between different experiments.

2. Perform fitting using "FLIM Image FIT" on LASX to produce two images containing the amplitudes of these two components: A_1 for the short lifetime and A_2 for the long lifetime.
3. To compute the weighted lifetime per pixel, the following equation was used:

$$\tau_{\text{amp weighted}} = (0.4 \times A_1 + 2.8 \times A_2)/(A_1 + A_2)$$

allowing pixel-based comparison between FRET-states.

4. Export the amplitude images as "raw ImageJ TIFF" with scaling 0.001 and enable the fixed range export option. These files can then be processed on your own dedicated Fiji/ImageJ macro or using the plugin FAST-HIPPOS.

3.7 FAST-HIPPOS Workflow and Custom Cell Segmentation

1. Open the ImageJ plugin and select "Confocal TCSPC": lifetime component images, fitted with a double-exponential and exported from LASX (as "raw ImageJ TIFF") or alternatively select "Ratio Imaging": intensity 2-channel TIFF files.
2. Use Cellpose (v2/3), integrated via Fiji, to generate cell masks. Alternatively, FAST-HIPPOS now supports segmentation using additional fluorescence channel, which can simplify the workflow; to use this feature, follow the detailed instructions provided on the plugin's official website.
3. Extract from LAS X as described for intensity export the NR12S stained membrane image.
4. Input this image to Cellpose to generate a label mask image. Manually review it by adjusting parameters according to cell boundaries and upload it into FAST-HIPPOS to override its native segmentation step.
5. FAST-HIPPOS computes average ratiometric intensity or lifetime values per cell over time.
6. Single-cell traces were plotted within the plugin. Export the data as .csv files for further analysis.
7. Import the files into R or any other software suitable for visualization, data representation, and statistical analysis.

4 Anticipated Results

4.1 Successful Ca^{2+} Biosensor Electroporation Confirmed by Confocal Fluorescence Microscopy

Confocal fluorescence microscopy verified the delivery, expression, and correct integration in the genome of Twitch-2B calcium biosensor in pig intestinal organoids (Fig. 1). For deeper and functional FRET dynamics studies, additional calcium imaging of both donor

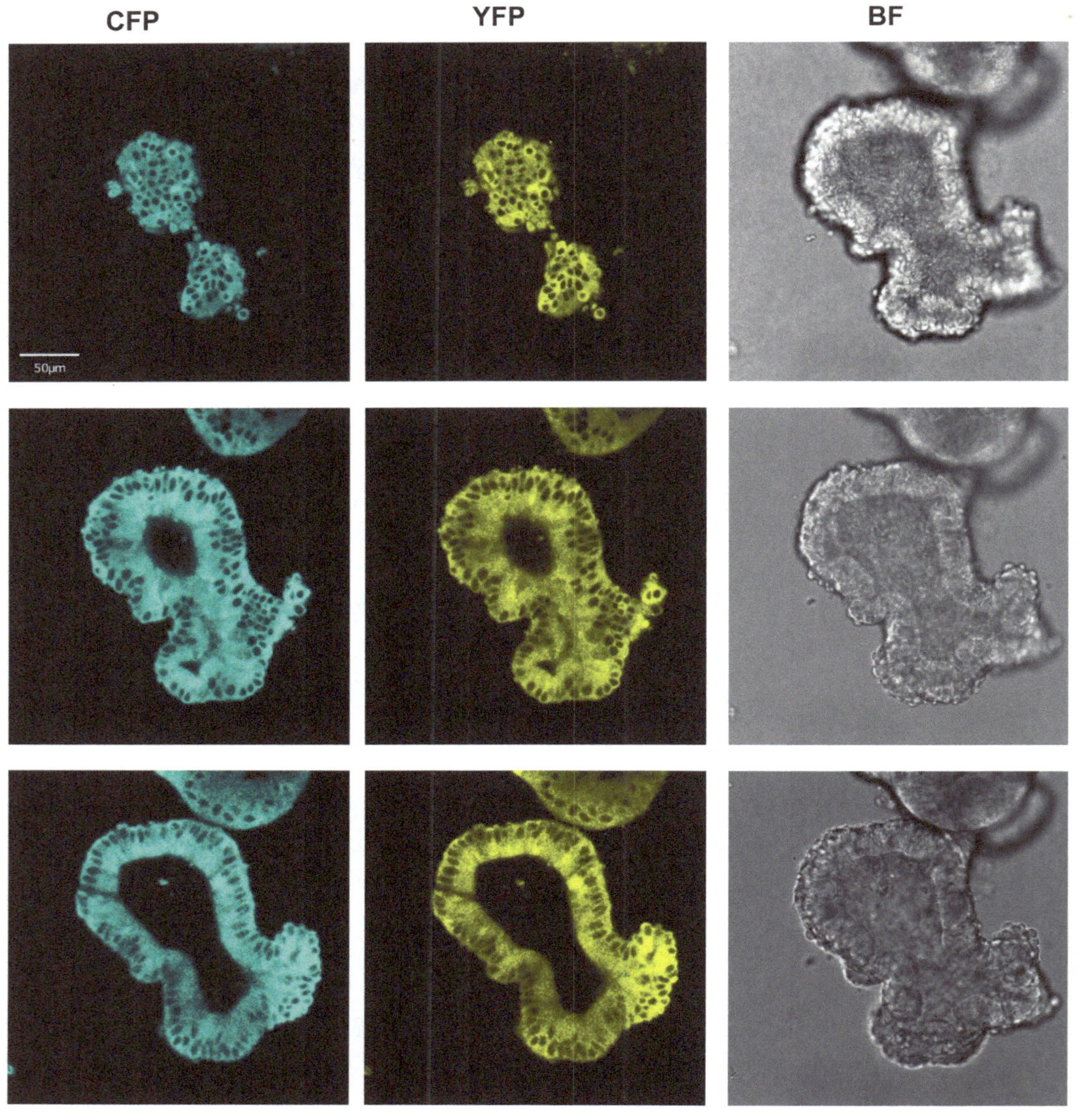

Fig. 1 Confocal microscopy of Twitch-2B biosensor expression in pig intestinal organoids showed at different z-stack (depth) projections. Left: CFP fluorescence following 405 nm excitation (emission 450–480 nm); Middle: YFP (Venus) fluorescence following 520 nm excitation (emission 530–550 nm); Right: corresponding brightfield images

and acceptor must be performed. Excitation of CFP only at 405 nm (Venus excitation efficiency = 0.006) and collection of emission with narrow bands between 450 and 480 (Venus emission efficiency = 0) showed expression of the blue fluorescent protein. Similarly, excitation of YFP was done at 520 nm (CFP excitation efficiency = 0) with collection of emission between 530 and 550 nm. Figure 1 illustrates cytoplasmic distribution through the volume of the organoid in both CFP and Venus channels together, indicating that the electroporation protocol resulted in uniform cellular expression of the biosensor.

4.2 Ratiometric Calcium Imaging of Pig Intestinal Organoid Under Bradykinin Stimulation

Live ratiometric imaging was performed as described previously following stimulation with 1 μM bradykinin (BDK) at 100 s time point (*see* **Note 2**) to study calcium dynamics in the electroporated organoids (Fig. 2). Membrane staining with NR12S was used to define cell boundaries. As shown in Fig. 2A, staining enabled generation of precise static segmentation masks, allowing extraction of single-cell data across the organoids.

Cells responded with a slightly asynchronized calcium rise, followed by a gradual decline, which indicates a partial recovery of the basal calcium levels. Differences in response amplitude and recovery kinetics were evident between individual cells. Four distinct phases can be distinguished and could be defined as resting baseline, rapid signal increase after stimuli exposure, peak of the response, and recovery among cells of the same organoid (Fig. 2B). This potentially reflects cell-specific differences in calcium buffering, receptor expression levels and allow for identification of different cell types based on their dynamic behavior. Beyond confirming sensor functionality, this method enables a spatially and temporally resolved analysis of calcium kinetics across a complex 3D structure of the organoids. It raises questions regarding spatial patterning in calcium signaling: by mapping onset timing or peak intensity, one could explain whether calcium influx is initiated from basal or apical sides, whether it propagates radially or with wave-like characteristics. Such information can be coupled to the organoid structure, shape and can also reflect age and stage of cell differentiation. Described method shows promise to study combined effects of different drugs or stimuli (mechanical, electrical, optical...) e.g., in studying potential paracrine effects or mechanical coupling of the Ca^{2+} signaling. In Fig. 2C (top), raw single-cell traces are shown, while in the bottom figure, the same traces are shown after LOESS smoothing (span = 0.1), which helps reducing high-frequency noise and highlights general trends in calcium dynamics. This visualization is particularly useful for comparing signal shape and temporal alignment across many cells and for detecting subtle features in response patterns.

4.3 FLIM-Based Calcium Imaging

FLIM was performed simultaneously to ratiometric imaging, with data collected under identical stimulation and acquisition conditions, so that direct comparison between intensity- and lifetime-based calcium readout was possible.

Figure 3A shows a clear drop in donor (CFP) lifetime, detectable upon binding with calcium and increase in FRET efficiency following

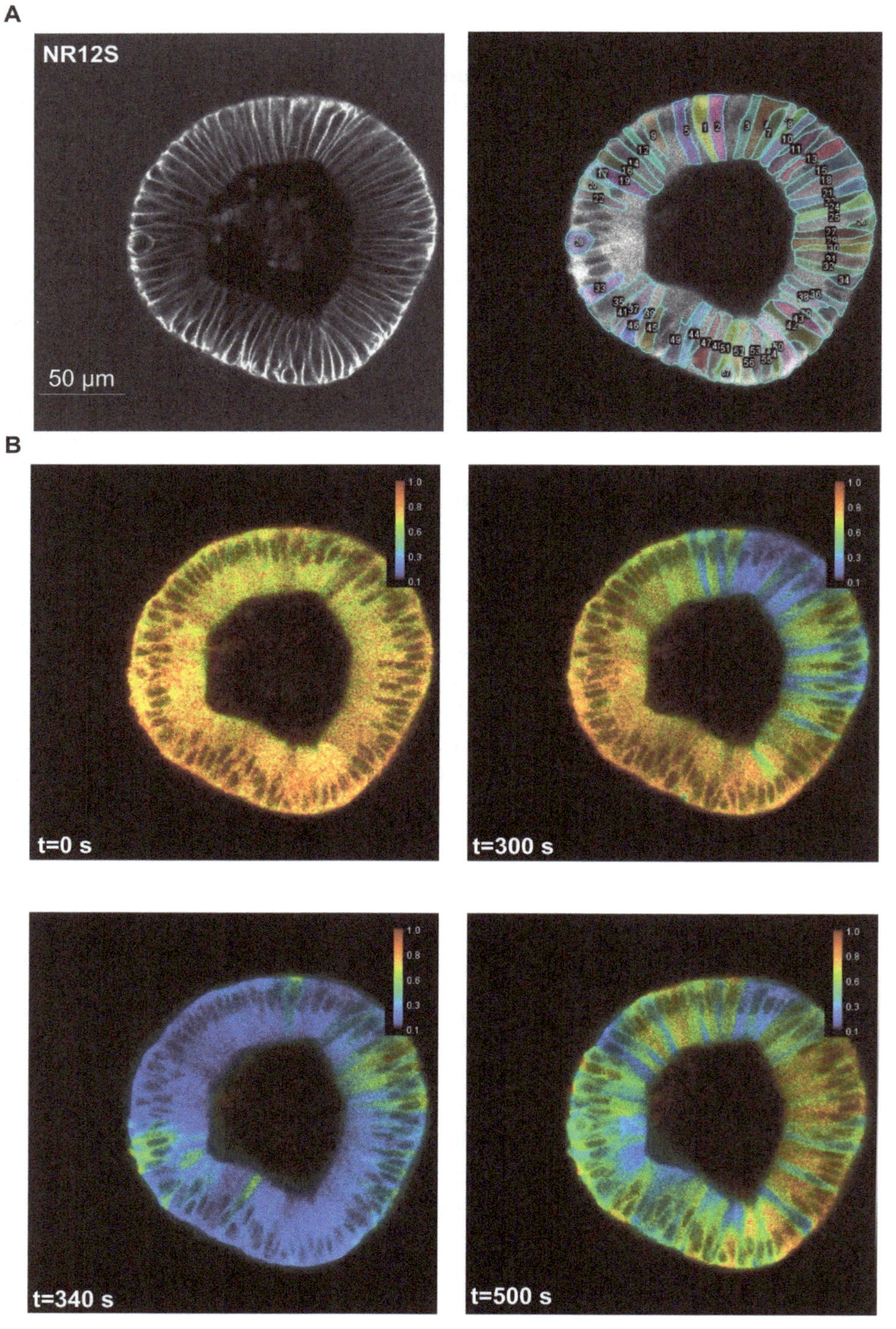

Fig. 2 Ratiometric CFP/YFP calcium imaging in intestinal organoids upon bradykinin stimulation. (**A**) NR12S membrane stain (left) and overlay of intensity, segmentation mask, and labels (right) showing identification of individual cells across the organoid through Cellpose. (**B**) Pseudocolor images of ratiometric values (CFP/YFP)

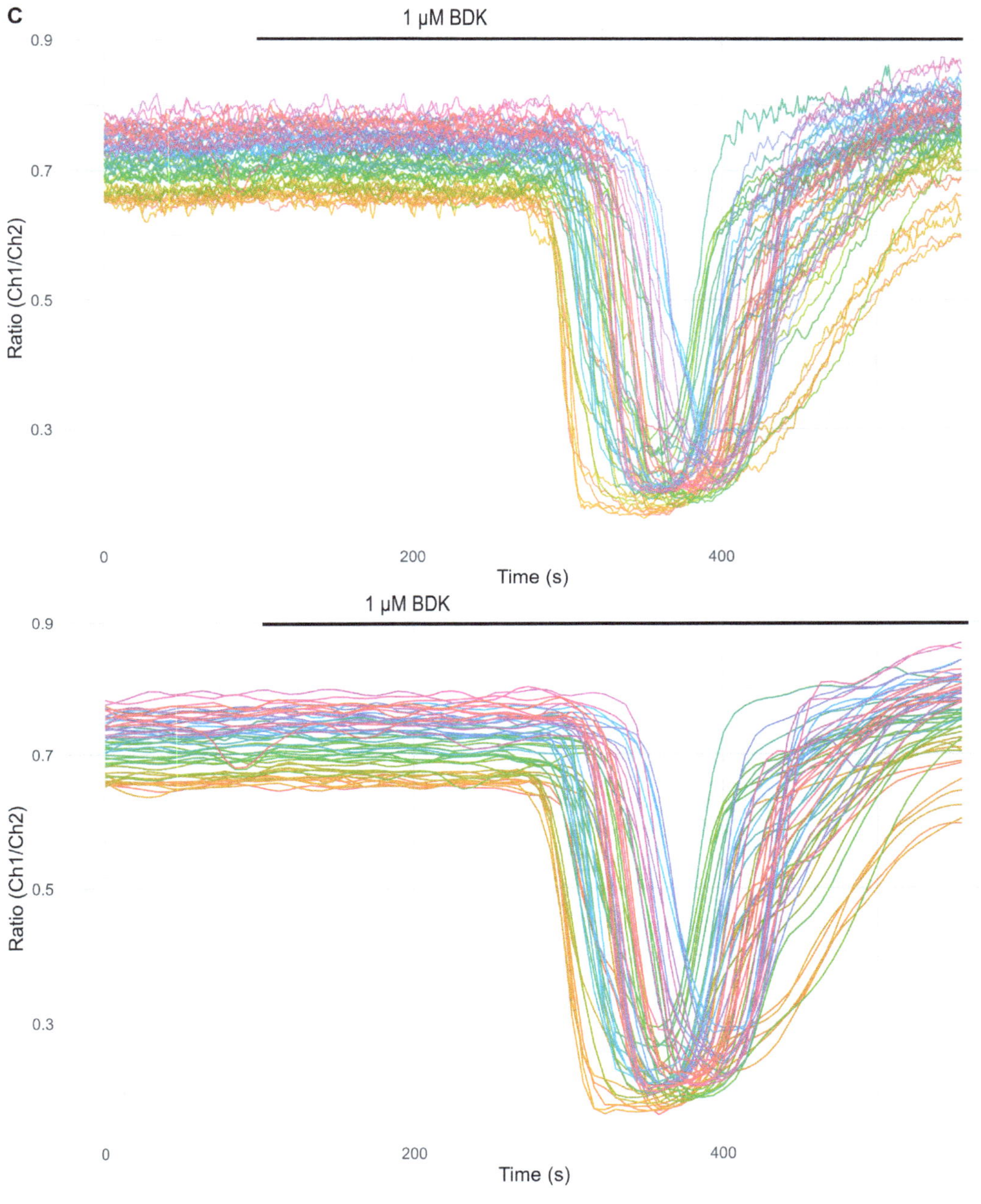

Fig. 2 (continued) at four timepoints ($t = 0$, 300, 340, and 500 s). (**C**) Raw single-cell traces (top) and smoothed traces (bottom) show distinct dynamic across cells, capturing changes in ratiometric changes response magnitude, activation onset timing, and recovery dynamics

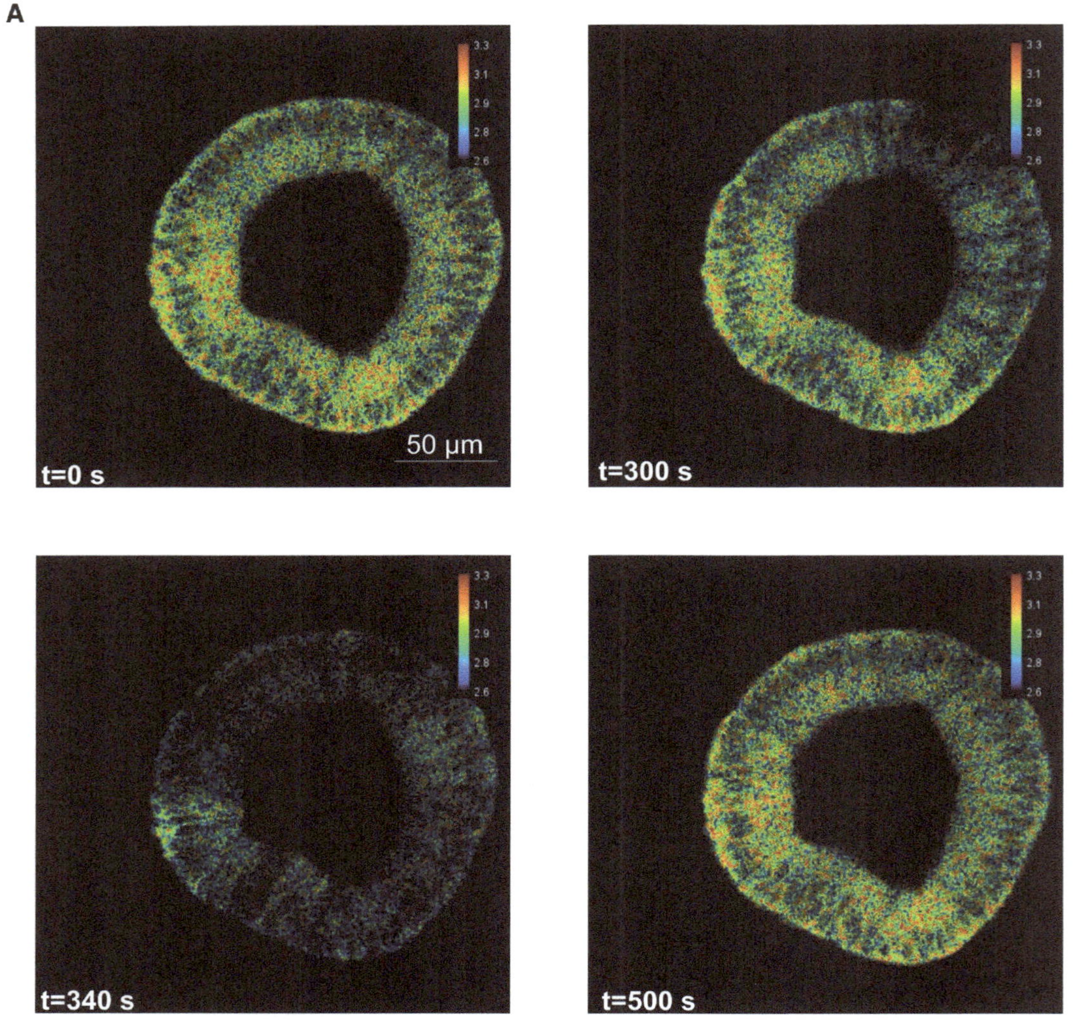

Fig. 3 FLIM microscopy (CFP or donor) of calcium dynamics in pig intestinal organoids expressing Twitch-2B biosensor. (**A**) Lifetime pseudo-colored images at four timepoints ($t = 0, 300, 340, 500$ s) show transient donor lifetime decrease following BDK stimulation. (**B**) Single-cell lifetime traces over time shown as raw (top) and smoothed (bottom). (**C**) Fitting of fluorescence decay curves extracted from the FLIM timelapse experiment, as global for overall time series and identification of a baseline and activation part. The global fit across all timepoints is shown in blue and IRF in grey. Exponential fit at two components (previously described) to two representative phases as baseline (pre-stimulation) and activation (post stimulation) are plotted in black and red determining two components lifetime fitting with $\tau_1 = 0.4$ ns and $\tau_2 = 2.8$ ns. The corresponding raw decay data are plotted in pink and orange. The observed shift in decay curves between baseline and activation reflects the expected shortening of donor lifetime due to increase in FRET efficiency upon calcium binding. Lower panel shows the fit residuals

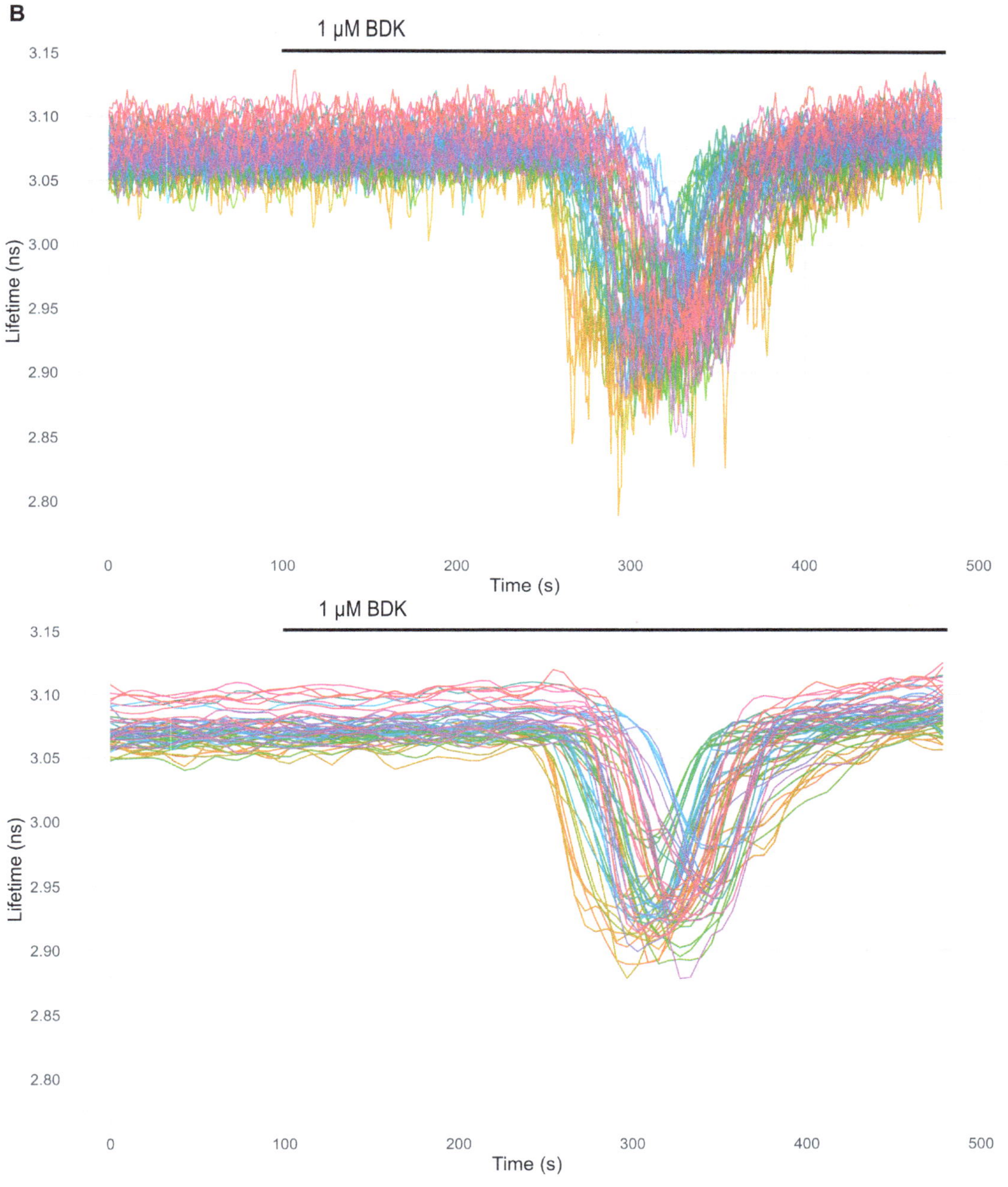

Fig. 3 (continued)

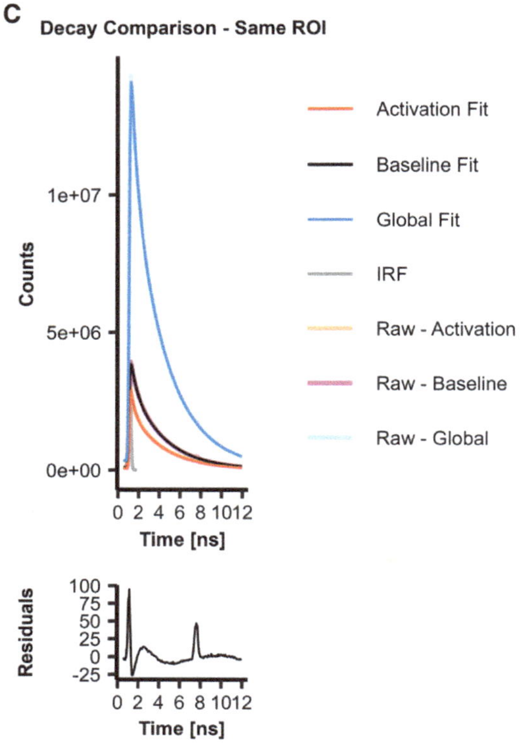

Fig. 3 (continued)

BDK stimulation. Subsequently, lifetime values return to baseline confirming a full functionality of the FRET biosensor. Single-cell traces (Fig. 3B) demonstrate a consistent but minor change in lifetime (~0.2 ns~), with dynamics similar to the ratiometric traces. However, the lifetime contrast between resting and activated states is limited. Such a narrow dynamic range likely results from a combination of various factors: very high FRET efficiency, which leads to a nearly complete quenching of the donor, reducing the measurable lifetime range. Additionally, photon yield is reduced, which makes lifetime fitting more challenging and less accurate in the low photon regions (as can be seen in Fig. 3, $t = 340$ s). To further validate the lifetime response, two-component global exponential decay fitting was applied to entire timelapse acquisition. Fitting global decay and extracted decays from intervals of baseline and activation showed a clear lifetime reduction, consistent with calcium-induced FRET (Fig. 3C). Collectively, these experiments confirm that Twitch-2B biosensor is functional under FLIM conditions, but its limited contrast and photon efficiency make it suboptimal for FLIM-based quantification. For experiments that prioritize lifetime sensitivity, sensors with broader dynamic range and tau shifts can offer better performance [35].

4.4 Ratiometric Calcium Imaging in Apical-Out Organoids

To verify functionality of calcium handling in the polarity-reversed organoids, we used apical-out organoids, where the apical membrane faces the surrounding growth medium [33].

These organoids are nonadherent and free floating, often resulting in rotation and other movements, which can affect quality of live imaging. In this context, rather than focusing on individual cells, we chose to visualize Ca^{2+} dynamics at the whole-organoid level, enabling robust analysis of collective responses despite segmentation challenges (*see* **Note 5**). Therefore, segmentation was performed using Cellpose, applying one mask per organoid and following ratiometric calcium over time for the three organoids present in the field of view. Following 1 μM BDK stimulation, the organoid responded with a clear drop in CFP/YFP ratio, followed by gradual recovery (Fig 4B, bottom). The temporal response matches dynamic profile of basal-out organoids, confirming the functionality of Twitch-2B biosensor under polarity reversion conditions. The movement seen in the image panels, particularly between $t = 0$ and $t = 100$ s, reflecting real translational and rotational motion, which may explain some drift in the extracted ratiometric traces. With free-floating organoids, manual drug application can also significantly affect baseline measurement and introducing shifts in the signal (noise), unrelated to calcium waves. To minimize this, we carefully optimized the stimulation protocol by using smallest volume possible and high concentration of the applied drugs. Described pipeline demonstrates the potential use of transfected organoids in apical-out reverted polarity in calcium imaging experiments. The future improvements, such as dynamic segmentation or tracking algorithms, can also help recovering and deepen the spatial information in these measurements (*see* **Note 5**).

5 Notes

1. The choice of 30% laser power was based on empirical optimization to achieve a sufficient photon counts per pixel per pulse for reliable fluorescence lifetime fitting, ensuring accurate lifetime determination, while preventing intensity saturation of the detectors, as it can distort both lifetime and intensity data [34]. Since the stably transfected cell lines often show lower expression levels compared to transiently transfected cells, such experiments may require moderately high laser power to achieve suitable signal levels. Control timelapse experiment must be performed to estimate bleaching of donor effect and stability of lifetime during the course of a timelapse

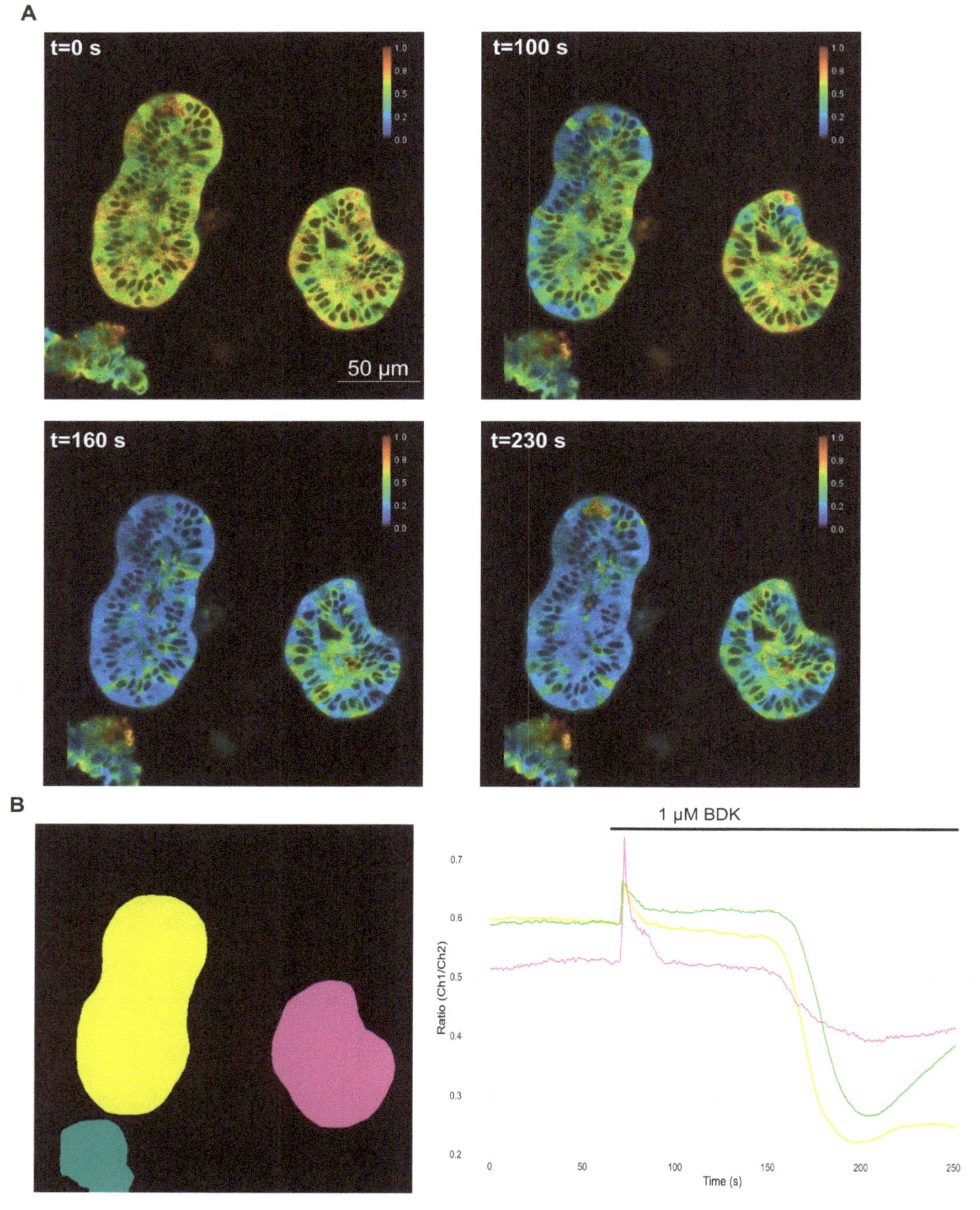

Fig. 4 Ratiometric calcium imaging in apical-out organoids. (**A**) Pseudocolor images of ratiometric values (CFP/YFP) at four timepoints ($t = 0$, 100, 160, and 230 s). (**B**) Segmentation masks (left), generated with Cellpose, show whole-organoid outlines for bulk signal extraction, and (right) whole-organoid ratiometric traces over time show robust calcium transients. Signal drift is likely due to organoid movement or bradykinin stimulation

measurement. Control experiments showed that under these settings during a 10 min timelapse with frame interval of about 800 ms bleaching remained under 5% with not changes in the lifetime values.

2. Usually, a highly concentrated (e.g., 1000×) reagent is used to prevent changes in volume, temperature, and/or osmolarity when adding the stimulus. We mix the reagent by first pipetting up a small volume with a tip on a P20. We then transfer the tip carefully to a P200, making sure to avoid losing the droplet by pressing the plunger beforehand, then gently releasing it when putting the tip on the p200. Thus, a small drop of reagent is present in the middle of the tip. To stimulate the cells, we carefully put the tip of the pressed pipette in the medium of the well, aspirate 50 μL of medium (which pre-dilutes the agonist), and gently mix by pipetting up and expelling the 50 μL medium for multiple times, while avoiding air bubbles.
3. FRET ratios have been used in this study without spectral correction. Such assessment reflects the absence of donor-only and acceptor-only cell lines, which are typically required to correct for bleed-through, crosstalk, and to deconvolve baseline FRET efficiencies. While this limits quantitative calibration of intracellular calcium levels, uncorrected FRET remains fully informative and adequate for tracking relative changes, response dynamics and stimulus-induced calcium fluctuations, resulting useful for assessing kinetics and spatial propagation of signals.
4. It is critical to establish a puromycin cell-killing curve for each organoid batch prior to applying selection, as the optimal concentration varies by species and culture density.
5. During experiments involving pharmacological stimulation, organoids often undergo substantial contraction and movement. These deformations interfere with cell tracking and segmentation over time. To preserve single-cell resolution, motion correction and/or tracking algorithms should be integrated in the image analysis workflow.

Acknowledgments

This research was supported by the European Union, fliM-AGIN3D-DN Horizon Europe-MSCA-DN No. 101073507 grant. We thank Dr. Irina A. Okkelman and Ghent Light Microscopy Core (GLiM) for support and supervision as well as Prof. Bert Devriendt for providing the organoid line.

References

1. Reynolds A, Parris A, Evans LA et al. (2007) Dynamic and differential regulation of NKCC1 by calcium and cAMP in the native human colonic epithelium. J Physiol 582:507–524. https://doi.org/10.1113/jphysiol.2007.129718
2. Engevik KA, Hanyu H, Matthis AL et al. (2019) Trefoil factor 2 activation of CXCR4 requires calcium mobilization to drive epithelial repair in gastric organoids. J Physiol 597:2673–2690. https://doi.org/10.1113/jp277259
3. Pauer SM, Buß B, Diener M, Ballout J (2024) Time-dependent effects of tumor necrosis factor α on Ca2+-dependent secretion in murine s mall intestinal organoids. Front Physiol 15: 1382238. https://doi.org/10.3389/fphys.2024.1382238
4. Zietek T, Rath E, Haller D, Daniel H (2015) Intestinal organoids for assessing nutrient transport, sensing and incretin secretion. Sci Rep 5:16831. https://doi.org/10.1038/srep16831
5. Mah AT, Yan KS, Kuo CJ (2016) Wnt pathway regulation of intestinal stem cells. J Physiol 594:4837–4847. https://doi.org/10.1113/jp271754
6. Vermeire B, Gonzalez LM, Jansens RJJ et al (2021) Porcine small intestinal organoids as a model to explore ETEC–host interactions in the gut. Vet Res 52(1):94. https://doi.org/10.1186/s13567-021-00961-7
7. Mussard E, Lencina C, Boudry G et al (2023) Culture of piglet intestinal 3D organoids from cryopreserved epithelial crypts and establishment of cell monolayers. J Vis Exp 10:192. https://doi.org/10.3791/64917
8. Kawasaki M, Goyama T, Tachibana Y et al (2022) Farm and companion animal organoid models in translational research: a powerful tool to bridge the gap between mice and humans. Front Med Technol 4:895379. https://doi.org/10.3389/fmedt.2022.895379
9. Berridge MJ, Bootman MD, Roderick HL (2003) Calcium signalling: dynamics, homeostasis and remodelling. Nat Rev Mol Cell Biol 4:517–529. https://doi.org/10.1038/nrm1155
10. Sanderson MJ, Charles AC, Boitano S, Dirksen ER (1994) Mechanisms and function of intercellular calcium signaling. Mol Cell Endocrinol 98:173–187. https://doi.org/10.1016/0303-7207(94)90136-8
11. Pelaez-Llaneza N, Jones V, Kam C et al (2022) Guardian crypt-base-goblet-cells protect the human colonic stem cell niche by triggering cholinergic calcium signal-dependent MUC2 secretion and luminal flushing. bioRxiv. https://doi.org/10.1101/2022.01.04.474646
12. Hofer M, Lutolf MP (2021) Engineering organoids. Nat Rev Mater 6:402–420. https://doi.org/10.1038/s41578-021-00279-y
13. Kim S, Choi YS, Lee JS et al. (2022) Intestinal extracellular matrix hydrogels to generate intestinal organoids for translational applications. J Ind Eng Chem 107:155–164. https://doi.org/10.1016/j.jiec.2021.11.044
14. Vanhove L, Van Gansbeke T, Devriendt B et al (2025) Lack of biochemical signalling in GelMA leads to polarity reversion in intestinal organoids independent from mechanoreciprocity. J Tissue Eng 16:20417314251345000. https://doi.org/10.1177/20417314251345000
15. Tsien RY, Rink TJ, Poenie M (1985) Measurement of cytosolic free Ca2+ in individual small cells using fluorescence microscopy with dual excitation wavelengths. Cell Calcium 6:145–157. https://doi.org/10.1016/0143-4160(85)90041-7
16. Grienberger C, Konnerth A (2012) Imaging calcium in neurons. Neuron 73:862–885. https://doi.org/10.1016/j.neuron.2012.02.011
17. McCombs JE, Palmer AE (2008) Measuring calcium dynamics in living cells with genetically encodable calcium indicators. Methods 46:152–159. https://doi.org/10.1016/j.ymeth.2008.09.015
18. Direnberger S, Mues M, Micale V et al (2012) Biocompatibility of a genetically encoded calcium indicator in a transgenic mouse model. Nat Commun 3:1031. https://doi.org/10.1038/ncomms2035
19. Thestrup T, Litzlbauer J, Bartholomäus I et al. (2014) Optimized ratiometric calcium sensors for functional in vivo imaging of neurons and T lymphocytes. Nat Methods 11:175–182. https://doi.org/10.1038/nmeth.2773
20. Reiff DF, Ihring A, Guerrero G et al. (2005) In vivo performance of genetically encoded indicators of neural activity in flies. J Neurosci 25:4766–4778. https://doi.org/10.1523/jneurosci.4900-04.2005
21. van der Linden FH, Mahlandt EK, Arts JJ, Beumer J, Puschhof J, de Man SM, Chertkova AO, Ponsioen B, Clevers H, van Buul JD, Postma M (2021) A turquoise fluorescence lifetime-based biosensor for quantitative imaging of intracellular calcium. Nat Commun 12:7159. https://doi.org/10.1038/s41467-021-27249-w
22. Horikawa K, Yamada Y, Matsuda T et al. (2010) Spontaneous network activity visualized by ultrasensitive Ca2+ indicators, yellow Cameleon-Nano. Nat Methods 7:729–732. https://doi.org/10.1038/nmeth.1488
23. Engevik KA, Karns RA, Oshima Y, Montrose MH (2020) Multiple calcium sources are

required for intracellular calcium mobilization during gastric organoid epithelial repair. Physiol Rep 8:e14384. https://doi.org/10.14814/phy2.14384

24. Waldeck-Weiermair M, Alam MR, Khan MJ et al. (2012) Spatiotemporal correlations between cytosolic and mitochondrial Ca2+ signals using a novel red-shifted mitochondrial targeted cameleon. PLoS ONE 7:e45917. https://doi.org/10.1371/journal.pone.0045917
25. Mank M, Santos AF, Direnberger S et al. (2008) A genetically encoded calcium indicator for chronic in vivo two-photon imaging. Nat Methods 5:805–811. https://doi.org/10.1038/nmeth.1243
26. Rinnenthal JL, Börnchen C, Radbruch H et al. (2013) Parallelized TCSPC for dynamic intravital fluorescence lifetime imaging: quantifying neuronal dysfunction in neuroinflammation. PLoS ONE 8:e60100. https://doi.org/10.1371/journal.pone.0060100
27. Micu I, Jiang Q, Coderre E et al. (2006) NMDA receptors mediate calcium accumulation in myelin during chemical ischaemia. Nature 439:988–992. https://doi.org/10.1038/nature04474
28. Zheng K, Jensen TP, Rusakov DA (2018) Monitoring intracellular nanomolar calcium using fluorescence lifetime imaging. Nat Protoc 13:581–597. https://doi.org/10.1038/nprot.2017.154
29. Skoufou-Papoutsaki N, Adler S, D'Santos P et al (2023) Efficient genetic editing of human intestinal organoids using ribonucleoprotein-based CRISPR. Dis Model Mech 16(10):dmm050279
30. Gaebler A-M, Hennig A, Buczolich K et al (2020) Universal and efficient electroporation protocol for genetic engineering of gastrointestinal organoids. J Vis Exp 156:e60704. https://doi.org/10.3791/60704
31. Fujii M, Matano M, Nanki K, Sato T (2015) Efficient genetic engineering of human intestinal organoids using electroporation. Nat Protoc 10:1474–1485. https://doi.org/10.1038/nprot.2015.088
32. Okkelman IA, Zhou H, Borisov SM, et al. (2025) Fluorescence lifetime imaging microscopy (FLIM) visualizes internalization and biological impact of nanoplastics in live intestinal organoids by Fluorescence Lifetime Imaging Microscopy (FLIM). Light Sci Appl 14:272. https://doi.org/10.1038/s41377-025-01949-0
33. Debruyne AC, Okkelman IA, Heymans N et al. (2024) Live microscopy of multicellular spheroids with the multimodal near-infrared nanoparticles reveals differences in oxygenation gradients. ACS Nano 18:12168–12186. https://doi.org/10.1021/acsnano.3c12539
34. Kakni P, López-Iglesias C, Truckenmüller R et al (2022) Reversing epithelial polarity in pluripotent stem cell-derived intestinal organoids. Front Bioeng Biotechnol 10:879024. https://doi.org/10.3389/fbioe.2022.879024
35. van der Linden, F.H., Mahlandt, E.K., Arts, J.J.G. et al. A turquoise fluorescence lifetime-based biosensor for quantitative imaging of intracellular calcium. Nat Commun 12, 7159 (2021).https://doi.org/10.1038/s41467-021-27249-w

Check for updates

Chapter 18

Methods for Processing Stem Cell-Derived Organoids for Histological and Immunofluorescence Staining

Samna Sagadevan and Paloma Ordóñez-Morán

Abstract

Intestinal stem cell (ISC)-derived organoids provide a physiologically relevant 3D culture system to model intestinal biology, regeneration, and disease mechanisms. Their complex architecture, recapitulating crypt structures, poses challenges for downstream histological and immunohistochemical analyses, necessitating optimized processing, and staining protocols. Here, we describe robust workflows for the fixation, embedding, sectioning, and staining of ISC-derived organoids. Each method is detailed stepwise, with emphasis on preserving morphology and antigenicity. We discuss reagent choices, critical technical considerations, and troubleshooting tips to maximize data reproducibility and quality. We have integrated recent advances and best practices, serving as a comprehensive guide for researchers working with intestinal organoids.

Keywords Stem cell-derived organoids, Paraffin embedding, Cryosectioning, Immunohistochemistry, Histology, Fixation, Staining

1 Introduction

The intestinal epithelium undergoes rapid turnover driven by intestinal stem cells (ISCs) located at the crypt base [1]. Understanding ISC biology is fundamental for elucidating intestinal homeostasis, regeneration, and disease pathogenesis, such as inflammatory bowel disease (IBD) and colorectal cancer. The advent of 3D ISC-derived organoid cultures has transformed in vitro modeling by faithfully recapitulating crypt-villus structures, cell diversity, and stem cell dynamics in a physiologically relevant microenvironment [2, 3].

Organoids enable functional studies of stem cell behavior, tissue regeneration, drug responses, and niche interactions [4, 5]. However, their complex 3D architecture requires meticulous processing to maintain morphology and antigenicity for histological and immunohistochemical analyses. The choice between paraffin embedding and cryosectioning depends on experimental goals, with paraffin embedding offering high-resolution morphology and

Paloma Ordóñez-Morán (ed.), *Intestinal Stem Cells*, Methods in Molecular Biology, vol. 3059,
https://doi.org/10.1007/978-1-0716-5412-5_18,

long-term storage, while cryosectioning preserves antigenicity better for immunostaining.

Recent advances in bioengineering and matrix scaffolding (e.g., HistoGel™) have improved organoid culture and handling. Nonetheless, robust protocols for fixation, embedding, sectioning, and staining remain essential to ensure reproducibility. Here, we present detailed stepwise protocols for processing ISC-derived organoids as cryosectioning supplemented with practical notes, reagent details, and troubleshooting guidance.

2 Materials

The following reagents, solutions, and antibodies (Table 1) and equipment (Table 2) were used for organoid processing, fixation, embedding, sectioning, and staining.

Table 1
Reagents and solutions

Reagent/solution	Purpose	Notes/source
HistoGel™	Embedding matrix for 3D support	HistoGel™ solidifies at 4°C
4% Paraformaldehyde (PFA)	Fixation	Higher quality/stability rather than regular histology grade preferred
25% Sucrose	Cryoprotection	Filter-sterilized; prevents ice crystal formation in cryosectioning
OCT (Optimal Cutting Temperature) Compound	Embedding medium for cryosectioning	Use fresh aliquots
Ethanol (graded series: 70%, 80%, 95%, 100%)	Dehydration	Molecular biology grade
Xylene	Clearing agent	Use under fume hood; toxic and flammable
PBS (pH 7.4)	Buffer	Sterile; for washing and dilutions
BSA (1.5% in PBS)	Blocking solution	Reduces nonspecific antibody binding
Primary antibodies	Target detection	Example: anti-Vimentin
Secondary antibodies	Detection	Fluorophore- or HRP-conjugated; species-specific
Citrate buffer (pH 6.0)	Antigen retrieval buffer	For heat-induced epitope retrieval (HIER) in paraffin sections
Fluoromount-G™ Mounting Medium	Mounting medium	Preserves fluorescence

Table 2
Equipment

Equipment	Purpose	Notes/source
Centrifuge	Pellet organoids	Tabletop, refrigerated recommended
Microcentrifuge tubes	Sample handling	RNase/DNase-free recommended
Paraffin embedding station	Paraffin infiltration and embedding	Heated, programmable
Cryostat	Cryosectioning	Adjustable thickness, maintain at −20°C
Microtome	Paraffin sectioning	Produces thin (5 μm) sections
Barrier pen (PAP pen)	Hydrophobic barrier on slides	Prevents reagent spread during staining
Glass slides	Mounting sections	For example: Superfrost Plus slides

3 Methods

3.1 Preparation of Intestinal Stem Cell-Derived Organoids

1. Gently harvest organoids from Matrigel domes by mechanical disruption using a P1000 pipette tip (*see* **Note 1**).
2. Transfer organoid suspension into a microcentrifuge tube.
3. Centrifuge at 300 × *g* for 5 min at 4°C to pellet organoids.
4. Carefully aspirate supernatant and wash pellet twice with cold PBS.

3.2 Paraffin Embedding and Sectioning

3.2.1 Paraffin Embedding (Fig. 1)

1. Resuspend organoid pellet in prewarmed HistoGel™ (~50 μL), transfer to embedding mold (*see* **Note 2**).
2. Place mold on ice for 5 min to solidify gel.
3. Fix embedded organoids in 4% PFA at room temperature for up to 12 h/overnight.
4. Dehydrate samples through graded ethanol series: 70%, 80%, 95%, 100% (1 h each).
5. Clear samples in xylene (2 × 30 min) under fume hood.
6. Infiltrate with molten paraffin wax at 60°C for 1 h, repeat twice.
7. Embed organoids in paraffin blocks.

3.2.2 Paraffin Sectioning

1. Cut paraffin sections at 4 μm thickness using a microtome.
2. Float sections on a 40°C water bath and mount on Superfrost Plus slides.

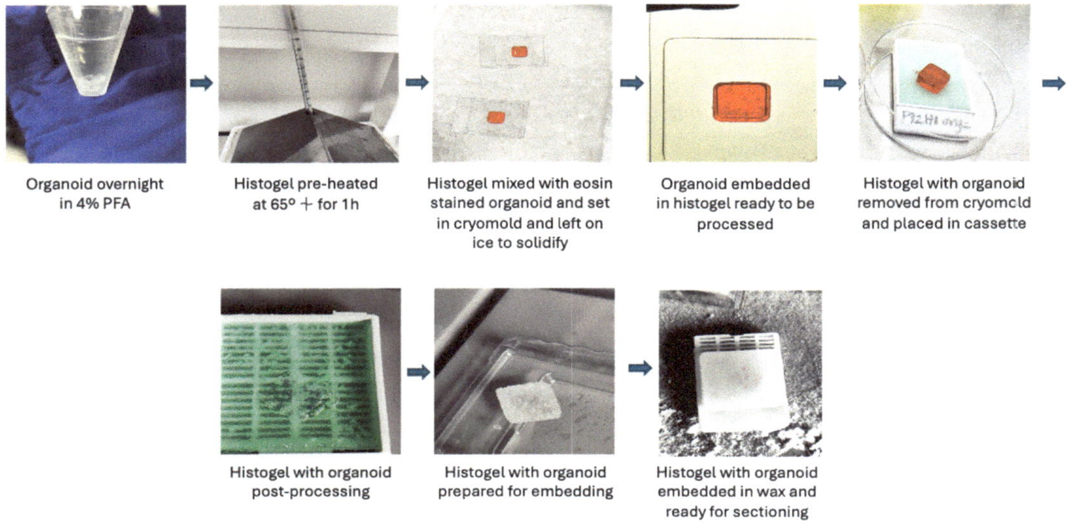

Fig. 1 A step-by-step schematic illustrates the preparation, casting, and processing of organoids for histological embedding. Organoids are first collected and pelleted, followed by careful removal of excess medium. The prewarmed HistoGel™ is then added, mixed, and dispensed onto chilled mold to solidify. Once polymerized, individual gel blocks containing organoids are trimmed and transferred to labeled tissue cassettes. The gel blocks undergo fixation, dehydration, clearing, and finally paraffin infiltration. The completed paraffin-embedded block is shown ready for downstream histological or immunohistochemical analyses

3. Dry slides overnight at room temperature or for 1 h at 60°C.
4. It can be stored at room temperature for long-term storage.

3.3 Cryosectioning

3.3.1 Fixation and Cryoprotection

1. Fix organoids in 4% PFA on ice for 1 h.
2. Wash twice with PBS.
3. Incubate organoids in 25% sucrose solution overnight at 4°C for cryoprotection.

3.3.2 Embedding and Sectioning (*Fig. 2*)

1. Embed organoids in OCT compound in cryomolds.
2. Freeze rapidly by placing molds on dry ice (*see* **Note 3**).
3. Store blocks at −80°C until sectioning.
4. Cut 5–10 μm sections on a cryostat set at −20°C.
5. Mount sections onto Superfrost Plus slides and air dry for 30 min at room temperature.

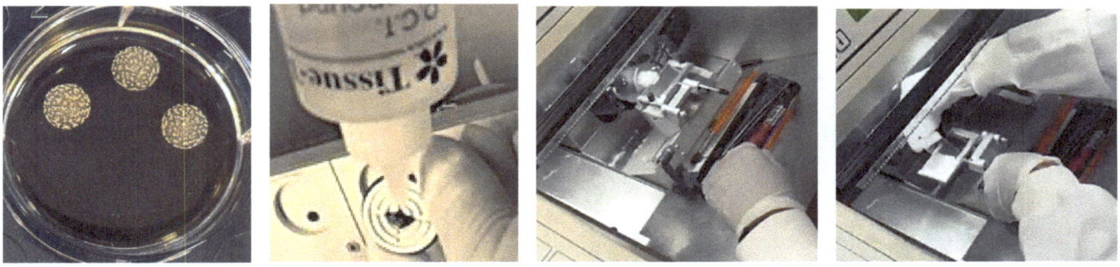

Wholemount gel fixed in 4% PFA

Embedding step: optimal cutting temperature (OCT) compound added to the cryostat chuck before and after placing the fixed wholemount gel.

Sectioning of OCT-embedded gels using a cryostat.

Collection of gel sections onto microscope slides, followed by storage at –80 °C until further use.

Fig. 2 Workflow for embedding organoids in OCT compound for cryosectioning. This figure illustrates the steps involved in preparing organoids for frozen sectioning. OCT is dispensed into a cryomold, and organoids are carefully transferred into the OCT to ensure proper positioning. The mold is then placed on the precooled stage of a cryostat to rapidly freeze the OCT and stabilize the sample. Once fully solidified, the embedded block is secured in the cryostat chuck, trimmed, and sectioned to generate thin cryosections suitable for downstream immunofluorescence or molecular analyses

3.4 Immunohistochemistry and Immunofluorescence Staining (Applicable for Both Paraffin and Cryosections) (see Note 4) (Fig. 3)

1. Deparaffinize paraffin sections with xylene and rehydrate through graded ethanol series and PBS (not required for cryosections).
2. Perform antigen retrieval in citrate buffer (pH 6.0) at 95°C for 20 min (paraffin sections only).
3. Block sections with 1% BSA in PBS for 1 h at room temperature.
4. Incubate with primary antibodies (e.g., anti-LGR5, anti-OLFM4) diluted in blocking buffer overnight at 4°C (*see* **Note 5**).
5. Wash sections 3 × 5 min with PBS.
6. Incubate with fluorophore- or HRP-conjugated secondary antibodies for 1 h at room temperature in the dark.
7. Wash 3 × 5 min with PBS.
8. Mount with Fluoromount-G™ and cover with coverslips.
9. Image using fluorescence or brightfield microscopy as appropriate.

4 Notes

1. Minimize mechanical stress to preserve crypt architecture; excessive pipetting or harsh centrifugation may disrupt morphology.
2. If organoids are small and require improved visualization, add a drop of eosin to the sample before embedding in HistoGel™.
3. Rapid freezing is essential to prevent ice crystal formation and preserve morphology and antigenicity.

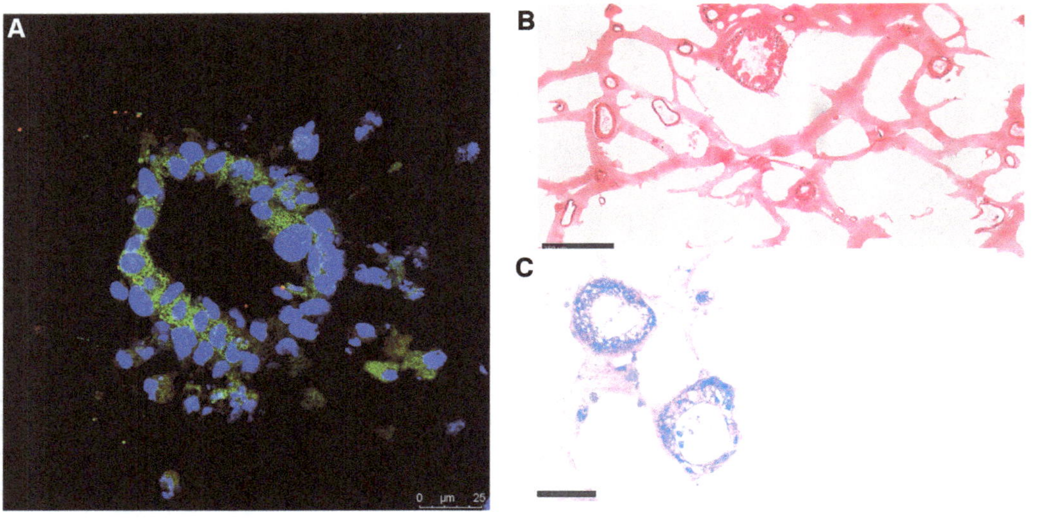

Fig. 3 Histological and immunofluorescence staining of organoids sections from paraffin- and OCT-embedded samples. (**A**) Immunofluorescence staining of paraffin-processed organoid sections cut on a microtome, showing expression of Vimentin (green) with nuclear counterstaining by DAPI (blue). (**B**) Hematoxylin and eosin (H&E) staining of cryosectioned OCT-embedded organoids, illustrating general morphology and architecture in Matrigel. (**C**) Alcian blue staining of OCT-embedded cryosections, highlighting mucin-producing cells within the organoid structures

4. Processing time: paraffin embedding requires ~2 days (including fixation and dehydration); cryosectioning can be completed in ~1 day.
5. Validate antibodies for specificity and optimize concentrations for each new batch.

Acknowledgments

This work was supported by MRC AIM iCASE Grant Ref: MR/W007002/1 and by MRC NIRG Grant Ref: MR/W01825X/1.

References

1. Zhao Q, Guan J, Wang X (2020) Intestinal stem cells and intestinal organoids. J Genet Genomics 47(6):289–299.
2. Tian C, Yang MF, Xu H et al. (2023) Stem cell-derived intestinal organoids: a novel modality for IBD. Cell Death Discov 9:310
3. Takahashi T, Fujishima K, Kengaku M (2021) Modeling intestinal stem cell function with organoids. Int J Mol Sci 22(19):10641
4. Gjorevski N, Nikolaev M, Brown T et al. (2022) Tissue geometry drives deterministic organoid patterning. Science. 375(6577):eabg3137
5. Takahashi T, Takase Y (2025) The intestinal stem cell niche: generation and utilization of intestinal organoids. Organoids 5(1):e00001

Part III

Stem Cells and Colon Cancer: In Vitro and In Vivo Models

Check for updates

Chapter 19

Stemness in Colon Cancer

Roei Cohen-Almagor, Samna Sagadevan and Paloma Ordóñez-Morán

Abstract

Cell plasticity is a major driver of colorectal cancer (CRC), as tumors hijack intestinal stem cell regenerative programs to create a dynamic, bidirectional axis between complementary stem-like states. Cancer stem-like cells (CSCs) display hallmark stem cell properties, including self-renewal, multipotency, controlled differentiation, and quiescence, and are now recognized as key orchestrators of CRC biology. They can sustain tumor initiation and long-term growth, seed metastases, enable periods of dormancy, and promote resistance to chemotherapy. As a result, therapeutically dismantling CSC populations and their supportive niches offers a compelling strategy to improve treatment responses and ultimately patient survival.

Keywords Colon cancer, Cancer-like stem cells, LGR5, Small intestine, Stemness, Metastasis

1 Introduction

Colorectal cancer (CRC) is a carcinoma of the large intestine and/or rectum, with over a million deaths reported annually [1]. It is the third-most prevalent cancer and the second leading cause of cancer-related mortality worldwide, highlighting its significance as a global health concern [2]. The 5-year survival is heavily dependent on disease stage, with a 90% survival rate for stage I patients, drastically dropping to 10% for stage IV patients [3].

Improvements in the understanding of the disease have led to the development of new screening techniques and therapeutic options. Alternative treatments include immunotherapy and targeted therapies, for example, against the endothelial growth factor receptor (EGFR) and vascular endothelial growth factor receptor (VEGFR) signaling pathways; however, clinical benefit has remained limited [4]. Surgery, radiotherapy, and chemotherapy, such as 5-fluorouracil (5-FU), have remained the backbone of CRC treatment for decades. Yet, only a fraction of advanced CRC patients respond to the routinely used chemotherapeutic agents [5].

Paloma Ordóñez-Morán (ed.), *Intestinal Stem Cells*, Methods in Molecular Biology, vol. 3059,
https://doi.org/10.1007/978-1-0716-5412-5_19,

The intra- and inter-tumor heterogeneity of CRC proves a considerable obstacle in achieving durable treatment responses [6]. Therapies impart a selective pressure on these cancer cells, often facilitating the generation of various plastic populations. Phenotypically diverse cancer cells can boost tumorigenic potential, drug resistance, and metastasis [7]. A deeper understanding of this type of cells and their molecular mechanisms' role in the progression of CRC is necessary. In this chapter, we review the role of these plastic populations that we term cancer-stem-like cells in colorectal cancer.

2 Colonic Stem Cells

In the adult intestine, several stem and progenitor populations have been described; however, the classically defined stem cells are the crypt base columnar (CBC) cells, which reside at the crypt base and generate transit-amplifying progenitors that differentiate as they migrate toward the villus apex. CBCs are reportedly marked by the leucine-rich repeat-containing G-protein-coupled receptor 5 (LGR5) [8]. Upon binding of R-Spondins, this receptor potentiates canonical Wnt/β-Catenin signaling, a key pathway in maintaining the pool of LGR5+ intestinal stem cells and thus the homeostatic renewal of the intestinal epithelium [9]. Disrupting the balance between stem cell and differentiation programs, such as Wnt signaling, is a hallmark of CRC [10]. Indeed, mouse studies showed that aberrant Wnt signaling through targeted deletion of *Apc* in Lgr5+ intestinal stem cells resulted in the rapid formation of an adenoma, also suggesting these cancer cells as the cell-of-origin for tumors [11]. Further discussion of colonic healthy stem cells appears in Chap. 1: "Intestinal Stem Cells: Regulation, Plasticity, and Therapeutic Potential."

3 Cancer Stem-Like Cells

Fundamental research from the 1990s defined CRC by the sequential accumulation of somatic mutations in oncogenes and tumor suppressors, such as *APC*, *KRAS*, and *TP53*, which disrupt epithelial homeostasis and drive malignant transformation [12]. Hereditary predispositions like familial adenomatous polyposis (FAP), caused by germline *APC* mutations, reinforce the importance of genetic drivers in CRC initiation [13]. Functional studies confirm that these oncogenic mutations confer a selective advantage and polarize epithelia toward a highly mitotic cancer stem cell-like (CSC) phenotype [14].

However, genetic similarity does not imply phenotypic uniformity; even tumors that share canonical driver mutations frequently display striking phenotypic heterogeneity and adaptability [15]. Recent work shows that dynamic nonmutational mechanisms complement canonical oncogenic programs to generate diverse tumor cell states [16]. This plasticity limits the utility of any singular molecular marker marker as a universal identifier of the CSC compartment. Indeed, commonly used surface markers (LGR5 [17], CD44 [18], CD133 [19], ALDH1 [20]) can enrich for these cells but none uniquely label the entire CSC population across tumors or assays. Instead, the CSC pool comprises multiple, coexisting states, and distinct marker panels preferentially detect particular functional subsets rather than a single homogenous population [21]. For these reasons, CSCs are more robustly defined by conserved behaviors, including long-term self-renewal, high tumor-initiating capacity and the ability to regenerate tumoral heterogeneity through multilineage differentiation [22, 23].

A major CSC phenotype in CRC is the proliferative cancer stem-like cell (proCSC) state. These cells are characterized by sustained cell-cycle activity, failure to differentiate into secretory and absorptive lineages, and upregulated oncogenic pathways, including Wnt/β-catenin, MYC, MAPK, and PI3K signaling [15]. High signaling entropy and mitogenic programs distinguish proCSCs [14]. Collectively, these features underpin rapid tumor expansion and recurrence.

Crucially, CSC identity in CRC has been shown to be dynamic. Experimental depletion of LGR5+ CSCs halts primary tumor growth in vivo models and patient-derived organoids, but isogenic LGR5− cells can dedifferentiate into LGR5+ populations to restore the stem cell pool and regenerate tumors [24, 25]. This plasticity mechanism mirrors normal intestinal repair. In healthy epithelium, slow-cycling stem cells at the crypts "+4 position" similarly replenish all intestinal lineages upon injury [26], suggesting CRC may co-opt this conserved regenerative program to sustain malignancy [27].

The dedifferentiation process triggered by the loss of LGR5+ CSCs was initially conceptualized as a passive model, wherein differentiated cells acquire stemness through exposure to Wnt and other signals in the vacated stem cell niche [28]. However, recent studies revealed that intestinal epithelia employ an active reprogramming response following injury [29], such as chemotherapy [30], irradiation [31], or parasitic infection [32]. This generates a revival stem cell population that can restore the entire epithelium [31]. Molecular profiling shows these cells bear transcriptional similarity to fetal intestinal stem cells [33].

The Hippo pathway is the most established inducer of the fetal stem cell programme and regulator of this intestinal regeneration programme, mediated through the activity of YAP/TAZ transcriptional co-activators [6]. YAP signaling has been shown to drive chromatin remodeling, promoting self-renewal and suppressing premature transition into functionally restricted adult intestinal stem cells [33]. This evolutionarily conserved regeneration programme is hijacked by CRC to balance clonogenic resilience with proliferative potential [15].

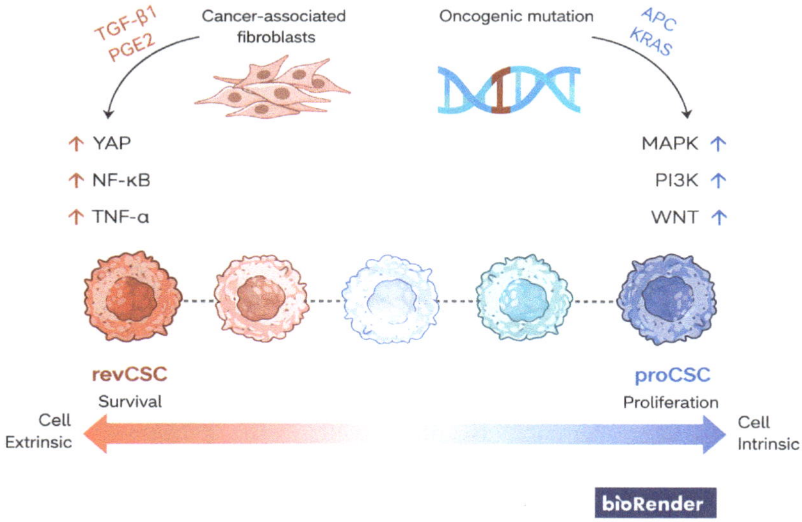

Fig. 1 Colonic cancer stem-like cell bidirectional axis from proCSC—revCSC phenotypes. proCSCs are often driven by cell-intrinsic oncogenic mutations such as *KRAS* and *APC* to activate mitogenic pathways such as MAPK. revCSCs are often driven by microenvironmental signals such as TGF-β1 and PGE2 to activate YAP signaling pathway (Image created using Biorender.com)

Within CRC tumors, revival cancer stem-like cells (revCSCs) coexist alongside LGR5+ proCSCs, forming a bidirectional plasticity axis regulated by environmental pressures and oncogenic mutations [34]. High-dimensional phenoscaping reveals that Wnt/β-catenin signaling, often driven by mutations, such as *APC* loss, leads to proCSC expansion and uncontrolled growth. Conversely, stromal stressors or therapy-induced damage activate Hippo and NF-κβ signaling, polarizing cells to enter a slow-cycling, stress-resistant revCSC state [14]. This dynamic interconversion (Fig. 1) underpins healthy mucosal regeneration but also fuels tumorigenesis, metastasis, and resistance to therapy.

4 Cancer Stem-Like Cells in Driving Tumorigenesis

The debate on the etiology of CRC centers two main hypotheses: "bottom-up," and "top-down." The former suggests that tumorigenesis initiates in CBCs or lineage-committed precursors and spreads up the crypts [35], and the latter suggests that differentiated cells reacquire stem-like characteristics, invading downwards and laterally [36]. However, both mechanisms converge on the principle: cancer arises when organismal fitness is dominated by cellular fitness. This is fueled by oncogenic mutations and hijacked regeneration programs to overcome homeostatic constraints.

CSCs exploit cell-autonomous and microenvironmental mechanisms to outcompete healthy stem cells in the crypt niche [37]. For example, recent studies have reported that *Apc*-mutant cells secrete Wnt antagonists to trigger differentiation in neighboring healthy stem cells, allowing the mutant cells to dominate the stem-cell niche and initiate tumor formation [38]. Indeed, aberrant Wnt signaling in sporadic CRC is a near-universal initiating step to enforce a stem-like state [10] while selecting for secondary mutations such as TGF-β/BMP pathway alterations to develop insensitivity to pro-differentiation signals [39]. This "double hit" drives proliferation and malignant transformation.

Stromal cells are also important in the initiation and acceleration of CRC tumorigenesis. Single-cell RNA sequencing analysis identified a rare population of fibroblasts, which secrete prostaglandin E2. This activates YAP signaling via the Ptger4 receptor to drive the expansion of revCSCs. *Apc-/Ptger4-* mouse models revealed attenuation of YAP target gene activation and revCSC expansion [40]. Human studies further corroborated these findings: immunostaining in colorectal tumor tissues revealed YAP nuclear localization but not in the neighboring normal areas of tissues across sporadic, hereditary, and inflammation-associated cases [40].

While YAP signaling in revCSCs is important for tumor initiation, their slow-cycling activity limits them from forming primary tumors [41]. Interestingly, YAP activation alone suppresses tumor growth by inhibiting Wnt/β-catenin signaling by sequestering Disheveled protein [42] and/or by binding to transcriptional co-repressors through Lats1/2 deletion and thus downregulate proCSCs [43]. These findings suggest that primary tumors maintain a hybrid population balancing the complementary roles of proCSCs with revCSCs acting as a dormant stem cell-like reservoir [15].

5 Cancer Stem-Like Cells in Metastasis

Metastasis is the leading cause of CRC-related mortality [44]. Conserved mutational patterns between matched suggest the main driving forces of metastasis may in part rely on non-genetic factors. Recent studies show that a pivotal step in the metastatic cascade is the disruption of epithelial integrity, enabling LGR5- cells to invade [45] and enter the circulation, adopting fetal-like gene signatures [46]. Following basal membrane invasion, metastatic progenitors upregulate the adhesion molecule L1CAM [47]. These cells employ L1CAM to spread on blood capillaries and interact with integrins to activate mechanotransduction transcription factors, YAP and myocardin-related transcription factor (MRTF). This L1CAM-YAP signaling axis licenses the growth of metastasis-initiating cells post infiltration and/or dormancy [48]. Whilst YAP is important in micrometastatic survival, its attenuation is required for cellular heterogeneity state to enable macrometastatic expansion [49]. Following primary tumor resection, high relapse cells (HRCs), an EMP1+ LGR5− subpopulation overlapping with revCSC signatures [14], were found to drive metastatic recurrence [4]. HRCs exhibit a hybrid, non-differentiated state marked by upregulated cell adhesion molecules, suggesting cluster-based migration and colonization to improve likelihood of survival. Post-dissemination, HRCs remain dormant until mircoenvironmental cues reactivate LGR5+ and/or L1CAM+ CSC programs, crucial for outgrowth into macrometastases (Fig. 2) [4, 47].

Recently, single-cell profiling of matched normal, primary, and metastatic tissue revealed PROX1 as a master regulator of stem cell plasticity, governing transitions between canonical intestinal states (LGR5+ stem cell-like, differentiated absorbative or secretory), and noncanonical states (fetal-like state, differentiated squamous-like or neuroendocrine-like) [16]. In the normal epithelium, PROX1 is induced in injury-responsive regenerative cells, where it enforces lineage fidelity by suppressing noncanonical differentiation and promoting redifferentiation into canonical intestinal lineages [50]. In CRC, however, this regulatory role becomes increasingly dysfunctional and context dependent [16]. Elevated PROX1 activity during early tumorigenesis reinforces ISC-like states [51]; however, at invasive fronts, metastatic cells escape PROX1-mediated repression, enter a

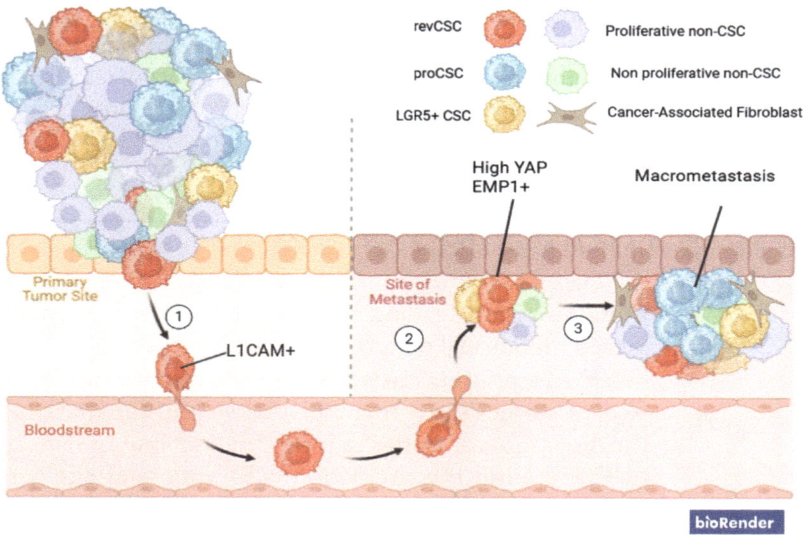

Fig. 2 Role of plastic populations in CRC metastasis. 1) Tumor expands to contain a heterogeneous mixture of stem cells. A locally invading tumor cell is reprogrammed into a revCSC state to break through the basement membrane and intravasate into the bloodstream. 2) The circulating revCSC extravasates and seeds into a new tissue site to form a micrometastasis. 3) Once the micrometasis survives in this new environment, adult stem cell programs are activated, and revCSCs can be polarized into proCSCs or differentiate to form a macrometastasis (Image created using Biorender.com)

fetal-like state, and adopt non-canonical states associated with poorer outcomes [16]. These metastatic-specific cell states underscore the importance of more widespread clinical cancer genomics studies of metastatic tissue to understand plasticity mechanisms.

6 Future Directions

Targeting CSCs is an attractive strategy to prevent drug resistance and relapse, but their plasticity and overlap with healthy stem cells have posed significant challenges. For instance, early attempts at targeting the cell surface receptors have yielded limited success [6]. More recently, however, a bispecific antibody, MCLA-158, co-engages EGFR and LGR5 to selectively deplete LGR5+ CSCs in *KRAS*-mutant organoids, with limited toxicity in healthy stem cells [52]. This agent is now entering early phase 1/2 clinical trials in solid tumors are currently underway (NCT03526835).

Another promising avenue for therapeutic intervention is to dismantle the revCSC survival programming. Small molecule inhibition of the YAP-TEAD interaction has been shown to repolarise revCSCs back into proCSCs and resensitise CAF-induced revCSCs to chemotherapy [53]. Conversely, revCSCs could be forced into the cell cycle by reactivating mitotic pathways using PI3K agonists [54]. Transient activation of these key mitotic pathways could potentially render these cells susceptible to cytotoxic agents [6]. Moreover, the relative scarcity of revCSCs in the healthy intestine makes the identification of revCSC-specific surface antigens attractive targets for CAR-T or antibody drug conjugate therapies with low off-target toxicity [15].

Going forward, successful therapies must exploit weaknesses that are genuinely cancer-specific. Examples include dependence on key epigenetic modifiers or rewired stemness circuits. These targeted interventions should be combined with standard of care treatments to reduce the likelihood of therapeutic escape. High-resolution time-course and single-cell profiling can further tease apart short-lived regenerative responses from persistent tumor stem cell states. This will ultimately provide a clearer roadmap for designing targeted, safer combination strategies.

7 Conclusion

Achieving durable drug responses in CRC will require combination regimens that simultaneously eliminate both proliferative and slow-cycling cancer stem-like cells, as well as the stromal niche supporting them. This goal necessitates the advancement of model systems to allow the deeper characterization of molecular pathways such as the switches governing the revCSC-proCSC bidirectional axis in individual patients and the discovery of CSC-specific markers. Improving the humanization of in vivo models through humanized patient-derived xenograft models [55] and the reproducibility of co-culture models will be indispensable for investigating tumor-stroma crosstalk in patients [56]. The development of more user-friendly "organ-on-chip" systems is another priority; implementing Artificial Intelligence to standardize their readouts will provide critical insights into the biomechanical interfaces that influence CSC behavior and drug response [57]. Together, multi-omic profiling integrated with these standardized, human-relevant models promise more accurate CSC definition. This will facilitate the design of therapies that effectively target both CSCs and their supportive niches with lower adverse effects.

References

1. Barker N, van Es JH, Kuipers J, Kujala P, van den Born M, Cozijnsen M, Haegebarth A, Korving J, Begthel H, Peters PJ, Clevers H (2007) Identification of stem cells in small intestine and colon by marker gene Lgr5. Nature 449(7165):1003–1007. https://doi.org/10.1038/nature06196
2. Rawla P, Sunkara T, Barsouk A (2019) Epidemiology of colorectal cancer: incidence, mortality, survival, and risk factors. Prz Gastroenterol 14(2):89–103. https://doi.org/10.5114/pg.2018.81072
3. Surveillance, Epidemiology, and End Results (SEER) Program. Cancer stat facts: colorectal cancer. National Cancer Institute. https://seer.cancer.gov/statfacts/html/colorect.html. accessed 2025
4. Cañellas-Socias A, Sancho E, Batlle E (2024) Mechanisms of metastatic colorectal cancer. Nat Rev Gastroenterol Hepatol 21(9):609–625. https://doi.org/10.1038/s41575-024-00934-z
5. Schütte M, Risch T, Abdavi-Azar N, Boehnke K, Schumacher D, Keil M, Yildiriman R, Jandrasits C, Borodina T, Amstislavskiy V, Worth CL, Schweiger C, Liebs S, Lange M, Warnatz H-J, Butcher LM, Barrett JE, Sultan M, Wierling C, Yaspo M-L (2017) Molecular dissection of colorectal cancer in pre-clinical models identifies biomarkers predicting sensitivity to EGFR inhibitors. Nat Commun 8(1):14262. https://doi.org/10.1038/ncomms14262
6. Fey SK, Vaquero-Siguero N, Jackstadt R (2024) Dark force rising: reawakening and targeting of fetal-like stem cells in colorectal cancer. Cell Rep 43(6):114270. https://doi.org/10.1016/j.celrep.2024.114270
7. Brockmueller A, Buhrmann C, Moravejolahkami AR, Shakibaei M (2024) Resveratrol and p53: how are they involved in CRC plasticity and apoptosis. J Adv Res 66:181–195. https://doi.org/10.1016/j.jare.2024.01.005
8. Barker N, van Es JH, Kuipers J, Kujala P, van den Born M, Cozijnsen M, Haegebarth A, Korving J, Begthel H, Peters PJ, Clevers H (2007) Identification of stem cells in small intestine and colon by marker gene Lgr5. Nature 449(7165):1003–1007. https://doi.org/10.1038/nature06196
9. Sphyris N, Hodder MC, Sansom OJ (2021) Subversion of niche-signalling pathways in colorectal cancer: what makes and breaks the intestinal stem cell. Cancers 13(5):1000. https://doi.org/10.3390/cancers13051000
10. Liang X, Duronio GN, Yang Y, Bala P, Hebbar P, Spisak S, Sahgal P, Singh H, Zhang Y, Xie Y, Cejas P, Long HW, Bass AJ, Sethi NS (2022a) An enhancer-driven stem cell-like program mediated by SOX9 blocks intestinal differentiation in colorectal cancer. Gastroenterology 162(1):209–222. https://doi.org/10.1053/j.gastro.2021.09.044
11. Barker N, Ridgway RA, van Es JH, van de Wetering M, Begthel H, van den Born M, Danenberg E, Clarke AR, Sansom OJ, Clevers H (2009) Crypt stem cells as the cells-of-origin of intestinal cancer. Nature 457(7229):608–611. https://doi.org/10.1038/nature07602
12. Fearon ER, Vogelstein B (1990) A genetic model for colorectal tumorigenesis. Cell 61(5):759–767. https://doi.org/10.1016/0092-8674(90)90186-I
13. Bodmer WF, Bailey CJ, Bodmer J, Bussey HJR, Ellis A, Gorman P, Lucibello FC, Murday VA, Rider SH, Scambler P, Sheer D, Solomon E, Spurr NK (1987) Localization of the gene for familial adenomatous polyposis on chromosome 5.

Nature 328(6131):614–616. https://doi.org/10.1038/328614a0
14. Qin X, Cardoso Rodriguez F, Sufi J, Vlckova P, Claus J, Tape CJ (2023) An oncogenic phenoscape of colonic stem cell polarization. Cell 186(25):5554–5568.e18. https://doi.org/10.1016/j.cell.2023.11.004
15. Tape CJ (2024) Plastic persisters: revival stem cells in colorectal cancer. Trends Cancer 10(3):185–195. https://doi.org/10.1016/j.trecan.2023.11.003
16. Moorman A, Benitez EK, Cambulli F, Jiang Q, Mahmoud A, Lumish M, Hartner S, Balkaran S, Bermeo J, Asawa S, Firat C, Saxena A, Wu F, Luthra A, Burdziak C, Xie Y, Sgambati V, Luckett K, Li Y, Ganesh K (2025) Progressive plasticity during colorectal cancer metastasis. Nature 637(8047):947–954. https://doi.org/10.1038/s41586-024-08150-0
17. Beumer J, Clevers H (2016) Regulation and plasticity of intestinal stem cells during homeostasis and regeneration. Development 143(20):3639–3649. https://doi.org/10.1242/dev.133132
18. Lee SY, Kim KA, Kim CH, Kim YJ, Lee J-H, Kim HR (2017) CD44-shRNA recombinant adenovirus inhibits cell proliferation, invasion, and migration, and promotes apoptosis in HCT116 colon cancer cells. Int J Oncol 50(1):329–336. https://doi.org/10.3892/ijo.2016.3801
19. Kazama S, Kishikawa J, Kiyomatsu T, Kawai K, Nozawa H, Ishihara S, Watanabe T (2018) Expression of the stem cell marker CD133 is related to tumor development in colorectal carcinogenesis. Asian J Surg 41(3):274–278. https://doi.org/10.1016/j.asjsur.2016.12.002
20. Li H-J, Reinhardt F, Herschman HR, Weinberg RA (2012) Cancer-Stimulated Mesenchymal stem cells create a carcinoma stem cell niche via prostaglandin E2 signaling. Cancer Discovery 2(9):840–855. https://doi.org/10.1158/2159-8290.CD-12-0101
21. Wang H, Gong P, Chen T, Gao S, Wu Z, Wang X, Li J, Marjani SL, Costa J, Weissman SM, Qi F, Pan X, Liu L (2021) Colorectal cancer stem cell states uncovered by simultaneous single-cell analysis of transcriptome and telomeres. Adv Sci 8(8):2004320. https://doi.org/10.1002/advs.202004320
22. Batlle E, Clevers H (2017) Cancer stem cells revisited. Nat Med 23(10):1124–1134. https://doi.org/10.1038/nm.4409
23. Plaks V, Kong N, Werb Z (2015) The cancer stem cell niche: how essential is the niche in regulating stemness of tumor cells. Cell Stem Cell 16(3):225–238. https://doi.org/10.1016/j.stem.2015.02.015
24. de Sousa e Melo F, Kurtova AV, Harnoss JM, Kljavin N, Hoeck JD, Hung J, Anderson JE, Storm EE, Modrusan Z, Koeppen H, Dijkgraaf GJP, Piskol R, de Sauvage FJ (2017) A distinct role for Lgr5+ stem cells in primary and metastatic colon cancer. Nature 543(7647):676–680. https://doi.org/10.1038/nature21713
25. Shimokawa M, Ohta Y, Nishikori S, Matano M, Takano A, Fujii M, Date S, Sugimoto S, Kanai T, Sato T (2017) Visualization and targeting of LGR5+ human colon cancer stem cells. Nature 545(7653):187–192. https://doi.org/10.1038/nature22081
26. Tian H, Biehs B, Warming S, Leong KG, Rangell L, Klein OD, de Sauvage FJ (2011) A reserve stem cell population in small intestine renders Lgr5-positive cells dispensable. Nature 478(7368):255–259. https://doi.org/10.1038/nature10408
27. Asfaha S, Hayakawa Y, Muley A, Stokes S, Graham TA, Ericksen RE, Westphalen CB, von Burstin J, Mastracci TL, Worthley DL, Guha C, Quante M, Rustgi AK, Wang TC (2015) Krt19+/Lgr5− cells are radioresistant cancer-initiating stem cells in the colon and intestine. Cell Stem Cell 16(6):627–638. https://doi.org/10.1016/j.stem.2015.04.013
28. Tetteh PW, Basak O, Farin HF, Wiebrands K, Kretzschmar K, Begthel H, van den Born M, Korving J, de Sauvage F, van Es JH, van Oudenaarden A, Clevers H (2016) Replacement of lost Lgr5-positive stem cells through plasticity of their enterocyte-lineage daughters. Cell Stem Cell 18(2):203–213. https://doi.org/10.1016/j.stem.2016.01.001
29. Yui S, Azzolin L, Maimets M, Pedersen MT, Fordham RP, Hansen SL, Larsen HL, Guiu J, Alves MRP, Rundsten CF, Johansen JV, Li Y, Madsen CD, Nakamura T, Watanabe M, Nielsen OH, Schweiger PJ, Piccolo S, Jensen KB (2018) YAP/TAZ-dependent reprogramming of colonic epithelium links ECM remodeling to tissue regeneration. Cell Stem Cell 22(1):35–49.e7. https://doi.org/10.1016/j.stem.2017.11.001
30. Solé L, Lobo-Jarne T, Álvarez-Villanueva D, Alonso-Marañón J, Guillén Y, Guix M, Sangrador I, Rozalén C, Vert A, Barbachano A, Lop J, Salido M, Bellosillo B, García-Romero R, Garrido M, González J, Martínez-Iniesta M, López-Arribillaga E, Salazar R, Espinosa L (2022) p53 wild-type colorectal cancer cells that express a fetal gene signature are associated with metastasis and poor prognosis. Nat Commun 13(1):2866. https://doi.org/10.1038/s41467-022-30382-9
31. Ayyaz A, Kumar S, Sangiorgi B, Ghoshal B, Gosio J, Ouladan S, Fink M, Barutcu S, Trcka D, Shen J, Chan K, Wrana JL, Gregorieff A

(2019) Single-cell transcriptomes of the regenerating intestine reveal a revival stem cell. Nature 569(7754):121–125. https://doi.org/10.1038/s41586-019-1154-y
32. Nusse YM, Savage AK, Marangoni P, Rosendahl-Huber AKM, Landman TA, de Sauvage FJ, Locksley RM, Klein OD (2018) Parasitic helminths induce fetal-like reversion in the intestinal stem cell niche. Nature 559(7712):109–113. https://doi.org/10.1038/s41586-018-0257-1
33. Pikkupeura LM, Bressan RB, Guiu J, Chen Y, Maimets M, Mayer D, Schweiger PJ, Hansen SL, Maciag GJ, Larsen HL, Lõhmussaar K, Pedersen MT, Teves JMY, Bornholdt J, Benes V, Sandelin A, Jensen KB (2023) Transcriptional and epigenomic profiling identifies YAP signaling as a key regulator of intestinal epithelium maturation. Sci Adv 9(28):eadf9460. https://doi.org/10.1126/sciadv.adf9460
34. Gil Vazquez E, Nasreddin N, Valbuena GN, Mulholland EJ, Belnoue-Davis HL, Eggington HR, Schenck RO, Wouters VM, Wirapati P, Gilroy K, Lannagan TRM, Flanagan DJ, Najumudeen AK, Omwenga S, McCorry AMB, Easton A, Koelzer VH, East JE, Morton D, Leedham SJ (2022) Dynamic and adaptive cancer stem cell population admixture in colorectal neoplasia. Cell Stem Cell 29(8):1213–1228.e8. https://doi.org/10.1016/j.stem.2022.07.008
35. Preston SL, Wong W-M, Chan AO-O, Poulsom R, Jeffery R, Goodlad RA, Mandir N, Elia G, Novelli M, Bodmer WF, Tomlinson IP, Wright NA (2003) Bottom-up histogenesis of colorectal adenomas: origin in the monocryptal adenoma and initial expansion by crypt fission. Cancer Res 63(13):3819–3825
36. Shih I-M, Wang T-L, Traverso G, Romans K, Hamilton SR, Ben-Sasson S, Kinzler KW, Vogelstein B (2001) Top-down morphogenesis of colorectal tumors. Proc Natl Acad Sci U S A 98(5):2640–2645. https://doi.org/10.1073/pnas.051629398
37. Che J, Yu S (2023) Ecological niches for colorectal cancer stem cell survival and thrival. Front Oncol 13:1135364. https://doi.org/10.3389/fonc.2023.1135364
38. van Neerven SM, de Groot NE, Nijman LE, Scicluna BP, van Driel MS, Lecca MC, Warmerdam DO, Kakkar V, Moreno LF, Vieira Braga FA, Sanches DR, Ramesh P, ten Hoorn S, Aelvoet AS, van Boxel MF, Koens L, Krawczyk PM, Koster J, Dekker E, Vermeulen L (2021) Apc-mutant cells act as supercompetitors in intestinal tumour initiation. Nature 594(7863):436–441. https://doi.org/10.1038/s41586-021-03558-4
39. Jung B, Staudacher JJ, Beauchamp D (2017) Transforming growth factor β superfamily signaling in development of colorectal cancer. Gastroenterology 152(1):36–52. https://doi.org/10.1053/j.gastro.2016.10.015
40. Roulis M, Kaklamanos A, Schernthanner M, Bielecki P, Zhao J, Kaffe E, Frommelt L-S, Qu R, Knapp MS, Henriques A, Chalkidi N, Koliaraki V, Jiao J, Brewer JR, Bacher M, Blackburn HN, Zhao X, Breyer RM, Aidinis V, Flavell RA (2020) Paracrine orchestration of intestinal tumorigenesis by a mesenchymal niche. Nature 580(7804):524–529. https://doi.org/10.1038/s41586-020-2166-3
41. Cheung P, Xiol J, Dill MT, Yuan W-C, Panero R, Roper J, Osorio FG, Maglic D, Li Q, Gurung B, Calogero RA, Yilmaz ÖH, Mao J, Camargo FD (2020) Regenerative reprogramming of the intestinal stem cell state via hippo signaling suppresses metastatic colorectal cancer. Cell Stem Cell 27(4):590–604.e9. https://doi.org/10.1016/j.stem.2020.07.003
42. Barry ER, Morikawa T, Butler BL, Shrestha K, de la Rosa R, Yan KS, Fuchs CS, Magness ST, Smits R, Ogino S, Kuo CJ, Camargo FD (2013) Restriction of intestinal stem cell expansion and the regenerative response by YAP. Nature 493(7430):106–110. https://doi.org/10.1038/nature11693
43. Li Q, Sun Y, Jarugumilli GK, Liu S, Dang K, Cotton JL, Xiol J, Chan PY, DeRan M, Ma L, Li R, Zhu LJ, Li JH, Leiter AB, Ip YT, Camargo FD, Luo X, Johnson RL, Wu X, Mao J (2020) Lats1/2 sustain intestinal stem cells and Wnt activation through TEAD-dependent and independent transcription. Cell Stem Cell 26(5):675–692.e8. https://doi.org/10.1016/j.stem.2020.03.002
44. Cheng X, Zhou J, Chen Y, Zhao Y, Zheng H, Wang Q, Li X, Jiang S (2024) Patterns and trends of mortality from metastatic colorectal cancer in Shanghai, China from 2005 to 2021: a population-based retrospective analysis. J Cancer Res Clin Oncol 150(2):68. https://doi.org/10.1007/s00432-023-05518-z
45. Fumagalli A, Oost KC, Kester L, Morgner J, Bornes L, Bruens L, Spaargaren L, Azkanaz M, Schelfhorst T, Beerling E, Heinz MC, Postrach D, Seinstra D, Sieuwerts AM, Martens JWM, van der Elst S, van Baalen M, Bhowmick D, Vrisekoop N, van Rheenen J (2020) Plasticity of Lgr5-negative cancer cells drives metastasis in colorectal cancer. Cell Stem Cell 26(4):569–578.e7. https://doi.org/10.1016/j.stem.2020.02.008
46. Mzoughi S, Schwarz M, Wang X, Demircioglu D, Ulukaya G, Mohammed K, Zorgati H, Torre D, Tomalin LE, Di Tullio F, Company C, Dramaretska Y, Leushacke M, Giotti B, Lannagan TR, Lozano-Ojalvo D, Karras P, Vermeulen PB, Hasson D, Guccione E (2025)

Oncofetal reprogramming drives phenotypic plasticity in WNT-dependent colorectal cancer. Nat Genet 57(2):402–412. https://doi.org/10.1038/s41588-024-02058-1

47. Ganesh K, Basnet H, Kaygusuz Y, Laughney AM, He L, Sharma R, O'Rourke KP, Reuter VP, Huang Y-H, Turkekul M, Er EE, Masilionis I, Manova-Todorova K, Weiser MR, Saltz LB, Garcia-Aguilar J, Koche R, Lowe SW, Pe'er D, Massagué J (2020) L1CAM defines the regenerative origin of metastasis-initiating cells in colorectal cancer. Nat Cancer 1(1):28–45. https://doi.org/10.1038/s43018-019-0006-x
48. Er EE, Valiente M, Ganesh K, Zou Y, Agrawal S, Hu J, Griscom B, Rosenblum M, Boire A, Brogi E, Giancotti FG, Schachner M, Malladi S, Massagué J (2018) Pericyte-like spreading by disseminated cancer cells activates YAP and MRTF for metastatic colonization. Nat Cell Biol 20(8):966–978. https://doi.org/10.1038/s41556-018-0138-8
49. Heinz MC, Peters NA, Oost KC, Lindeboom RGH, van Voorthuijsen L, Fumagalli A, van der Net MC, de Medeiros G, Hageman JH, Verlaan-Klink I, Borel Rinkes IHM, Liberali P, Gloerich M, van Rheenen J, Vermeulen M, Kranenburg O, Snippert HJG (2022) Liver colonization by colorectal cancer metastases requires YAP-controlled plasticity at the micrometastatic stage. Cancer Res 82(10):1953–1968. https://doi.org/10.1158/0008-5472.CAN-21-0933
50. Yan KS, Gevaert O, Zheng GXY, Anchang B, Probert CS, Larkin KA, Davies PS, Cheng Z, Kaddis JS, Han A, Roelf K, Calderon RI, Cynn E, Hu X, Mandleywala K, Wilhelmy J, Grimes SM, Corney DC, Boutet SC, Kuo CJ (2017) Intestinal enteroendocrine lineage cells possess homeostatic and injury-inducible stem cell activity. Cell Stem Cell 21(1):78–90.e6. https://doi.org/10.1016/j.stem.2017.06.014
51. Petrova TV, Nykänen A, Norrmén C, Ivanov KI, Andersson LC, Haglund C, Puolakkainen P, Wempe F, von Melchner H, Gradwohl G, Vanharanta S, Aaltonen LA, Saharinen J, Gentile M, Clarke A, Taipale J, Oliver G, Alitalo K (2008) Transcription factor PROX1 induces colon cancer progression by promoting the transition from benign to highly dysplastic phenotype. Cancer Cell 13(5):407–419. https://doi.org/10.1016/j.ccr.2008.02.020
52. Herpers B, Eppink B, James MI, Cortina C, Cañellas-Socias A, Boj SF, Hernando-Momblona X, Glodzik D, Roovers RC, van de Wetering M, Bartelink-Clements C, Zondag-van der Zande V, Mateos JG, Yan K, Salinaro L, Basmeleh A, Fatrai S, Maussang D, Lammerts van Bueren JJ, Throsby M (2022) Functional patient-derived organoid screenings identify MCLA-158 as a therapeutic EGFR × LGR5 bispecific antibody with efficacy in epithelial tumors. Nat Cancer 3(4):418–436. https://doi.org/10.1038/s43018-022-00359-0
53. Ramos Zapatero M, Tong A, Opzoomer JW, O'Sullivan R, Cardoso Rodriguez F, Sufi J, Vlckova P, Nattress C, Qin X, Claus J, Hochhauser D, Krishnaswamy S, Tape CJ (2023) Trellis tree-based analysis reveals stromal regulation of patient-derived organoid drug responses. Cell 186(25):5606–5619.e24. https://doi.org/10.1016/j.cell.2023.11.005
54. Gong GQ, Bilanges B, Allsop B, Masson GR, Roberton V, Askwith T, Oxenford S, Madsen RR, Conduit SE, Bellini D, Fitzek M, Collier M, Najam O, He Z, Wahab B, McLaughlin SH, Chan AWE, Feierberg I, Madin A, Vanhaesebroeck B (2023) A small-molecule PI3Kα activator for cardioprotection and neuroregeneration. Nature 618(7963):159–168. https://doi.org/10.1038/s41586-023-05972-2
55. Kanikarla Marie P, Sorokin AV, Bitner LA, Aden R, Lam M, Manyam G, Woods MN, Anderson A, Capasso A, Fowlkes N, Overman MJ, Menter DG, Kopetz S (2022) Autologous humanized mouse models to study combination and single-agent immunotherapy for colorectal cancer patient-derived xenografts. Front Oncol 12:994333. https://doi.org/10.3389/fonc.2022.994333
56. Wang J, Tao X, Zhu J, Dai Z, Du Y, Xie Y, Chu X, Fu G, Lei Z (2025) Tumor organoid-immune co-culture models: exploring a new perspective of tumor immunity. Cell Death Discov 11(1):195. https://doi.org/10.1038/s41420-025-02407-x
57. Avula LR, Grodzinski P (2024) How organ-on-a-chip is advancing cancer research and oncology - a cancer hallmarks' perspective. Front Lab Chip Technol 3:1487377. https://doi.org/10.3389/frlct.2024.1487377

Chapter 20

Deep Insights Without Clearing: An Optimized Protocol for Patient-Derived Whole-Organoid Immunofluorescence Imaging

Laura Gull, Mireia Sueca-Comes and Ana Pires

Abstract

Here, we present an optimized whole-organoid immunofluorescence staining protocol that preserves morphology, achieves uniform antibody penetration, and enables deep-tissue imaging without the need for tissue clearing. Organoids are three-dimensional, self-organizing structures that replicate many of the cellular and architectural features of their tissue of origin, making them powerful preclinical models for studying development, disease, and therapeutic responses. Most organoid research is performed on formalin-fixed, paraffin-embedded (FFPE) samples, in which organoids are sectioned across the Z-planes. While this method facilitates conventional histological analysis, it limits spatially resolved visualization of protein expression and tissue organization. The inherent density and abundant extracellular matrix of organoids present substantial barriers to reagent penetration and imaging depth, further constraining the ability to capture their three-dimensional architecture in full. Our new workflow includes steps for fixation, permeabilization, and antibody incubation and has been validated on intestinal organoids using markers for different membrane-associated proteins. By providing high-quality staining through the full organoid depth without clearing, this protocol streamlines sample preparation, reduces processing artifacts, and allows more rapid, accurate 3D mapping of cell types and structures in organoid-based research.

Keywords Organoids, Whole-mount immunostaining, 3D cell culture, Deep-tissue imaging, Antibody penetration, No tissue clearing, Confocal microscopy

1 Introduction

Deriving from stem cells or tissue Progenitor, organoids self-organize into three-dimensional (3D) structures recapitulating key aspects of organ architecture, function, and cellular diversity. Their spatial organization and cellular heterogeneity that closely resemble in vivo tissues are a clear advantage over the classic monolayer cultures, making them invaluable tools in fields, such as developmental biology, oncology, neurobiology, and regenerative medicine [1, 2].

Over the past decade with the increasing use of organoids has created an equivalent demand for techniques that increase the

Paloma Ordóñez-Morán (ed.), *Intestinal Stem Cells*, Methods in Molecular Biology, vol. 3059,
https://doi.org/10.1007/978-1-0716-5412-5_20,

high-resolution analysis potential of their complex internal structures and phenotypic variation. Although transcriptomic and proteomic methods can provide molecular insight at the cell level, they lack spatial context. To overcome this gap, immunohistochemistry permits visualization of specific markers within the 3D architecture of the organoids, providing information on cell identity and tissue organization, in a spatial perspective [3].

However, immunostaining of whole organoids poses unique technical challenges, since the penetration of reagents and antibodies is restricted by the dense cell and extracellular structure, particularly in larger and/or more complex structures. Standard protocols optimized for thin tissue sections or 2D cultures are insufficient to obtain uniform staining throughout entire organoids. Additionally, poor preservation of morphology during processing, nonspecific background staining, and optical opacity impeding deep-tissue imaging are common complications observed in this type of assays [4, 5]. Consequently, careful optimization of steps like fixation, permeabilization, and antibody incubation is essential to achieve consistent and reliable immunolabeling in whole organoids.

Here, we describe an optimized protocol for immunofluorescence staining of whole organoids, which is suitable for different downstream imaging approaches, including confocal and light-sheet microscopy. This method is designed to ensure uniform antibody penetration and high signal-to-noise ratio, while preserving organoid's primary structure. Importantly, it achieves these outcomes without the need for tissue clearing, reducing processing time, avoiding potential artefacts from clearing agents, and maintaining native optical properties. Developed using intestinal models, this method can be adapted to different tissue types and marker panels. By addressing the specific challenges of 3D immunostaining, this method allows for accurate and reproducible visualization of tissue architecture and localization of specific targets within whole organoids.

2 Materials

2.1 Equipment and Consumables

1. P1000, P200, P20, and P10 pipettes and respective tips.
2. Centrifuge for Eppendorfs.
3. Shaker suitable for incubations in cool room.
4. 1.5 mL Eppendorfs.
5. Foil.
6. SuperFrost Plus slides.
7. Square coverslips.
8. Double-side sticky tape.

2.2 Solutions

1. 4% Paraformaldehyde (PFA) in PBS.
2. Cell Recovery Solution.
3. Hoechst 33342.
4. Prolong Gold Mounting Medium.

2.3 Buffers Composition

1. PBS-BSA (1% wt/vol): PBS 100 mL, BSA 1 g.
2. PBS-T (0.1% vol/vol): PBS 500 mL, Tween-20 0.5 mL.
3. PBS-TX (0.1% vol/vol): PBS 500 mL, Triton X-100 0.5 mL.
4. PBS-TX (0.5% vol/vol): PBS 100 mL, Triton X-100 0.5 mL.
5. Antibody Dilution Buffer (ADB): PBS-TX (0.1%) 100 mL, BSA 2 g.

3 Method

3.1 Organoid Recover

This step allows for disruption of the 3D matrix whilst preserving the organoid structure (see Fig. 1A for a visual reference of the starting point).

1. Using a P1000 pipette, remove the culture medium from the wells and add 1 mL of PBS without disrupting the matrix. Remove the PBS to clear the remaining culture medium.
2. With a P1000 pipette, add 1 mL of ice-cold cell recovery solution (from freezer) to each well and incubate on a shaker at 4°C for 30–60 min, depending on how long it takes for the domes to dissolve.

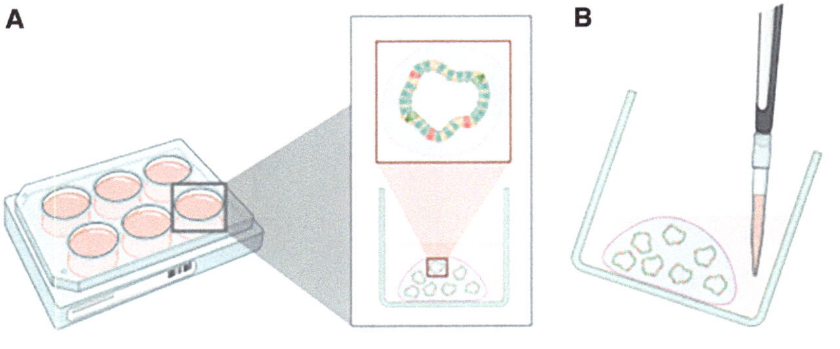

Fig. 1 Representation of organoids in cell culture plates. (**A**) Organoids presented in a 3D matrix dome inside the well. The dome will be submerged in cell culture medium (represented in pink), and each dome will contain multiple organoids. (**B**) To pipette contents in the well, tilt the plate in an angle, and place the pipette on a 45° angle, against the well wall. This will avoid any disruption of the 3D matrix or the organoids. Created with BioRender.com

3. Gently resuspend the contents of each well five to ten times by using a P1000 (Fig. 1B).
4. Transfer the organoids to a 1.5 mL Eppendorf tube.
5. Add 0.5 mL of ice-cold 1% PBS-BSA to the well, and transfer to the 1.5 mL Eppendorf tube to ensure all organoids are collected.
6. Centrifuge at 70×*g* for 3 min. Remove the supernatant by pipetting carefully and slowly without disrupting the pellet. The pellet should be tight without a visible layer of 3D matrix.

3.2 Fixation

This step preserves protein location and structure.

1. Gently resuspend the pellet of organoids in 1 mL of PFA (Fig. 2) (*see* **Notes 1 and 2**).
2. Incubate at 4°C for 2 h (halfway through the incubation period, gently resuspend the organoids).
3. Fill the Eppendorf tube up to 1.5 mL with cold (4°C) PBS-T, gently mix, incubate for 10 min at 4°C.
4. Centrifuge at 70×*g* for 5 min. Remove the supernatant by pipetting carefully and slowly without disrupting the pellet. Resuspend the pellet in 1 mL of cold PBS-T.

Pause Point: Organoids can be stored in PBS-T at 4°C for at least 4 weeks.

3.3 Permeabilization

This step is carried out to allow access of antibodies for intracellular staining.

1. Aspirate off the supernatant, paying careful attention to not disrupt the pellet. Resuspend the pellet in 1 mL of 0.1% PBS-TX and repeat centrifugation and aspiration.

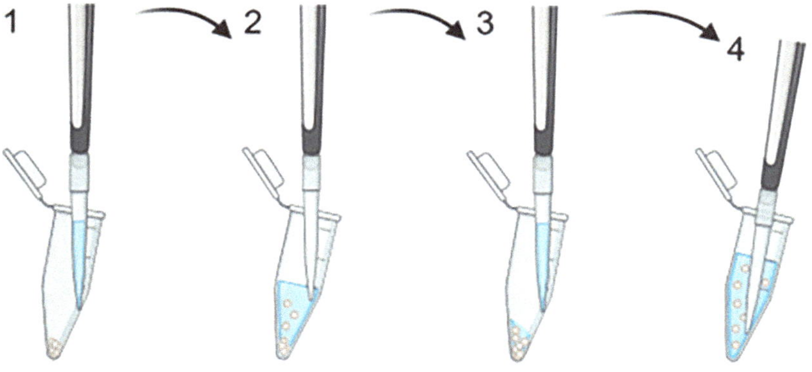

Fig. 2 Step by step on how to resuspend organoids in a pellet. Dispense the liquid slowly on the side of the tube. Organoids should visibly start to detach from the pellet. Then aspirate the liquid, avoiding touching the organoids, then dispense the liquid slowly again, next to the pellet. Created with BioRender.com

2. Resuspend each pellet in 1.5 mL 0.5% PBS-TX (*see* **Note 3**).
3. Place tubes on shaker, lying down horizontally, at room temperature (20°C–22°C) to permeabilize for 30 min (*see* **Note 4**).

3.4 Blocking

Blocking is performed to prevent nonspecific staining.

1. Centrifuge at 70×*g* for 5 min, aspirate off the supernatant, paying careful attention to not disrupt the pellet. Resuspend the pellet in 1 mL of 0.1% PBS-TX and repeat centrifugation and aspiration.
2. Resuspend pellets in lying down 1.5 mL of ADB.
3. Place on shaker, lying down, for 1 h at room temperature (20°C–22°C) to block.

<u>**Pause Point:**</u> Organoids can be left in ADB for blocking overnight at room temperature (12–16 h).

3.5 Primary Antibody Incubation

1. Add all primary antibodies needed, at the appropriate dilution, directly to the ADB.
2. Place on shaker, lying down, at 4°C (in cool room) overnight.

3.6 Secondary Antibody Incubation

1. Centrifuge at 70×*g* for 5 min, aspirate off the supernatant, paying careful attention to not disrupt the pellet. Resuspend the pellet in 1 mL of 0.1% PBS-TX and repeat centrifugation and aspiration.
2. Resuspend pellets in 1.5 mL of ADB.
3. Add relevant secondary antibodies at the appropriate dilution directly to the ADB.
4. Place on shaker, in the dark (wrapped in foil), at RT for 3 h.

3.7 Counterstain

1. Centrifuge at 70×*g* for 5 min, aspirate off the supernatant, paying careful attention to not disrupt the pellet. Resuspend the pellet in 1 mL of 0.1% PBS-TX and repeat centrifugation and aspiration.
2. Resuspend pellets in 1.5 mL 0.1% PBS-TX.
3. Add Hoechst at a 1:1000 dilution.
4. Place on shaker, in the dark (wrapped in foil), for 20 min.
5. Centrifuge at 70×*g* for 5 min, aspirate off the supernatant, paying careful attention to not disrupt the pellet. Resuspend the pellet in 1 mL of 0.1% PBS-TX and repeat centrifugation and aspiration.

3.8 Optional Phalloidin Stain

1. Resuspend pellets in 1.5 mL of 0.1% PBS-TX.
2. Add Phalloidin at a 1:200 dilution.
3. Place on shaker, in the dark (wrapped in foil), for 20 min.

4. Centrifuge at 70×*g* for 5 min, aspirate off the supernatant, paying careful attention to not disrupt the pellet. Resuspend the pellet in 1 mL of 0.1% PBS-TX with a P1000 and repeat centrifugation and aspiration.

3.9 Slide Preparation

(See Fig. 3 for guidance on slide mounting).

1. Remove as much of the supernatant as possible without disturbing the pellet.
2. Place a 1-cm-long piece of sticky tape, creating a middle area for sample deposition. Use one, two, or three layers, depending on the organoid size.
3. Resuspend organoids in 20 μL Prolong Gold Antifade per layer of tape and pipette the organoid suspension onto the slide, in between the tape layers.
4. Place a coverslip on top by placing the left side of the coverslip on the left stack of sticky tape and then slowly lowering the

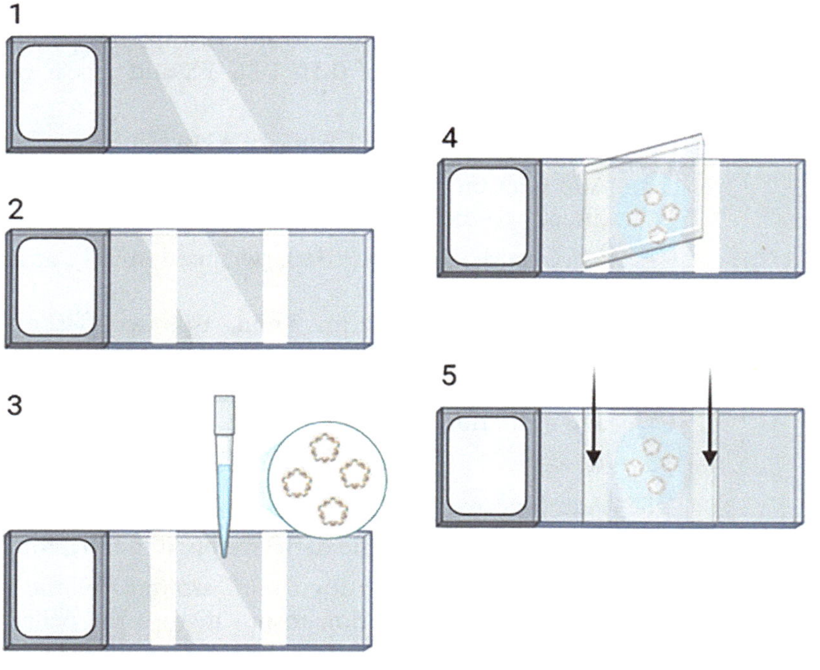

Fig. 3 Whole organoid slide preparation. **1**, label a clean slide; **2**, place sticky tape on slide to create an area for sample deposition. Place one to three layers of sticky tape on each side, depending on the size of the organoids; **3**, dispense the organoids resuspended in mountant in between the layers of tape; **4**, place coverslip on top of layers of sticky tape, starting from the left side, without applying pressure; **5**, let the liquid spread for 1 min and then gently apply pressure on the coverslip, on top of the sticky tape areas. Allow to cure and store slides. Created with BioRender.com

right side of the coverslip until it touches the right stack of sticky tape.

5. Wait for 1 min to allow for the liquid to spread out.
6. Gently apply pressure to both sides of the coverslip to firmly attach it to the double-sided sticky tape.
7. Allow slides to cure.
8. Slides are ready for imaging or storage. Slides can be stored indefinitely.

3.10 Validation of the Immunostaining Protocol

To assess the performance of the whole-organoid immunostaining protocol, we applied it to intact intestinal organoids and labeled three representative markers: phalloidin, tubulin, and E-cadherin (see Table 1 for full antibody details). These targets were selected to evaluate antibody penetration and specificity across the whole structure. Organoids varied in size and were processed according to the optimized fixation, permeabilization, and antibody incubation conditions described above.

Following immunostaining, samples were imaged in 3D using confocal microscopy. The protocol yielded bright, specific labelling throughout the entire organoid, with no detectable loss of fluorescence intensity from the outer layers to the central core (Fig. 4). Orthogonal views confirmed uniform penetration of primary and secondary antibodies. Staining patterns matched the expected distribution of each marker, indicating high specificity.

Morphological integrity was preserved in all samples, with no evidence of collapse, shrinkage, or surface distortion during staining or clearing. Together, these results demonstrate that the protocol allows robust and specific immunolabeling of whole organoids, producing image quality suitable for high-resolution volumetric analysis.

Table 1
Primary and secondary antibodies

Target	Host species	Supplier	Cat. number	Dilution
Phalloidin (AF568)	NA	ThermoFisher	A12380	1:200
Tubulin	Rabbit	Abcam	Ab52866	1:200
E-cadherin	Mouse	Agilent Technologies	M36121	1:1000
Go a-Mouse AF647	Goat	ThermoFisher	A32728	1:200
Go a-Rabbit AF790	Goat	ThermoFisher	A11369	1:200

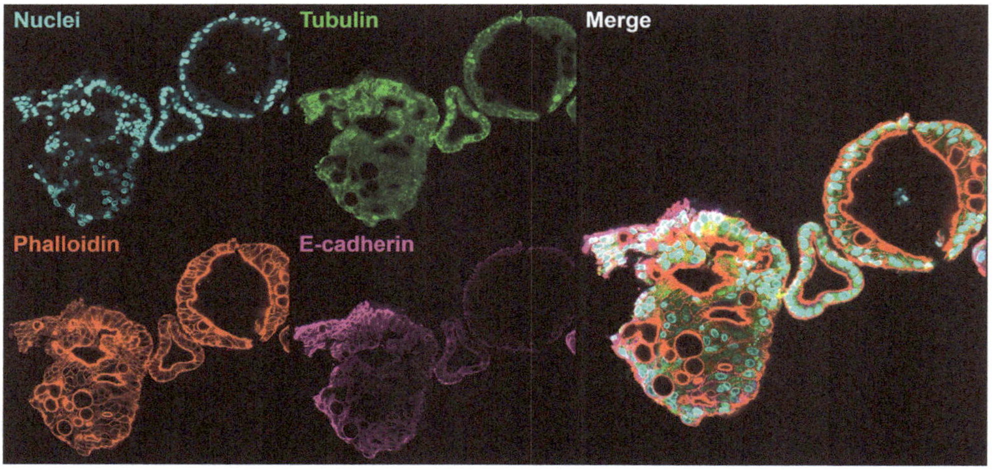

Fig. 4 A representative whole-mount confocal image of mouse intestinal organoid immunofluorescently labeled. A three-dimensional intestinal organoid imaged by whole-mount immunofluorescent staining and confocal microscopy. Nuclei (blue), tubulin (green), phalloidin (red), and E-cadherin (magenta)

4 Notes

1. From this point forward, to resuspend organoids in a pellet (see Fig. 2 for reference), dispense the liquid slowly on the side of the tube (Fig. 2-1,2). Organoids should visibly start to detach from the pellet (Fig. 2-2). Then aspirate the liquid avoiding touching the organoids (Fig. 2-3), then dispense the liquid slowly again, next to the pellet (Fig. 2-4).
2. This protocol was optimized using <100 μm thick organoids, and no shredding or disruption was observed. However, larger organoids might need to be handled with cut P1000 tips, so the opening is larger.
3. For longer incubations in shaker, tubes need to be completely full to guarantee that organoids are completely submerged in solution and do not dehydrate.
4. From this point forward, all steps requiring incubation of samples in Eppendorf tubes on a shaker, place the tube horizontally on the surface of the shaker. This is to increase the surface area between organoids and solution, to allow for a better mix and incorporation of reagents into the sample. If needed, secure the tube with tape.
5. If organoids are lost during the experiment, this may be because they stick to pipettes or tubes, stuck to 24-well plates, or were aspirated during washing. To avoid this, carefully precoat tubes and tips with 1% (wt/vol) PBS-BSA, use

low-adherence suspension plates, increase the time allowed for organoids to settle, and remove buffers more gently.

6. If organoids collapse during fixation, this can be due to suboptimal fixation conditions. To address this, try using a different fixative such as formalin or 2% (vol/vol) PFA-glutaraldehyde.
7. If organoids are no longer intact, they may have broken during mixing or transferring. In this case, using a P1000 tip that has been cut to make the opening larger can help reduce mechanical damage.
8. If high background staining is observed, this may be caused by incomplete dissociation of matrigel or by inadequate blocking and/or primary antibody incubation conditions. A longer fixation time, the use of cold reagents for fixation, or blocking followed by overnight incubation with primary antibodies at 4°C can improve the outcome.
9. If weak signal is obtained when using directly conjugated antibodies, this can be due to a low abundance of the target epitope. To improve detection, amplify the signal with a secondary antibody step
10. If the signal from a fluorescent reporter is weak, fixation time may have been too long or too short, or the reporter itself may have been weak from the start. Optimizing fixation time and performing immunolabelling of the fluorophore can help to amplify the signal
11. If weak signal is observed for immunolabelled nuclear proteins, this can result from inefficient antibody penetration into the nucleus. In this case, use ADB supplemented with 0.02% (wt/vol) SDS to enhance penetration and improve labelling
12. If organoids move while imaging, often seen as suboptimal results after image stitching, this may be due to suboptimal fixation of organoids between the slide and coverslip. Using fewer layers of sticky tape during slide preparation can help secure the organoids in place.
13. If organoids clump together, this may be because they are of a stickier nature. Ensuring organoids are properly mixed before transferring them to the slide can reduce clumping.

Acknowledgments

The authors would like to thank Magda Martínez-Espuga (University of Nottingham) and Paloma Ordóñez-Morán (University of Nottingham) for providing the organoids; and Robert Kasper (Evident) for imaging the slides, using the FluoView FV4000 Confocal Laser Scanning Microscope.

References

1. Clevers H (2016) Modeling development and disease with organoids. Cell 165(7):1586–1597
2. Lancaster MA, Knoblich JA (2014) Generation of cerebral organoids from human pluripotent stem cells. Nat Protoc 9(10):2329–2340
3. Dekkers JF et al (2019) High-resolution 3D imaging of fixed and cleared organoids. Nat Protoc 14(6):1756–1771
4. Schwarz N et al (2015) Immunofluorescence staining of whole mount zebrafish embryos and organoids. Cold Spring Harb Protoc 2015(8):pdb.prot083030
5. Renner H et al (2017) Self-organized developmental patterning and differentiation in cerebral organoids. EMBO J 36(10):1316–1329

Chapter 21

Comprehensive Isolation Strategy for Intestinal Epithelial and Fibroblast Cells from Healthy and Tumor Human Tissues

Samna Sagadevan and Paloma Ordóñez-Morán

Abstract

Organoids and fibroblasts derived from patient tissue serve as physiologically meaningful in vitro models to investigate tissue biology, diseases, and treatment responses. In this study, we present a robust protocol for the simultaneous isolation and long-term culture of patient-derived organoids (PDOs) and fibroblasts from both healthy and colorectal cancer samples. The workflow uses enzymatic dissociation followed by efficient separation into epithelial and stromal cell fractions. Organoids are embedded in Matrigel or Matrigel-Collagen I mixtures to support three-dimensional growth, whereas fibroblasts are maintained on conventional two-dimensional culture dishes. This dual-culture approach facilitates a broad range of downstream applications.

Keywords Extracellular matrix, Patient-derived organoids, Fibroblasts, Tumor microenvironment, Matrigel-Collagen I

1 Introduction

Patient-derived organoids (PDOs) are three-dimensional epithelial cultures that closely mimic the architecture, heterogeneity, and genetic landscape of the tissue of origin. PDO technology has been widely applied to cancer biology, regenerative medicine, and personalized medicine studies [1–3]. In contrast to traditional two-dimensional (2D) cell cultures, 3D organoid cultures better recapitulate the complex cell–cell and cell–matrix interactions, tissue architecture, and physiological functions of native tissues. While 2D cultures are easier to establish and manipulate, they lack the spatial organization and microenvironmental cues crucial for accurate modeling of tissue biology and drug responses.

Fibroblasts, including cancer-associated fibroblasts (CAFs), are key stromal cells within the tumor microenvironment, influencing tumor progression, immune responses, and extracellular matrix

Paloma Ordóñez-Morán (ed.), *Intestinal Stem Cells*, Methods in Molecular Biology, vol. 3059,
https://doi.org/10.1007/978-1-0716-5412-5_21,

(ECM) remodeling [4, 5]. The co-isolation and parallel culture of epithelial and stromal populations from the same patient sample enables the study of epithelial-stromal interactions under controlled conditions [6]. Liberase™, a blend of collagenases and neutral protease, combined with DNase I, provides efficient and gentle dissociation of human tissues, preserving cell viability and minimizing clumping from released DNA [5].

Matrigel is a solubilized basement membrane extract derived from the Engelbreth–Holm–Swarm mouse sarcoma, enriched in laminin, type IV collagen, entactin/nidogen, and heparan-sulfate proteoglycans. It provides a complex array of biochemical signals and growth factors that support cell attachment, survival, proliferation, and 3D tissue organization in organoid cultures [7]. Collagen I, the primary fibrillar collagen secreted by fibroblasts, imparts mechanical strength and influences the spatial organization of the extracellular matrix, thereby playing a key role in cell migration, morphogenesis, and tissue homeostasis [8]. When used together, Matrigel and Collagen I more closely mimic the multifaceted composition and mechanical characteristics of the native tissue microenvironment, enhancing the physiological relevance of in vitro models and supporting diverse cellular behaviors, such as proliferation, polarization, and tissue-specific differentiation. This combinatorial matrix approach enables more faithful recapitulation of in vivo conditions for both healthy and tumor-derived organoids, increasing the robustness and versatility of downstream applications including drug testing, mechanistic studies, and regenerative medicine. This chapter outlines a reproducible dual-culture protocol for PDOs and fibroblasts, adaptable to both normal and malignant tissue sources (Fig. 1).

2 Materials

2.1 Reagents

Reagents are listed in Tables 1–3.

2.2 Equipment and Supplies

1. Class II biological safety cabinet.
2. Petri dishes.
3. Sterile scalpels.
4. Tweezers.
5. 15 mL conical tubes, pipettes, sterile tips (low binding are recommended).
6. 70 μm cell strainers.
7. Benchtop centrifuge (swinging-bucket rotor preferred).
8. CO_2 incubator set at 37°C, 5% CO_2.

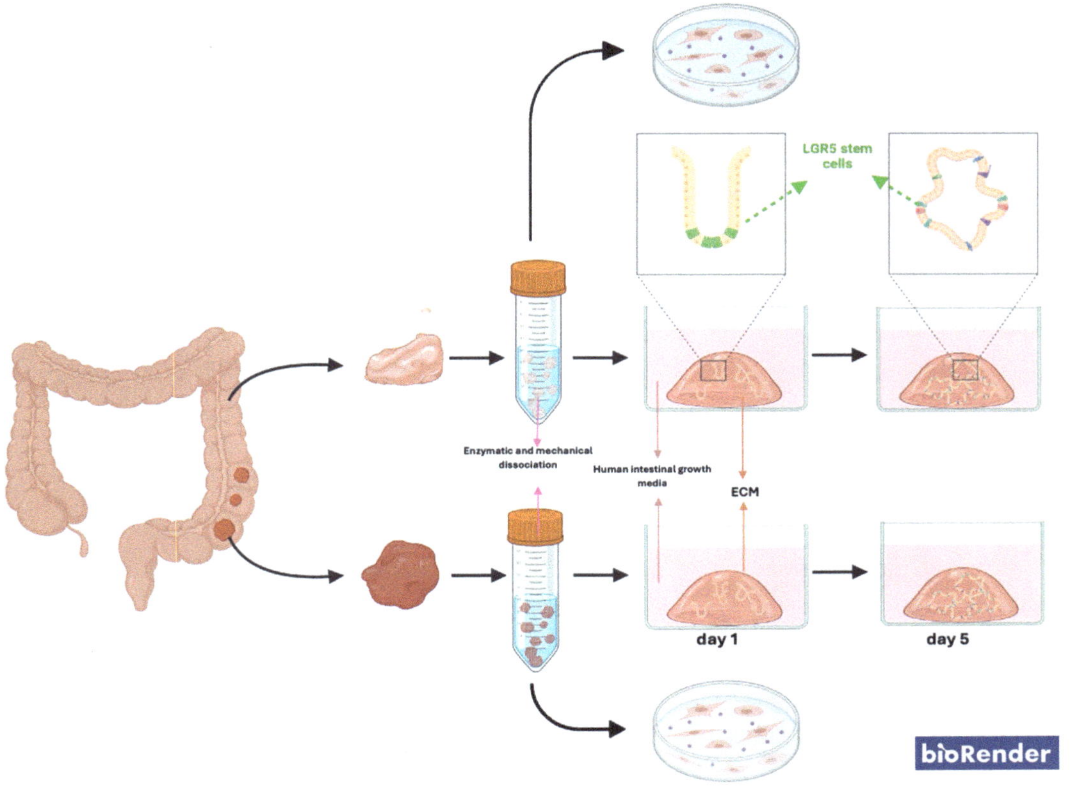

Fig. 1 Tissue processing workflow. Representative image showing freshly resected healthy and tumor tissue. Figure created with BioRender.com

3 Methods

3.1 Preparation of Tissue Samples

1. Perform all procedures under sterile conditions in a Class II biosafety cabinet.
2. Transfer freshly collected tissue into a sterile Petri dish containing 1 mL of complete growth factor (GF) medium (see Fig. 2 and *see* **Note 1**).
3. Using sterile scalpels, mince the tissue into small fragments approximately 1–2 mm^3 in size.
4. Carefully transfer the minced tissue fragments into sterile 15 mL conical tubes for subsequent processing.

3.2 Enzymatic Digestion with Liberase™ and DNase I

1. Add 1 mL complete medium to tissue fragments followed by Liberase™ (20 μL of 50 μg/mL) and DNase I (10 μL of 20 μg/mL) (*see* **Note 2**).
2. Vortex briefly and incubate at 37°C for 1 h (Fig. 2).

Table 1

Reagents used for patient-derived organoid (PDO) and fibroblast culture, with their main purpose or functional role in tissue dissociation, extracellular matrix support, growth factor supplementation, and antimicrobial protection

Reagent/material	Purpose/notes
Liberase™	Collagenase blend for efficient tissue dissociation
DNase I	Degrades free DNA to reduce clumping
Advanced DMEM/F12	Basal medium for PDOs
Matrigel	ECM for organoid embedding
VitroCol	Type I human collagen for organoid embedding
GlutaMAX	Glutamine source for PDO medium
Primocin	Broad-spectrum antibiotic
Antibiotic–Antimycotic (Anti-Anti)	Broad-spectrum antibiotic and antifungal
N21, B27 supplements	Serum-free supplements
CHIR99021, A83-01, SB431542, Y-27632	Compounds for pathway modulation
PGE_2	Prostaglandin for stemness maintenance
Noggin, EGF, R-spondin 1, Wnt3A	Growth factors essential for PDO culture
Advanced DMEM	Basal fibroblast medium
Gastrin	Peptide hormone that promotes epithelial proliferation and survival
Fetal calf serum (FCS)	Fibroblast growth supplement
L-glutamine, nonessential amino acids	Fibroblast culture supplements
Penicillin/Streptomycin, Gentamicin	Antibiotics for fibroblast medium

3. Every 15 min, gently pipette to aid dissociation and prevent clumping, or place the sample on a rotating incubator.

3.3 Separation of Epithelial Cells and Fibroblasts

1. Add an equal volume of Advanced DMEM (without growth factors) to the digestion mixture to dilute enzyme activity and prevent over-digestion.
2. Filter through a 70 μm cell strainer into a clean tube (Fig. 2) (*see* **Note 3**).
3. Centrifuge at 1600 rpm for 10 min at room temperature.

Epithelial fraction:

- After centrifugation, carefully aspirate the supernatant and resuspend the pellet in cold Matrigel or a 1:1 mixture of

Table 2
Medium composition of patient-derived organoid (PDO) (Reagents and concentration)

Reagent	Final concentration
Basal medium	
Advanced DMEM/F12	To final volume
Gastrin	10 nM
N-acetyl-L-cysteine (NAC)	1 mM
GlutaMAX	1×
Antibiotic-Antimycotic (Anti-Anti)	1×
Primocin	1×
N21	1×
B27	1×
Nicotinamide	10 mM
CHIR99021	3 μM
A83-01	500 nM
SB431542	3 μM
Y-27632	10 μM
PGE_2	1 μM
Supplements (fresh)	
Noggin	100 ng/mL
EGF	50 ng/mL
R-spondin 1	100 ng/mL
Wnt3A[a]	50 ng/mL

[a]Wnt3A is only added for healthy PDOs; omit for tumor PDOs

Table 3
Medium composition of fibroblasts (Reagents and concentration)

Reagent	Final concentration
Advanced DMEM	To final volume
Fetal calf serum (FCS)	10% (v/v)
L-glutamine	2 mM
Nonessential amino acids	1×
Penicillin/Streptomycin	1×
Gentamicin	50 μg/mL

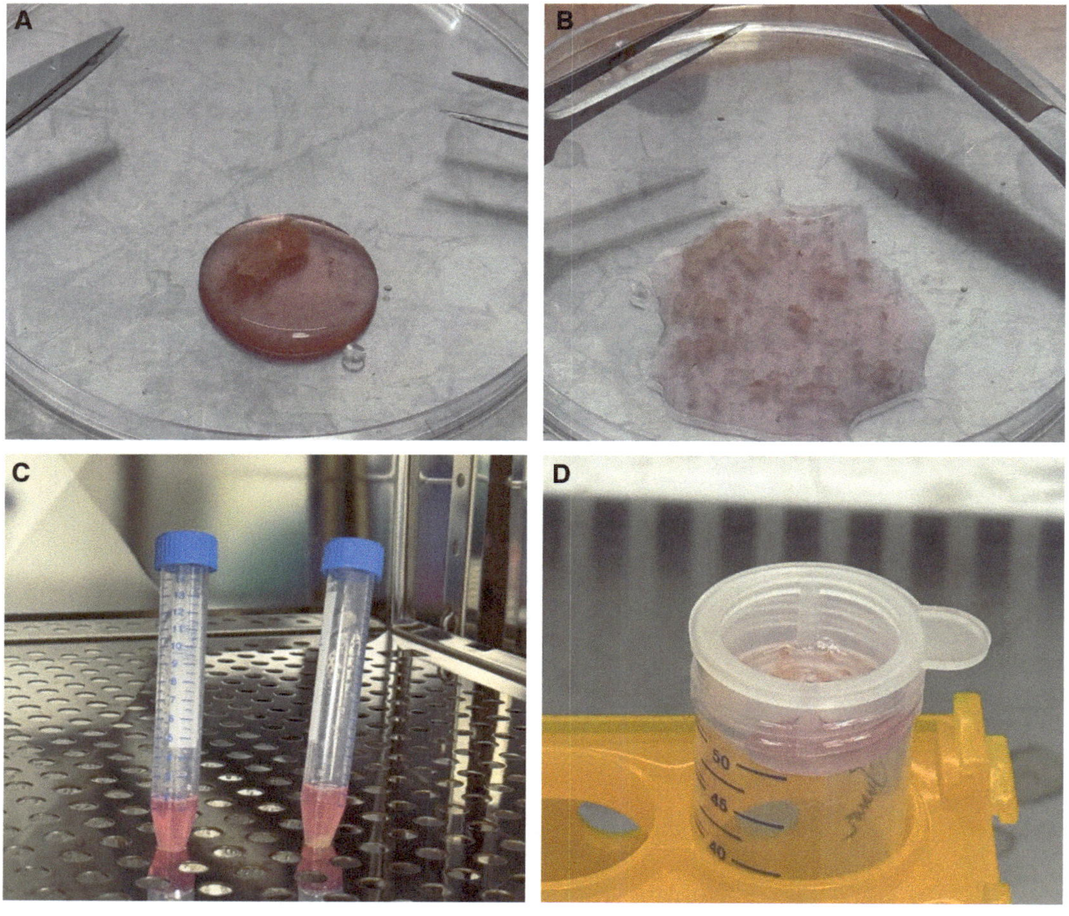

Fig. 2 (**A**, **B**) Images of tissue fragments placed in sterile Petri dishes containing complete growth factor medium before and after mechanical mincing. (**C**) Enzymatic digestion step. Tissue fragments incubated with Liberase™ and DNase I at 37°C with intermittent pipetting to aid dissociation. (**D**) Enzymatic digested tissue step filtered through a 70 μm cell strainer into a clean tube

Matrigel and Collagen I to achieve a final concentration of approximately $1–2 \times 10^4$ cells per 30 μL dome.

- Pipette 4–5 30 μL domes of the cell-Matrigel mixture in pre-warmed 24-well culture plates. Avoid introducing bubbles during pipetting.
- Allow the domes to solidify at 37°C for 30–60 min, depending on gel composition, before gently adding the organoid culture medium along the side of the wells to prevent disruption of the dome (*see* **Note 4**).
- Incubate at 37°C with 5% CO_2, changing the medium every other day. Organoid growth should be visible within 3–5 days (see Fig. 3 and see **Note 5**).

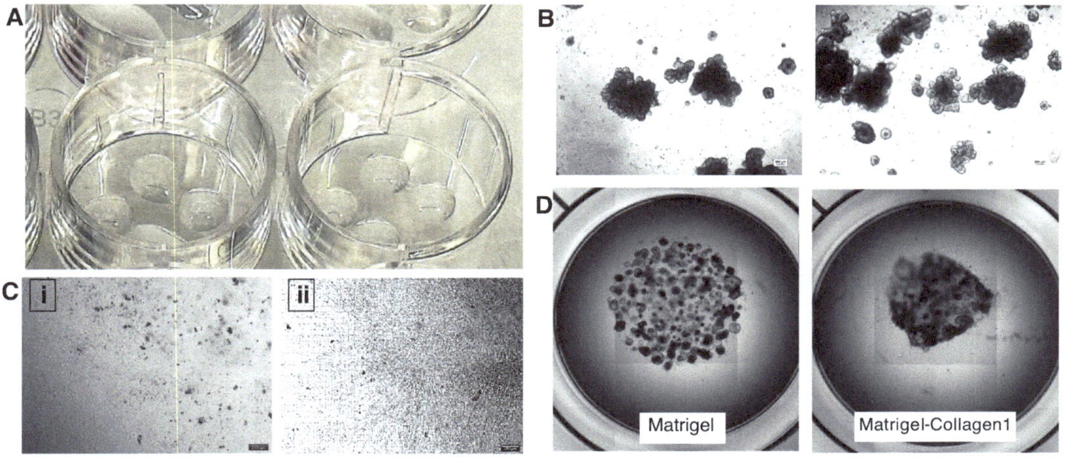

Fig. 3 (**A**) Matrigel domes with embedded organoids in a 12-well plate. (**B**) Patient-derived colorectal organoids cultured in Matrigel for 14 days (scale bar, 100 μm). (**C**) (i) Healthy and (ii) cancer-associated fibroblasts adhered to the culture plate and maintained in culture for 5 days. (**D**) Patient-derived colorectal organoids cultured in Matrigel and Matrigel-Collagen I for 30 days in a 48-well plate

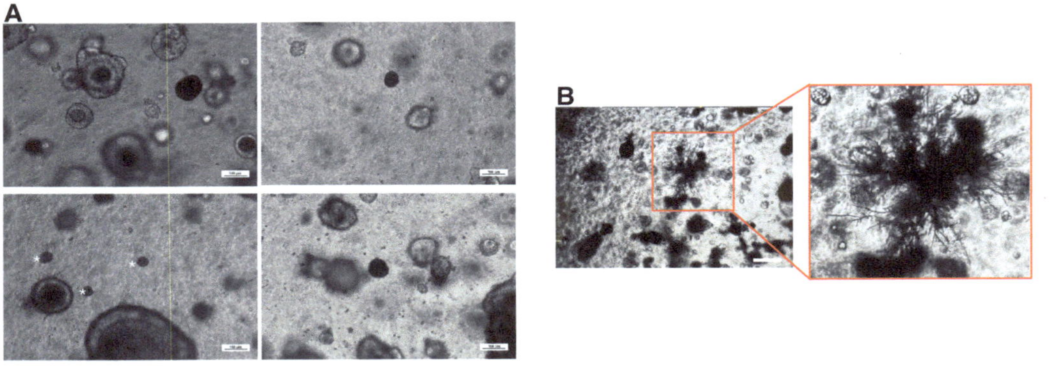

Fig. 4 (**A**) Bacterial contamination can manifest as colonies within the Matrigel domes, occasionally mimicking organoid morphology and thus complicating identification. (**B**) Fungal contamination may also arise, presenting as either filamentous molds or unicellular yeasts. Scale 100 μm

Fibroblast fraction:

- Collect the remaining cells after separation of the organoid fraction and resuspend in fibroblast growth medium (e.g., DMEM supplemented with 10% FBS and antibiotics).
- Plate cells in a 6-well plate at a density of ~1–2 × 10^5 cells per well.
- Incubate at 37°C with 5% CO_2, changing medium every 2–3 days. Fibroblasts typically adhere within 24 h and proliferate to confluence within 5–7 days (see Fig. 3). Passage cells when they reach 80%–90% confluence to maintain healthy growth (*see* **Notes 6 and 7**). For troubleshooting see Table 4.

Table 4

Common observations encountered during organoid and fibroblast culture, together with likely underlying causes and practical troubleshooting steps to restore growth, differentiation, and viability

Observation	Possible cause(s)	Suggested solution(s)
Organoids fail to establish	Low tissue quality or prolonged storage	Minimize the time between biopsy and culture initiation
Organoids fail to establish	Excessive digestion time	Shorten enzymatic digestion and monitor progress microscopically
Organoids stop growing	Poor-quality media components or old media	Test new batches before use and replace with fresh medium
Cell loss at passage	Foam formation during pipetting	Pipette gently against the tube wall and avoid creating bubbles
Organoids fail to differentiate	Incomplete removal of stem cell niche factors	Increase the number and duration of wash steps before induction
Fibroblasts fail to attach	Suboptimal Collagen I coating	Prepare a fresh coating solution and ensure uniform coverage of the culture surface
Fibroblasts senesce early	Over-confluence or late passage	Split cultures before 80% confluence and use early-passage cells

4 Notes

1. Always keep tissue moist; drying reduces cell viability.
2. Over-digestion may damage epithelial structures; check under a microscope after 30 min.
3. Filtering enzymatically digested tissue through a 70 μm strainer removes large debris and clumps, yielding a uniform single-cell suspension for improved culture viability and downstream processing.
4. Matrigel alone typically gels within ~30 min; however, Matrigel-Collagen I mixtures require a full 60 min incubation at 37°C to ensure complete polymerization and dome integrity.
5. Microbial contamination is frequently observed in organoid cultures derived from patient tissues (representative examples are shown in Fig. 4). Inclusion of Primocin and Anti–Anti in the culture medium has been demonstrated to reduce both bacterial and fungal contamination, thereby improving culture integrity.
6. To prepare a fibroblast-derived Collagen I gel, mix eight parts of chilled Collagen I solution with 1 part of chilled 10× PBS. Adjust the pH of the mixture carefully to 7.0–7.5 using sterile 0.1 M NaOH while monitoring pH closely. Finally, bring the

total volume to ten parts by adding sterile PBS. To prevent premature gelation, keep the mixture cold at 2°C–10°C until ready to form the gel.

7. Fibroblasts adhere within 6–12 h; changing medium after 24 h removes non-adherent cells.

5 Troubleshooting guide for PDO and fibroblast culture

Troubleshooting guide for PDO and fibroblast culture (see Table 4).

Acknowledgments

This work was supported by Medical Research Council (MRC) grants MRC AIM iCASE (MR/W007002/1). MRC NIRG (MR/W01825X/1), Royal Society (RGS\R2\222116), and GUTS UK (ECR2021_1). Ethics license: 17/EM/0126. We would like to thank patient donors and Nottingham QMC NHS team.

References

1. Zhang S et al (2024) Protocol for the generation, characterization, and functional assays of organoid cultures from normal and cancer-prone human esophageal tissues. STAR Protoc 5(4):103316
2. Dekkers J et al (2021) Long-term culture, genetic manipulation and xenotransplantation of human normal and breast cancer organoids. Nat Protoc 16:1936–1965
3. Guo L et al (2024) Toward reproducible tumour organoid culture: focusing on primary liver cancer. Front Immunol 15:1290504
4. Eenjes E et al (2023) Functional characterization and visualization of esophageal fibroblasts using organoid co-cultures. J Vis Exp 191:e64905
5. Högström J, Muranen T (2025) An optimized protocol for simultaneous propagation of patient- derived organoids and matching CAFs. Bio-protoc 15(2):e5160
6. Padmanaban V et al (2020) Organotypic culture assays for murine and human primary and metastatic-site tumours. Nat Protoc 15:2413–2442.
7. Passaniti A, Kleinman HK, Martin GR (2022) Matrigel: history/background, uses, and future applications. J Cell Commun Signal 16(4):621–626.
8. Pajic-Lijakovic I, Milivojevic M, Clark AG (2022) Collective cell migration on Collagen-I networks: the impact of matrix viscoelasticity. Front Cell Dev Biol 10:901026.

Check for updates

Chapter 22

A Tissue Explant Model for Colon Cancer

Justine Epiney, Stefano Gussago, Pénélope Saint-Amour, Joe Khallouf, Teresa Torres, Aurélie Vuagniaux, Nicola Colucci, Fabio Butti, Frédéric Bretagnol, Thibaud Koessler, Pierre Fournier and Albert Santamaria-Martínez

Abstract

Colorectal cancer (CRC) remains a leading cause of cancer-related morbidity and mortality worldwide. Recent advancements in cancer research emphasize the importance of personalized medicine, necessitating robust preclinical models that closely mimic the tumor microenvironment. Colorectal cancer organoids have emerged as a pivotal tool in this domain, providing a platform that preserves the cellular heterogeneity and allows high-throughput drug testing. However, organoids do not recapitulate the complexity of the tumor ecosystem, its architecture, and stromal interactions characteristic of native tumors. Here, we describe a model to maintain CRC tissue explants in culture. These explants enable the assessment of therapeutic responses ex vivo. Furthermore, they may offer insights into tumor-immune interactions and facilitate the evaluation of novel immunotherapeutic agents. The continued refinement of CRC tissue explant models promises to enhance our understanding of tumor biology and improve the predictive accuracy of preclinical studies, ultimately contributing to more effective and personalized treatment strategies for colorectal cancer patients.

Keywords Explants, Tumour microenvironment, ALI

1 Introduction

Colorectal cancer (CRC) is one of the most common malignancies worldwide and a leading cause of cancer-related death. Although advances in surgery, chemotherapy, targeted agents, and immunotherapy have improved outcomes, responses remain highly variable, particularly in advanced disease. This variability reflects the high degree of tumor heterogeneity, driven by intrinsic genetic and epigenetic differences that not only determine treatment sensitivity but also influence the complex crosstalk between cancer cells and their surroundings. The tumor microenvironment (TME) in colorectal cancer is exceptionally complex, shaped not only by interactions

Paloma Ordóñez-Morán (ed.), *Intestinal Stem Cells*, Methods in Molecular Biology, vol. 3059,
https://doi.org/10.1007/978-1-0716-5412-5_22,

among malignant, stromal, and immune cells but also by continuous exposure to nutrients, toxicants, and microbial metabolites from the intestinal lumen [1]. Altogether, these factors create a fine equilibrium that contributes to patient heterogeneity and underscores the need for personalized, physiologically relevant tumor models.

Preclinical models are indispensable for studying tumor biology and testing new therapies, yet reproducing the complexity of human colorectal cancer remains challenging. Traditional two-dimensional cell cultures have long been used because they are simple and scalable, but they fail to capture the spatial organization, cellular diversity, and intricate interactions of real tumors [2]. In vivo models address some of these limitations by preserving tissue architecture and systemic interactions, yet they are costly, rely on cross-species similarity, and cannot reproduce the heterogeneity observed among patients.

Three-dimensional spheroids and organoids have advanced cancer modeling by enabling cells to self-organize into structured tissues that better mimic native architecture due to the presence of cancer stem cells [3, 4]. They are scalable, genetically tractable, and useful for drug screening [5], and while cancer stem cells can sustain tumor growth and give rise to a differentiated progeny in these models, organoids lack the stromal and immune components essential to reproduce tumor-microenvironment interactions [6, 7]. Organ-on-chip systems further improve physiological relevance by integrating multiple cell types within microfluidic platforms that recreate biochemical and mechanical cues [8, 9]. These models allow controlled analysis of cancer-immune interactions but remain technically demanding and rely on the artificial reconstruction of tissues outside their native context. As a result, both organoids and organ-on-chip systems only partially capture the spatial and cellular complexity of human tumors.

Tissue explant models have recently been developed to bridge these limitations by preserving the native tumor architecture, stromal framework, and immune composition of patient-derived samples [10]. Among them, air-liquid interface (ALI) cultures have emerged as a robust approach to maintain tissue viability and microenvironmental integrity ex vivo [11, 12]. In this setup, small tumor fragments are embedded in extracellular matrix and cultured at the interface between air and medium, which ensures optimal oxygenation while preserving the spatial organization of the tissue. ALI explants retain endogenous stromal and immune components, enabling functional testing of therapeutic agents, including immunotherapies, in a setting that closely mirrors patient biology. This physiologically faithful model provides a powerful tool for translational studies and personalized medicine in colorectal cancer. In this chapter, we present a simple patient-derived tissue explant model based on the ALI system that reproduces the architecture and composition of the original tumor and that can be used to test therapies ex vivo for a short period of time.

2 Materials

2.1 Equipment

1. McIlwain Tissue chopper.
2. BSL-2 laminar flow hood.
3. Cell culture refrigerated centrifuge.
4. CO_2 incubator set to 5% CO_2 and 37°C.

2.2 Tissue Dissociation

1. Tweezers.
2. Disposable sterile safety scalpels.

2.3 Plastics

1. Tissue culture dishes (10 cm).
2. 12-well tissue culture plates.
3. Inserts with permeable, membranous bottom (0.4 μm), ⌀ 12 mm.
4. Polypropylene conical tubes (50 mL and 15 mL).
5. Serological disposable pipettes (5, 10, and 25 mL).
6. Pipettes (2, 200, and 1000 μL).
7. Pipette controller.
8. Speci-bag for transporting samples and substances of category B (UN3373).
9. Polysterene box.

2.4 Reagents

1. Growth medium: Advanced DMEM/F-12 supplemented with GlutaMAX™, 1% Penicillin/Streptomycin, 100 μg/mL Primocin, 10 mM HEPES, 1× B27 supplement without vitamin A, 1.25 mM N-acetylcysteine, 50 ng/mL EGF, 100 ng/mL Noggin, 0.5 μg/mL R-Spondin 1, 10nM Gastrin, 50 ng/mL FGF-2, 0.2% Lipid mixture 1, 1 μM PGE2.
2. Collection medium: Advanced DMEM/F-12 supplemented with GlutaMAX™, 10 mM HEPES, 1% Penicillin/Streptomycin, 100 μg/mL Primocin, 10 μM Y-27632.
3. Wash buffer: Sterile Phosphate Buffered Saline 1× (pH 7.4) supplemented with 1% Penicillin/Streptomycin, 100 μg/mL Primocin.
4. Cultrex Basement Membrane Extract (BME).
5. VitroGel® Hydrogel Matrix.
6. Formaldehyde.

3 Methods

3.1 Sample Collection and Transportation

1. The collection of human biological specimens is governed by both international and national regulations. Compliance with established guidelines, such as the Declaration of Helsinki, is essential to ensure ethical standards. This includes strict adherence to protocols concerning informed consent from patients when handling human-derived materials.
2. Collect sample in a 50 mL Falcon tube containing 25 mL of collection medium and transport on ice (4°C). Generally, untested oncological patient samples are typically transported as Biological Substance, Category B (UN 3373) under Packing Instruction P650, using triple packaging that includes an absorbent material to contain potential leaks and a rigid outer layer with at least one side measuring 100 mm × 100 mm (*see* **Notes 1–6**).

3.2 Before Starting

1. Confirm that all necessary materials, reagents, and equipment are in place. Prior to initiating the processing, irradiate the BSL-2 laminar flow hood for a minimum of 30 min to maintain sterility.
2. Confirm that the incubator is clean and has undergone routine maintenance, including the autoclave sterilization protocol.
3. Equilibrate Vitrogel to room temperature (RT) for 30 min prior to use.
4. Thaw Cultrex Basement Membrane Extract slowly at 4°C overnight and maintain in ice until use.
5. Prepare culture plates with Millipore inserts within the laminar flow hood under sterile conditions.
6. Prepare fresh CRC medium and store at 4°C.
7. Prepare the wash buffer and store at 4°C.

3.3 Preprocessing

1. Place the sample in a 10 cm dish under the P2 hood.
2. If not already performed, excise a portion of the sample using a clean safety scalpel and immediately immerse it in 4% formaldehyde. ⚠ Formaldehyde is volatile and toxic, follow the precautions described in *Guidelines for Safe Work Practices in Human and Animal Medical Diagnostic Laboratories*, 2012).
3. Carefully collect the remaining tissue, avoiding any mechanical trauma, and place it on the tissue chopper disk. Ensure the sample remains moist; if needed, add a small amount of medium to the plastic disk. Chop the tissue in four different orientations until most fragments are smaller than 1.5 mm^3. ⚠ The tissue chopper uses steel blades and needs to be

operated carefully, following the manufacturer's instructions. Dispose of the blades in designated sharps containers for safe handling.

4. Transfer the tissue fragments to a 50 mL falcon tube, wash them at least eight times in 40 mL of cold CRC wash buffer, and centrifuge at 300 g. ⚠ CRC samples are highly susceptible to microbial contamination; inadequate washing significantly increases the likelihood of contamination. If the sample is large, divide it into multiple tubes to facilitate washing and minimize contamination risk.
5. During the washing steps, prepare 1.7 mL tubes, each containing 200 μL of growth medium.
6. Remove the supernatant, add 1 mL of complete growth medium, and gently resuspend the pellet of tissue fragments. Decant the medium with the tissues in a sterile 10 cm dish.

3.4 Seeding of Tissue Explants

1. Using a low-retention filter tip, add 340 μL of Vitrogel Matrix to the 1.7 mL tube containing 200 μL of growth medium and then add 60 μL of Cultrex BME.
2. Gently mix the matrix solution, making sure to avoid introducing bubbles. After mixing, carefully and slowly pipette 600 μL of the prepared mixture into the insert.
3. Using a 1 mL pipette, carefully place tissue explants one at a time into the upper half of the matrix (*see* **Notes 7–9**).
4. Repeat Steps 1–3 for each additional insert.
5. Once all the inserts have been filled, place the plate in a 5% CO_2 incubator at 37°C for 15 min.
6. Retrieve the plate from the incubator and add 1.6 mL of full prewarmed CRC growth medium to the designated well.
7. If applicable, add the required drugs directly into the medium at the desired concentration.
8. Gently pipette the medium up and down ten times to mix, taking care to avoid forming bubbles. Likewise, ensure that no bubbles are trapped on the porous bottom of the insert.
9. Return the plate to the incubator (*see* **Note 10** and schematic diagram in Fig. 1).

3.5 Fixation and Analysis

1. Part of the specimen should be fixed in formaldehyde immediately before obtention or preprocessing to be used as a reference (control) as described in Sect. 3.2.2.
2. For the explants, fix the entire gel in 5 mL of 4% formaldehyde in a 15 mL polypropylene tube without disturbing the tissues.
3. Samples to be embedded in paraffin should be fixed in 4% formaldehyde for at least 24 h and a maximum of 72 h.

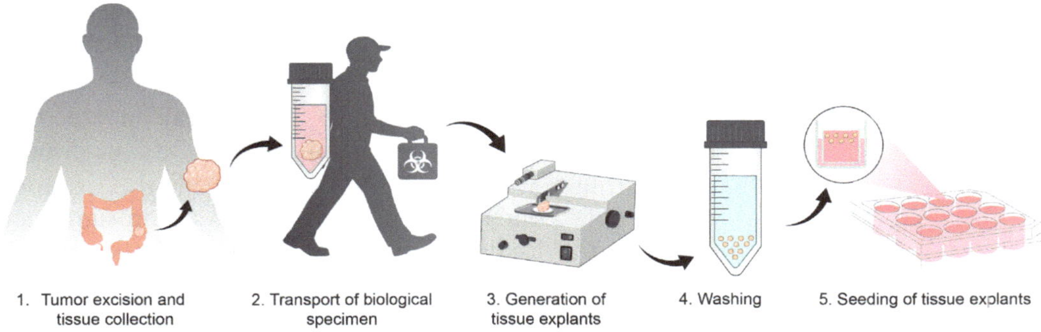

Fig. 1 Schematic diagram of the protocol workflow

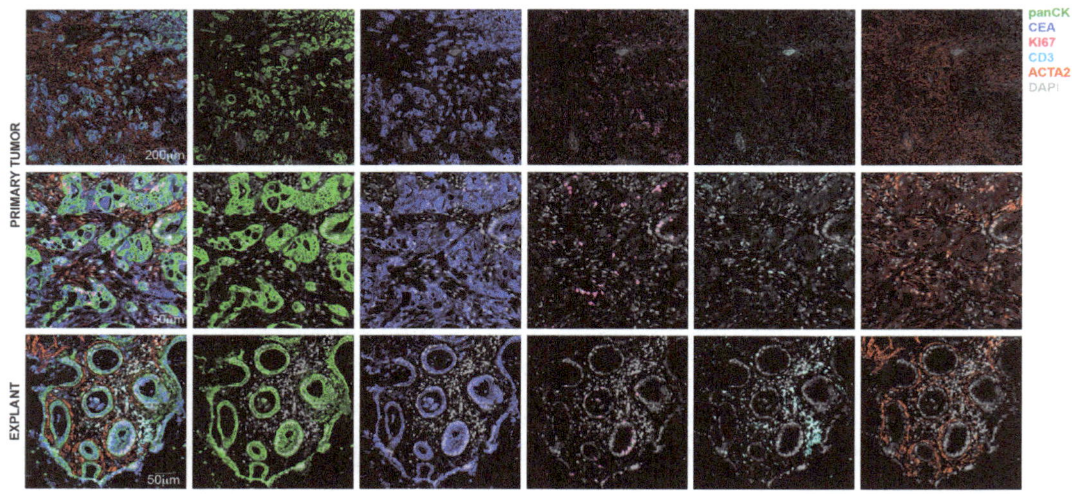

Fig. 2 Multiplex immunofluorescence staining of a primary tumor and a corresponding explant derived from the same tumor, performed using a 5-plex antibody panel (pancytokeratin, carcinoembryonic antigen, KI67, CD3, α-smooth muscle actin) with DAPI nuclear counterstaining (scale bars 200 μm and 50 μm) on a fully automated Ventana system

4. The samples are then placed in labeled histology cassettes and transferred to 70% ethanol prior to paraffin embedding.
5. The Formalin-Fixed Paraffin-Embedded (FFPE) blocks can be then sectioned and stained (see example in Fig. 2).

4 Notes

1. The sample should be transported on ice, for example, placing a sheet of aluminum foil on top of the ice (lower temperatures may cause a decrease in viability).

2. Untested human samples are to be considered P2. Always use PPE, such as nitrile gloves (double-gloving for high-risk samples), lab coats/gowns, eye protection, and face masks or shields during sample manipulation, and use screw-cap tubes for collection.
3. During collection and pathological examination, the sample must not be allowed to air-dry (e.g., after excision and before immersion in medium, saline solution, or fixative). Avoid placing the sample directly on gauze, absorbent paper, or similar materials, as this can lead to tissue dehydration.
4. Mechanical trauma caused by improper handling, such as the use of surgical instruments like tweezers or applying excessive compression to the specimen, can damage the sample, significantly reduce its viability, and compromise the tissue architecture.
5. After excision, the samples should be placed immediately at 4°C in collection medium and pre-processed in sterile conditions as soon as possible. Extended periods at RT may result in the loss of essential components of the tumor microenvironment (e.g., certain subsets of immune cells) and the loss of the tissue architecture.
6. The time between transportation and processing should not exceed 1.5 h.
7. When seeding the explants in the matrix, avoid adding extra medium to the gel (carry-over), as it can alter its strength and prevent direct cell contact with the functional hydrogel.
8. When pipetting the tissue fragments, apply just enough pressure to maintain the tissue fragment attached to the tip of the pipette to avoid damaging the explants.
9. To maximize the chances to capture the expected heterogeneity among tissue fragments, add six to nine explants per well. Note that increasing the number of explants per well also raises the risk of contamination.
10. For this protocol, 24–72 h of culture is what works best in terms of viability and tissue architecture.

Acknowledgments

The authors would like to thank all the patients who generously agreed to participate in the study CER-VD, project-ID 2020-02532, approved by the *Commission cantonale d'éthique de la recherche sur l'être humain*. J.E. was supported by the Fondation Bryn Turner-Samuels via Swiss Cancer Research (MD-PhD-5595-06-2022).

This project was supported by the Fondation Panacée with a grant to T.K. and A.S.M.

References

1. Kennel KB, Greten FR (2025) The immune microenvironment of colorectal cancer. Nat Rev Cancer 25(12):945–964. https://doi.org/10.1038/s41568-025-00872-1
2. Ben-David U et al (2018) Genetic and transcriptional evolution alters cancer cell line drug response. Nature 560:325–330
3. Kim S-Y, van de Wetering M, Clevers H, Sanders K (2025) The future of tumor organoids in precision therapy. Trends Cancer 11:665–675
4. Clevers H (2016) Modeling Development and Disease with Organoids. Cell 165:1586–1597
5. Tuveson D, Clevers H (2019) Cancer modeling meets human organoid technology. Science 364(6444):952–955. https://doi.org/10.1126/science.aaw6985
6. Polak R, Zhang ET, Kuo CJ (2024) Cancer organoids 2.0: modelling the complexity of the tumour immune microenvironment. Nat Rev Cancer 24:523–539
7. Recaldin T et al (2024) Human organoids with an autologous tissue-resident immune compartment. Nature 633:165–173
8. Huh D et al (2010) Reconstituting Organ-Level Lung Functions on a Chip. Science 328:1662–1668
9. Mitrofanova O et al (2024) Bioengineered human colon organoids with in vivo-like cellular complexity and function. Cell Stem Cell 31:1175–1186.e7
10. Voabil P et al (2021) An ex vivo tumor fragment platform to dissect response to PD-1 blockade in cancer. Nat Med 27:1250–1261
11. Neal JT et al (2018) Organoid Modeling of the Tumor Immune Microenvironment. Cell 175:1972–1988.e16
12. Santamaria-Martínez A et al (2024) Development of patient-derived lymphomoids with preserved tumor architecture for lymphoma therapy screening. Nat Commun 15: 10650

Chapter 23

A Straightforward and Reliable Mouse Model of Colorectal Cancer by Orthotopic Transplantation

Cewen Chen, Qiaochu Fu, Chao Gao, Shinya Tanaka and Masamichi Imajo

Abstract

Colorectal cancer (CRC) is a major malignancy with significant global implications for human health. The development of a reliable and pathologically relevant orthotopic CRC model is essential for advancing our understanding of its molecular mechanisms and for developing more effective therapeutic interventions. However, the construction of such models is fraught with challenges primarily because of the technical complexities involved in the transplantation of CRC cells into the intestinal epithelium. In this chapter, we describe a recently developed method for generating an orthotopic CRC model within the mouse cecal epithelium. This model enables tumor development and progression within a natural tissue microenvironment and facilitates complex interactions between tumor cells and surrounding normal cells, thereby replicating the intratumor heterogeneity of CRC. All procedures in this method are visualized under stereomicroscopic observation, leading to the efficient engraftment of CRC cells and subsequent tumor development. The establishment of this reliable orthotopic CRC model, which mimics tumor development in a more natural microenvironment, offers new opportunities to explore the molecular mechanisms underlying CRC and to assess novel anticancer therapies in pathologically relevant contexts.

Keywords Transplantation, Colorectal cancer, Orthotopic cancer model

1 Introduction

Colorectal cancer (CRC) is a major malignancy that poses a global threat to human health [1, 2]. It ranks third in both incidence and cancer-related mortality among all cancers [2]. As a result, there is a growing need to develop new therapeutic strategies that can effectively improve patient outcomes. Creating an appropriate in vivo model of CRC not only aids in understanding the molecular mechanisms behind CRC initiation and progression but also serves as an essential platform for the development of new anticancer drugs in preclinical research [3, 4]. To date, subcutaneous xenograft models have been extensively utilized in cancer research because of their ease of cancer cell transplantation and tumor growth monitoring [5, 6]. However, subcutaneous tumor models lack a suitable cancer

Paloma Ordóñez-Morán (ed.), *Intestinal Stem Cells*, Methods in Molecular Biology, vol. 3059,
https://doi.org/10.1007/978-1-0716-5412-5_23,

microenvironment resembling the original organs or tissues, limiting the applicability of the results to human cancers [3, 7]. Therefore, orthotopic cancer models that more faithfully replicate clinically relevant biological contexts are necessary for a better understanding of cancer development and the advancement of new therapies.

In this chapter, we describe a step-by-step protocol for establishing an orthotopic CRC model that can replicate cancer development in the natural tissue microenvironment [8]. In this method, CRC cells are transplanted into the luminal surface of the cecum under the guidance of a stereoscopic microscope, where all procedures for surgical transplantation of CRC cells are visualized to achieve efficient engraftment of the transplanted cells and subsequent tumor initiation. Tumors originate as small foci surrounded by normal epithelial cells and later expand into larger, homogeneous tumors while suppressing the proliferation of the adjacent normal epithelial cells. Since tumors develop in the natural tissue microenvironment resembling that of the original tumors in this model, complex interactions between tumor cells and adjacent normal epithelial cells, submucosal stromal cells, and immune cells are faithfully replicated. The establishment of an orthotopic CRC model mimicking tumor development within a natural microenvironment would provide valuable tools for studying the molecular mechanisms underlying CRC and for evaluating novel anticancer therapies in more pathologically relevant contexts.

2 Materials

1. Immunodeficient mice (BALB/c athymic nude mice are routinely used, but other severely immunocompromised mice, such as NOG (NOD/Scid/IL2Rgmma null) mice, can also be used for this protocol.
2. Small-animal anesthetizer.
3. Isoflurane.
4. Surgical instruments (dissecting scissors and forceps).
5. Stereomicroscope (ideally equipped with filters and light sources suitable for observation of red fluorescence).
6. Phosphate-buffered saline (PBS):137 mM NaCl, 2.7 mM KCl, 10 mM Na_2HPO_4, 1.8 mM KH_2PO_4, pH 7.4.
7. Sterile cotton swab.
8. Surgical suture (6-0 nylon).
9. Fast Green FCF (Fujifilm).
10. Dulbecco's modified Eagle's medium (DMEM) (SHIMADZU).

11. Plasticware: sterile 15 mL conical centrifuge tubes, 1.5 mL microcentrifuge tubes, P200 pipette tips, uncut P20, P200, and P1000 pipette tips, and 100 mm polystyrene dishes.
12. Inverted microscope (for cell culture).
13. CO_2 incubator (for cell culture).
14. Refrigerated microcentrifuge (for microcentrifuge tubes).
15. Caco-2 cells (ATCC).
16. Caco-2 cell culture medium: DMEM supplemented with 10% fetal bovine serum (FBS) and penicillin/streptomycin (1×).
17. 0.25% Trypsin-EDTA (Gibco).
18. Centrifuge for 15 mL tubes.
19. Sucrose.
20. Optimal Cutting Temperature Compound (O.C.T. compound) (Sakura Finetek Japan).
21. Cryostat microtome.
22. Hoechst33342 (Thermo Fisher Scientific).
23. Mowiol (Sigma-Aldrich).
24. Epifluorescence microscope equipped with filters and illuminators suitable for observation of blue and red fluorescence.

3 Methods

Here, we describe a method for creating an orthotopic CRC model by transplanting CRC cells into the intestinal mucosa of immunodeficient mice. In this surgical cell transplantation, the cecum offers several advantages: close similarity of the tissue environment between the cecum and colon, low risk of fatal intestinal obstruction even after invasive surgical procedures, and ease of externalizing the luminal surface to directly observe the mucosal layer. Additionally, even when tumors grow large within the intestinal lumen, intestinal obstruction is less likely to occur, making the cecum an optimal site for establishing an orthotopic CRC model. The step-by-step protocol below includes all procedures, such as removal of the epithelium from the intestinal mucosa prior to cell transplantation (Sect. 3.1), transplantation of cancer cells (Sect. 3.2), and methods for analyzing the formed tumors (Sect. 3.3), enabling even readers without much experience in mouse xenograft experiments to easily establish an orthotopic CRC model. Each step of these transplantation methods can be visualized under a microscope, which ensures the efficient engraftment of transplanted cells and subsequent tumor formation.

3.1 Removal of Cecal Epithelium

Before cell transplantation, it is necessary to remove the epithelium from the cecal mucosa and expose the underlying extracellular matrix in order to create a scaffold for colon cancer cells to engraft. Since the cecum can be easily everted to expose its luminal surface, the epithelial layer can be mechanically removed by direct brushing while observing under a stereomicroscope. The detailed protocol for these procedures is provided below.

1. After anesthesia with isoflurane, place the mouse in a supine position on the surgical table, and disinfect the abdominal region with 70% ethanol. We routinely use the female BALB/c athymic nude mice (8–12 weeks old). (For non-nude mice, the hair on the abdominal surface should be removed prior to disinfection.)
2. To expose the abdominal viscera, make a longitudinal incision (through the skin (including subcutaneous tissue and fat), the superficial muscle layer (typically the rectus abdominis or adjacent ventral muscles), and the peritoneum), approximately 1.5–2 cm in length, along the midline of the lower abdomen using surgical scissors.
3. By using dissection scissors and forceps, expose and free the cecum from the surrounding connective tissue. Gently exteriorize the cecum through the incision and place on a sterile platform using forceps.
4. Gently ligate the cecum at both ends with non-traumatic surgical sutures to occlude luminal flow. Make a 0.2-cm incision on the anti-mesenteric side near the cecal tip (or blind-ended pouch), avoiding visible vessels (Fig. 1a) (*see* **Note 1**).
5. Gently push the intestinal contents from the ligated end toward the incision using a sterile cotton swab moistened with PBS, ensuring minimal contact between the extruded contents and the serosal surface (Fig. 1b, c).
6. Gently evert the inner wall of the cecum through the incision to allow direct visualization of the mucosal surface (Fig. 1d). Carefully clear residual luminal contents with PBS and sterile cotton swabs and disinfect the surgical platform. (A small blunt instrument can be gently applied to the outer surface of the intestinal wall near the ligated end and used to push toward the incision to facilitate eversion.)
7. Under a stereomicroscope, stain the luminal surface of the cecum to facilitate visualization of the epithelial structures by dropping 10 μL of Fast Green FCF solution (Fig. 1e). Gently and repeatedly brush the cecal surface using a trimmed hard-bristled toothbrush to detach and remove the epithelial layer. Repeat the above procedures until the extracellular

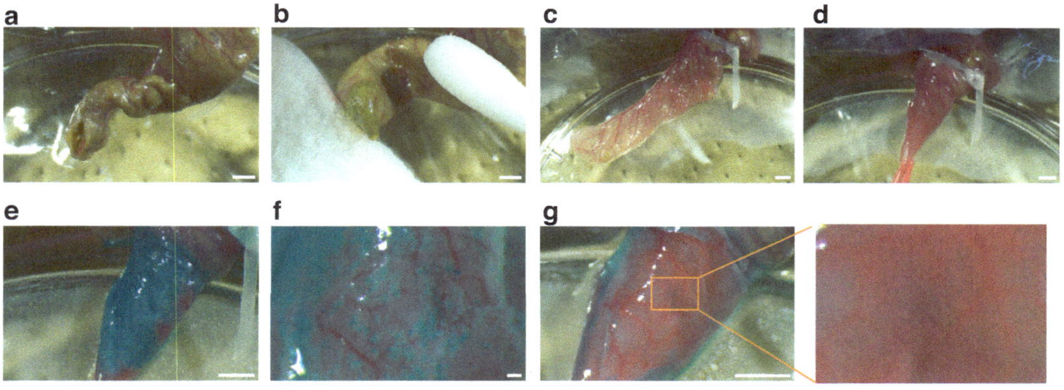

Fig. 1 Surgical procedures for externalization and removal of the cecal epithelium. Gross view of the mouse cecum during surgical procedures. (**a**) A small incision was made at the cecal tip. (**b**) The luminal contents were expelled from the incision by pushing the cecum with a cotton swab moistened with PBS. (**c**) Gross view of the cecum after the removal of the luminal contents. (**d**) Gross view of the externalized luminal surface of the cecum. (**e**) The cecal surface stained with Fast Green solution. (**f**) A magnified view of the cecal surface stained with Fast Green. Individual crypts were identified as small white spots. (**g**) The cecal surface after the removal of the epithelium. Crypts, devoid of epithelial cells, appeared as black holes. Scale bars, 2 mm in (**a–e, g**) and 200 μm in (**f**)

matrix layer is successfully exposed to facilitate the adhesion of transplanted CRC cells (Fig. 1f) (*see* **Note 2**).

3.2 Transplantation of Caco-2 Cells into Cecal Epithelium

After removing the epithelium from the cecum, transplantation can be performed by directly seeding CRC cells cultured under standard conditions onto the transplantation sites. Fluorescent markers are typically used to identify transplanted cancer cells. Fluorescent markers not only help confirm successful seeding of cells during transplantation but also make it easier to distinguish developed tumors during dissection, and are useful for observing interactions between tumor cells and surrounding nontumor cells in tumor sections. The method for generating CRC cells that stably express fluorescent markers is beyond the scope of this chapter and will not be described here; however, conventional methods using viral vectors or transfection reagents can be applied. The protocol used for transplanting Caco-2 CRC cells expressing the red fluorescent protein mScarlet is described below. In principle, this protocol can also be applied, with some modifications, to the transplantation of other CRC or normal intestinal epithelial cells.

1. Caco-2 cells expressing mScarlet should be cultured in DMEM supplemented with 10% FBS and antibiotics. For cell seeding, aspirate the culture medium from the 100 mm dish culturing the cells. Briefly rinse the cell layer by adding and then aspirating PBS.

2. Add 1.0 mL of 0.25% trypsin-EDTA solution to the dish and incubate at 37°C until the cells are dispersed (usually within 3 min).
3. Add 9 mL of the culture medium to the dish. Dissociate the cells into a single-cell suspension by pipetting up and down.
4. Centrifuge the cells at 1500×*g* for 3 min and discard the supernatant. Resuspend the cell pellet in PBS.
5. Count and dilute the cell suspension to a final density of 1.25 × 10^4 cells/μL, with a total volume of approximately 8–10 μL.
6. Under a stereomicroscope, carefully apply the cell suspension onto the cecal surface using a 2–20 μL micropipette. Allow cells to adhere for 10 min. A stereomicroscope equipped with filter units containing excitation and emission filters for red fluorescence [e.g., BP530-550 and BA575IF (SZX2-FRFP1)] and appropriate objective lenses should be used to better assess the adhesion status of the cells (Fig. 2).
7. Gently reposition the cecum to the original orientation using forceps, and suture the incision site using a 6–0 nylon suture.
8. To restore the luminal passage within the cecum, release non-traumatic surgical sutures ligated to occlude luminal flow at both ends of the cecum (Step 4 in the Sect. 3.1).
9. Rinse and moisten the cecum with PBS, gently return cecum to the abdominal cavity to avoid detachment of transplanted cells.
10. Suture the abdominal wall and skin separately using 6-0 nylon sutures, and disinfect the incision area. Monitor mice continuously until consciousness and mobility fully recover.

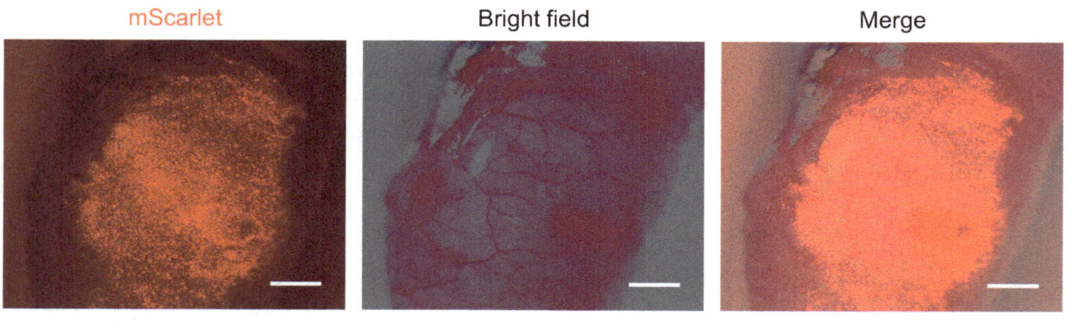

Fig. 2 Transplantation of CRC cells labeled with a red fluorescent protein mScarlet onto the epithelium-depleted surface of the cecum. After seeding on the cecal surface, Caco-2 cells expressing mScarlet are readily distinguishable under a fluorescence stereomicroscope. The attachment of Caco-2 cells to the tissue surface should be confirmed before proceeding to the next step in the surgical procedures. Scale bars, 1 mm

3.3 Analysis of the Developed CRC Xenografts

By 2 weeks after transplantation, transplanted CRC cells (Caco-2 cells expressing mScarlet) should form small tumor foci in the cecum of immunodeficient mice, which grow into small and large tumors by 1 and 3 months, respectively. Observation of fluorescent markers labeling transplanted cells is instrumental in identifying the developed tumors in the cecum, which are characterized by numerous surface irregularities. Here, we outline the comprehensive procedures involved in mouse dissection and tumor identification using fluorescent markers, followed by the preparation and examination of tissue sections for in-depth tumor analysis.

3.3.1 Mouse Dissection and Identification and Fixation of CRC Xenografts

1. Prior to sacrifice, disinfect the abdominal surface of the mouse with 70% ethanol. Prepare a 100 mm polystyrene dish and fill it with ice-cold PBS. All tissue dissection procedures should preferably be performed under low-temperature conditions.
2. Sacrifice mice in accordance with approved institutional guidelines. The cecum (from the terminal ileum to the proximal colon) should be immediately dissected using dissection scissors and placed in a 100 mm dish containing cold PBS. In our experience, the efficiency of transplanted CRC cell engraftment can be evaluated 1 week after transplantation, while tumor formation can be analyzed 1 month or later. The proper timing of these analyses can vary depending on CRC cell lines used, mouse strain, gender and age, and the number of transplanted cells, and should be appropriately adjusted as needed.
3. Observe the outer surface of the cecum directly under a stereomicroscope equipped with filters suitable for the observation of red fluorescence. Typically, in successfully transplanted cecum specimens collected one month or more after surgery, red fluorescence emitted from Caco-2 cell tumor foci expressing mScarlet can be clearly identified (Fig. 3a).
4. Flush the cecal lumen with cold PBS several times using a 100–1000 μL pipette to remove as much luminal content inside the cecum as possible.
5. Longitudinally open the cecum using micro-scissors. To avoid direct contact with the tumor foci, incise the cecum along the mesenteric line, whenever possible.
6. Rinse the luminal surface with cold PBS to remove any residual luminal content. Directly observe the luminal surface of the cecum to locate the xenograft region highlighted with red fluorescence (Fig. 3b).
7. Fix the tissue overnight in 10% neutral-buffered formalin at 4°C with gentle agitation. Avoid the exposure to light throughout the following procedures to prevent quenching of mScarlet fluorescence.

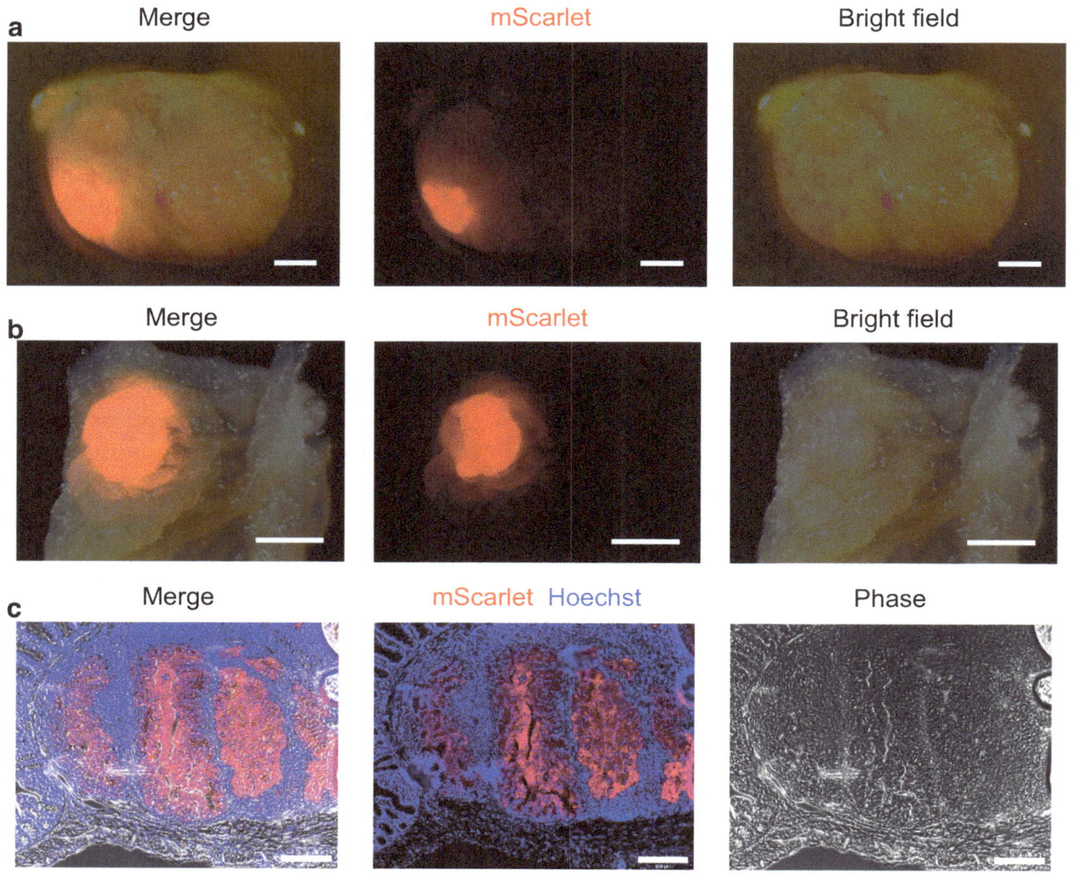

Fig. 3 Analysis of the xenografts developed after CRC cell transplantation into the cecum epithelium. (**a**) Gross view of the cecum developing the Caco-2-mScarlet xenografts. Tumor foci expressing mScarlet were visible from the outside of the cecum. (**b**). Caco-2-mScarlet cell tumors observed from the luminal side after dissection. (**c**) Fluorescence images of Caco-2-mScarlet xenografts stained with Hoechst33342. Caco-2 cells formed tumor foci surrounded by numerous normal host cells. Scale bars, 2 mm in (**a**,**b**) and 100 μm in (**c**)

3.3.2 Preparation, Staining, and Analysis of Xenograft Sections

1. After fixation, discard the fixative and wash the tissue with PBS at 4°C three times (1-h each).
2. Discard PBS. Subsequently, equilibrate the samples in a series of sucrose/PBS solutions (6%, 10%, 15%, 20%, 25%, and 30%) at 4°C with gentle agitation at 2-h intervals for each concentration.
3. Embed and rapidly freeze the samples in O.C.T. compound.
4. Cut sections of the frozen cecum specimen with a thickness of 4–8 μm by using a cryostat microtome and mount the sections onto a microscope slide. Allow sections to air dry.
5. Wash sections in PBS at room temperature for 5 min to remove the residual O.C.T. compound.

6. To stain the nuclei of cells, incubate the sections with Hoechst33342 diluted in PBS (1/500) for 30 min at room temperature.
7. Discard Hoechst33342 solution and wash sections in PBS three times for 5 min each. Rinse the slide briefly in DDW, and then seal the sections with the Mowiol mounting medium and coverslips.
8. Observe mScarlet and Hoechst33342 fluorescence in the sections under an epifluorescence microscope. For reference, typical images of the CRC xenograft stained by this protocol are shown in Fig. 3c. After transplantation into the cecal epithelium, CRC cells gradually grow into large tumors, while interacting with the surrounding normal epithelial cells, mesenchymal cells, and immune cells [8]. The distribution of these normal cells can be examined by performing conventional immunostaining analysis of cell type-specific markers on the sections prepared using this protocol [8]. The orthotopic CRC model described here enables the replication of CRC development in the native tissue microenvironment and provides new opportunities to analyze the roles of cancer cell-microenvironment interactions in tumor initiation, growth, invasion, and drug resistance.

4 Notes

1. Intestinal injury can cause severe adhesions in the abdomen, which may induce organ deformation and life-threatening intestinal obstruction. To reduce these risks, it is critical to minimize injury and bleeding during the surgery. Especially, care should be taken to avoid damaging visible blood vessels when making an incision at the distal edge of the cecum and ligating the cecum with nylon sutures to occlude luminal flow.
2. After staining with Fast Green solution, individual crypts are identifiable as small white spots under stereomicroscopic observation. Mechanical brushing can efficiently detach surface epithelial cells, leaving empty holes at crypt locations. These crypts, devoid of epithelial cells, appear as black holes under stereomicroscopic observation.

Acknowledgments

This work was supported by JSPS KAKENHI (Grant Numbers 22K06874 and 25K10197). This work was also supported by a Research Grant from HFSP (Ref.-No: RGP004/2026) (https://doi.org/10.52044/HFSP.RGP0042026.pc.gr.243776). We thank all members of the Bio Mix Laboratory at ICReDD and the Department of Cancer Pathology at the Faculty of Medicine for active discussion and helpful suggestions.

References

1. Abedizadeh R, Majidi F, Khorasani HR, Abedi H, Sabour D (2024) Colorectal cancer: a comprehensive review of carcinogenesis, diagnosis, and novel strategies for classified treatments. Cancer Metastasis Rev 43:729–753
2. Siegel RL, Miller KD, Wagle NS, Jemal A (2023) Cancer statistics, 2023. CA Cancer J Clin 73:17–48
3. Stribbling SM, Beach C, Ryan AJ (2024) Orthotopic and metastatic tumour models in preclinical cancer research. Pharmacol Ther 257:108631
4. Liu Y, Wu W, Cai C, Zhang H, Shen H, Han Y (2023) Patient-derived xenograft models in cancer therapy: technologies and applications. Signal Transduct Target Ther 8:160
5. Stribbling SM, Ryan AJ (2022) The cell-line-derived subcutaneous tumor model in preclinical cancer research. Nat Protoc 17:2108–2128
6. Schmidt KM, Geissler EK, Lang SA (2016) Subcutaneous murine xenograft models: a critical tool for studying human tumor growth and angiogenesis in vivo. Methods Mol Biol 1464:129–137
7. Zhao X, Li L, Starr TK, Subramanian S (2017) Tumor location impacts immune response in mouse models of colon cancer. Oncotarget 8:54775–54787
8. Chen C, Fu Q, Wang L, Tanaka S, Imajo M (2025) Establishment of a novel mouse model of colorectal cancer by orthotopic transplantation. BMC Cancer 25:405

INDEX

Paloma Ordóñez-Morán (ed.), *Intestinal Stem Cells*, Methods in Molecular Biology, vol. 3059, https://doi.org/10.1007/978-1-0716-5412-5,

FSC
www.fsc.org
MIX
Papier | Fördert
gute Waldnutzung
FSC® C083411

Kolibri 360 GmbH
Ferdinand-Jühlke-Straße 7
99095 Erfurt, Deutschland
produktsicherheit@kolibri360.de